Advanced Research in Chromatography

Advanced Research in Chromatography

Edited by **Carlos Dayton**

R CALLISTO REFERENCE

New York

Published by Callisto Reference,
106 Park Avenue, Suite 200,
New York, NY 10016, USA
www.callistoreference.com

Advanced Research in Chromatography
Edited by Carlos Dayton

International Standard Book Number: 978-1-63239-023-3 (Hardback)

Printed in the United States of America.

Contents

Preface

Chromatography was first experimented upon and invented by a Russian botanist. Since then it has evolved from a mere experiment to a technique, which has been improvised with different applications in industries as wells as academics. Chromatography is a versatile chemical analysis methodology, used for the separation of chemical compounds from complex mixtures in accordance to their distribution between two phases, mobile phase and stationary phase. Chromatography plays a vital role in a large number of detection systems, such as electrochemical, photometric and mass spectrometry. In 1952, James and Martin first used gas chromatography for analyzing fatty acid mixtures. The different techniques which have been developed in Chromatography are Column chromatography, High performance liquid chromatography (HPLC), Gas chromatography, Size exclusion chromatography, Ion exchange chromatography etc.

I am happy to have received contributions from internationally renowned contributors from different parts of the world. They joined hands with us to report on traditional and innovative approaches and reviewed some of the most relevant and impacting aspects of chromatography in this book. This effort would not have been possible without the kind cooperation of our contributors who patiently went through revisions of their chapters.

Instead of organizing the book into a pre-formatted table of contents with chapters, sections and then asking the authors to submit their respective chapters based on this frame, the authors were encouraged by the publisher to submit their chapters based on their area of expertise. The editor was then commissioned to examine the reading material and put it together as a book.

I convey my heartfelt thanks to all the contributors and to the team at the publishing house for their encouragement and excellent technical assistance as and when required.

Editor

Fingerprinting of Natural Product by Eastern Blotting Using Monoclonal Antibodies

Hiroyuki Tanaka,[1] Waraporn Putalun,[2] and Yukihiro Shoyama[3]

[1] *Faculty of Pharmaceutical Science, Kyushu University, 3-1-1 Maidashi, Higashiku, Fukuoka 812-0855, Japan*
[2] *Faculty of Pharmaceutical Science, Khon Kaen University, Khon Kaen 40002, Thailand*
[3] *Faculty of Pharmaceutical Science, Nagasaki International University/2825-7 Huis Ten Bosch, Sasebo, Nagasaki 859-3298, Japan*

Correspondence should be addressed to Yukihiro Shoyama, shoyama@niu.ac.jp

Academic Editor: Irena Vovk

We succeeded in developing the fingerprint of natural product by eastern blotting using monoclonal antibodies. After developing and separating them on a TLC plate, solasodine glycosides are oxidized by $NaIO_4$ and reacted with a protein to give conjugates which are recognized with anti-solamargine monoclonal antibody (MAb). Anti-solamargine MAb having wide cross-reactivity can stain and detect all solasodine glycosides by fingerprint. Different sensitivity between solamargine and solasonine was observed. The detection limit was 1.6 ng of solasonine. The hydrolysed products of solamargine were determined by fingerprint of eastern blotting compared to their Rf values depending on the sugar number. Fingerprint by eastern blotting using anti-ginsenoside Rb1 MAb distinguished the formula containing ginseng prescribed in traditional Chinese medicine. By double-staining of ginsenosides it is possible to suggest that the staining color shows the pharmacological activity, such as the purple bands indicate ginsenosides having stimulation activity, and the blue color indicated compound like ginsenosides possessed the depression affect for the central nervous system (CNS), respectively.

1. Introduction

In the recent rapid development of the molecular biosciences and their biotechnological applications, immunoassay systems using monoclonal antibody (MAb) against drugs and small molecular weight bioactive compounds have become an important tool for studies on receptor binding analysis, enzyme assay, and quantitative and/or qualitative analytical techniques in animals or plants; owing to their specific affinity. Previously we prepared various kinds of MAb against natural products like forskolin [1], solamargine [2], crocin [3], marihuana compound [4], opium alkaloids [5], ginsenosides [6, 7], berberine [8], sennosides [9], paeoniflorin [10], glycyrrhizin [11, 12], ginkgolic acid [13], aconitine alkaloid [14], baicalin [15], and so on, and developed individual competitive enzyme-linked immunosorbent assay (ELISA) as a high sensitive, specific, and simple methodology.

Western blotting is widely used as an immunostaining technique to detect high-molecular compounds like peptides and proteins based upon an antigen-antibody reaction. However, low-weight molecular compounds had not been previously analyzed by western blotting. We succeeded the blotted staining of solasodine glycoside on PVDF membrane using MAb after developed and separated solasodine glycosides on TLC and called as new-western blotting [16]. From this evidence, we applied this new methodology to licorice glycoside, glycyrrhizin, and named it as eastern blotting [12] following ginsenosides [17–19] and saikosaponin [20], and so on. In this paper, we introduce the fingerprint of a steroidal alkaloid glycoside, solasodine glycoside and ginseng saponin, ginsenosides by eastern blotting using MAbs.

2. Materials and Methods

2.1. Preparation of Anti-Solamargine MAb. A hybridoma producing MAb reactive to solamargine was obtained by the general procedure in our laboratory and classified into IgG2b which had k light chains [2]. The reactivity of

IgG type MAb, SMG-BD9, was tested by varying antibody concentration and by performing a dilution curve. The antibody concentration was selected for competitive ELISA. The MAb following competition was bound to polystyrene microtiter plates precoated with solamargine HSA. Under these conditions, the full measuring range of the assay extended from 20 to 400 ng/mL [2].

Anti-ginsenoside Rb1 [6] and Rg1 [7] were also prepared by the same ways in our laboratory.

2.2. Fingerprint by Eastern Blotting [16]. Solasodine glycosides were developed on a TLC plate. The TLC plate was covered with the PVDF membrane and blotted by short heating. The blotted PVDF membrane was dipped in water containing NaIO$_4$ under stirring at room temperature for 1 hr. After washing with water, a carbonate buffer containing BSA was added and stirred at room temperature for 3 hr. The PVDF membrane was washed twice with phosphate buffer for 5 min, and then washed with water. The PVDF membrane was immersed in anti-solamargine MAb, stirred at room temperature for 1 hr. After washing the PVDF membrane twice with phosphate buffer and water, a 1000-fold dilution of peroxidase-labeled goat anti-mouse IgG in phosphate buffer (pH 7.2) was added and stirred at room temperature for 1 hr. The PVDF membrane was washed twice with phosphate buffer and water, then exposed to 4-choloro-1-naphthol (1 mg/mL)—H$_2$O$_2$ (0.03%) in phosphate buffer (pH 7.2), and eastern blotting was stopped by washing with water. The immunostained PVDF membrane was allowed to dry.

2.3. Hydrolysis of Solasodine Glycosides and Stained by H$_2$SO$_4$, Dragendorff, and Eastern Blotting. Solamargine and solasonine were hydrolyzed by using 1 M HCl and heated for 10, 20, 30, 60, and 90 minutes, respectively. The hydrolyzed products were applied on three silica gel TLC plates. Developments were made by using CHCl$_3$/MeOH/NH$_4$OH (7 : 2.5 : 1) as mobile phase. Two plates were sprayed with H$_2$SO$_4$ and Dragendorff reagent, respectively. But the third TLC plate was transferred to a PVDF membrane by heating and stained as indicated above.

3. Results and Discussion

3.1. Fingerprint of Solasodine Glycosides by Eastern Blotting. The natural resources of adrenocortical hormones and sex hormones, which are mainly obtained from diosgenin, are becoming rare. The most important feature of solasodine is that it can be converted to dehydropregnenolone. Solasodine is found with a series of sugar residues attached to the oxygen at the C-3 position. By far the most common forms are the triglycosides, solamargine being predominant. Therefore, the steroidal alkaloid glycosides of the solasodine type like solamargine have become important as a starting material for the production of steroidal hormones in the pharmaceutical and medicinal areas. Rapid, simple, highly sensitive, and reproducible assay systems are required for a large number of plants, and a limited small amount of samples. In order to select the strain of higher yielding steroidal alkaloid

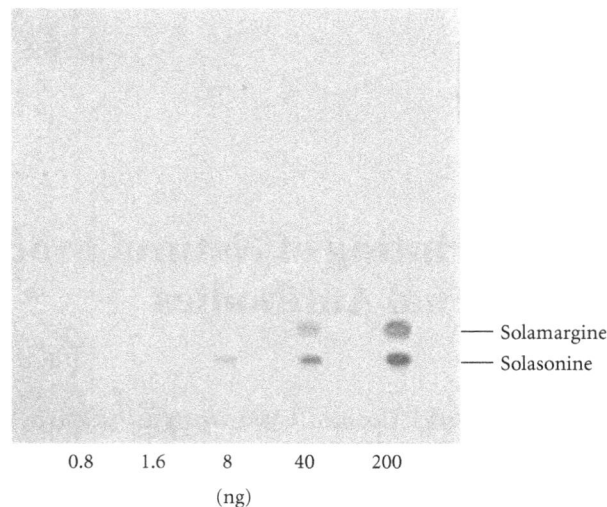

FIGURE 1: Fingerprint of solamargine and solasonine by eastern blotting.

glycosides in the resources of *Solanum* species we prepared an anti-solamargine MAb which was unique having a wide cross-reactivity [2].

Clear blue spots appeared on PVDF membrane as shown in Figure 1, the immunostains of solamargine and solasonine were observed. Lane 1–5 demonstrated the concentrations of both alkaloids, 0.8, 1.6, 8, 40, and 200 ng, respectively. Different sensitivities between solamargine and solasonine were observed in individual concentrations, and the sensitivity of solasonine was somewhat higher than that of solamargine. The detection limit was 1.6 ng of solasonine. We succeeded to separate the function of small molecule compounds such as solasodine glycosides into a part of epitope and fixing on the membrane [16]. When glycosides are treated by NaIO$_4$, the sugar part is cleaved to release an aldehyde group which is prepared a Schiff base with protein and then fixed to PVDF membrane. On the other hand, an epitope can be detected by MAb as described in Figure 2.

In order to confirm the preparation of solamargine-BSA conjugate on the PVDF membrane, the band corresponding to the solamargine-BSA conjugate was assessed by MALDI mass spectrometry. A broad peak [M+H]+ of solamargine-BSA conjugate appeared at around m/z 69043 in MALDI mass spectrometry demonstrating that at least 1-2 molecules of solamargine had combined [6] (data not shown). Therefore, it became clear that the sugar moiety which was conjugated with BSA is necessary in this staining system. Moreover, we confirmed the above evidence that solasodine glycoside-protein conjugate is necessary for the eastern blotting because solasodine which has no sugar moiety in a molecule cannot be stained by eastern blotting although solasodine can be stained by H$_2$SO$_4$ and Dragendorff reagent (see Figures 3(b) and 3(c)).

As an application of finger printing by newly established eastern blotting, the hydrolysis pathway of solamargine was surveyed by eastern blotting using anti-solamargine MAb. Solamargine was hydrolyzed using 1 M HCl for 10, 20,

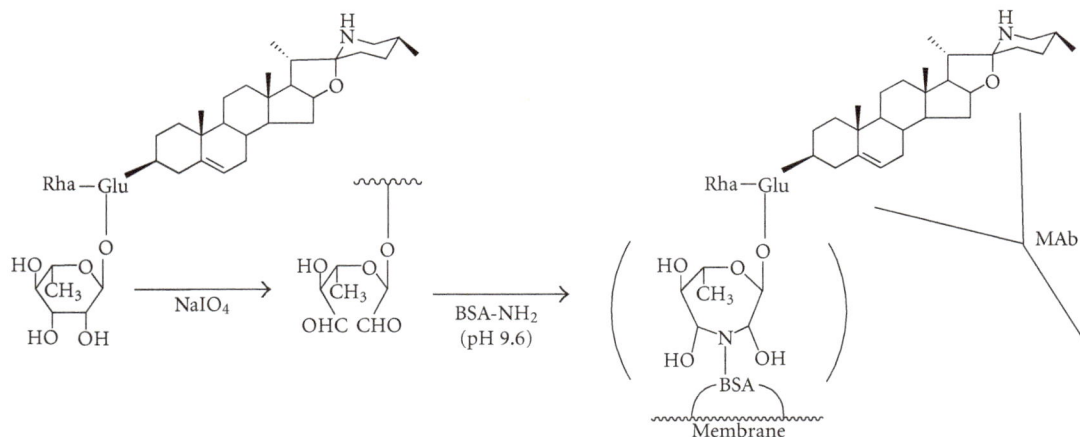

FIGURE 2: Staining mechanism of eastern blotting.

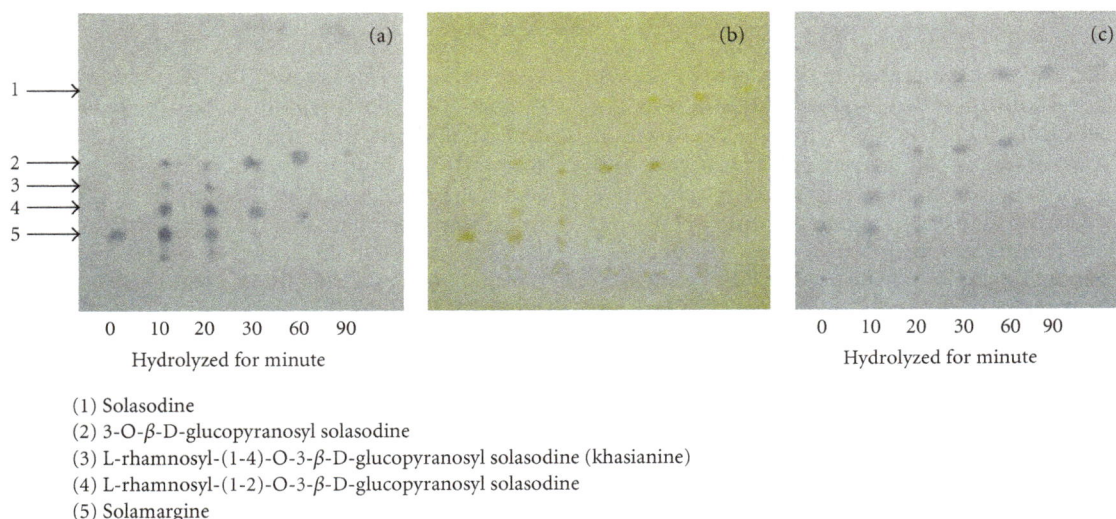

(1) Solasodine
(2) 3-O-β-D-glucopyranosyl solasodine
(3) L-rhamnosyl-(1-4)-O-3-β-D-glucopyranosyl solasodine (khasianine)
(4) L-rhamnosyl-(1-2)-O-3-β-D-glucopyranosyl solasodine
(5) Solamargine

FIGURE 3: Fingerprint of solasodine glycosides by eastern blotting (a), Dragendorff reagent (b), and sulfuric acid (c).

30, 60, and 90 minutes, respectively. The hydrolyzed products were developed by three silica gel TLC plates using $CHCl_3$/MeOH/NH_4OH solvent system. Two plates were sprayed and colored with H_2SO_4 and Dragendorff reagent, individually. One TLC plate was blotted to a PVDF membrane and stained by eastern blotting. Clear blue spots appeared as shown in Figure 3. Figure 3(a) shows the eastern blotting of solasodine glycosides. Lanes 1–6 demonstrate solamargine, 10, 20, 30, 60, and 90 minutes hydrolyzed products, individually, solasodine, 3-O-β-D-glucopyranosyl solasodine, L-rhamnosyl-(1-4)-O-3-β-D-glucopyranosyl solasodine, L-rhamnosyl-(1-2)-O-3-β-D-glucopyranosyl solasodine, and solamargine, respectively. This fingerprinting phenomenon agreed with the structural confirmation of solasodine glycosides elucidated by $_{13}$C NMR [21]. When compared with the sensitivity of fingerprinting regarding solasodine glycosides, the eastern blotting was more sensitive than those stained by H_2SO_4 (Figure 3(b)) and Dragendorff (Figure 3(c)).

We also analyzed the fingerprinting of *Solanum* species as shown in Figure 4. The profile stained by H_2SO_4 (Figure 4(b)) indicated the complicated pattern which is not limited to solasodine glycosides compared to the eastern blotting profile (Figure 4(a)). Major solasodine glycoside are solasonine, solamargine, and khasianine as indicated in Figure 4. From the survey of solasodine glycoside we decided that *S. khasianum* is an important resource of adrenocortical hormones and sex hormones.

Moreover, we succeeded to breed the strain of higher yielding solasodine glycoside by inducing the single chain Fv gene cloned from anti-solamargine MAb into *Solanum khasianum* plant [22].

3.2. Fingerprint of Ginseng Saponin, Ginsenoside, by Eastern Blotting. Ginseng, the crude drug of *Panax ginseng* is one of the most important natural medicine in many countries. It has been used to enhance stamina and capacity to cope with fatigue and physical stress, and as a tonic against

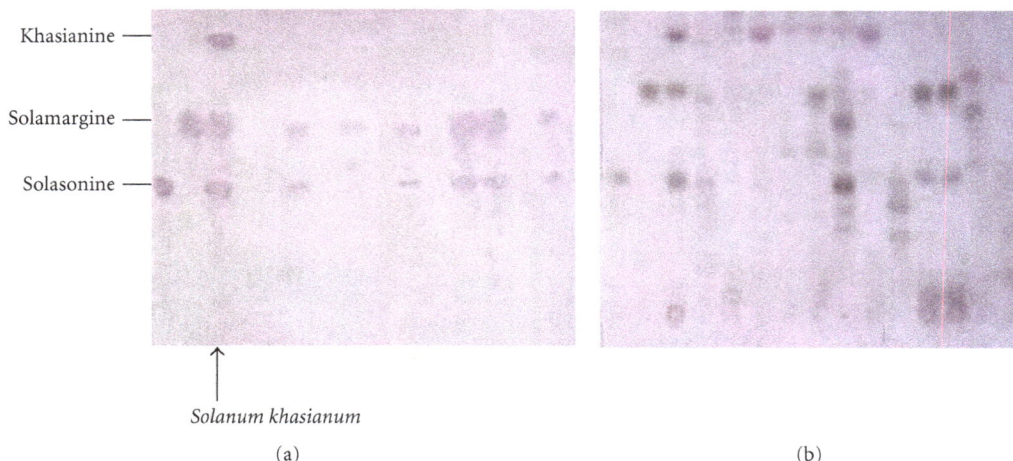

Khasianine —

Solamargine —

Solasonine —

↑
Solanum khasianum

(a)   (b)

FIGURE 4: The eastern blotting (a) and TLC stained with sulfuric acid (b) of solasodine glycosides from *Solanum* spp.

cancers, disturbances of the central nervous system (CNS), hypothermia, carbohydrate and lipid metabolism, immune function, the cardiovascular system, and radioprotection [23]. They contain more than 50 kinds of dammarane and oleanane saponins considered to be pharmacologically active components. Ginsenoside Rb1 is a main saponin in ginseng. However, since the concentration varies in the ginseng root or the root extract depending on the method of extraction, subsequent treatment, or even the season of its collection, standardization of quality is required. For this purpose we have prepared anti-ginsenoside Rb1 [6] and Rg1 MAbs [7].

Cross-reactivity is the most important factor in determining the value of an antibody. Since the ELISA for ginsenoside Rb1 was established for phytochemical investigations involving crude plant extract, the assay specificity was checked by determining the cross-reactivity of the MAb with various related compounds. The cross-reactivity data of MAb that was obtained were examined by competitive ELISA and calculated using picomole of ginsenoside Rb1. The cross-reactivity of ginsenoside Rc and Rd which possess a diglucose moiety attached to the C-3 hydroxy group were weak compared to ginsenoside Rb1, 0.024 and 0.020%, respectively. Ginsenoside Re and Rg1 showed no cross-reactivity (less than 0.005%). It is evident that the MAb reacted only with a small number of structurally related ginsenoside Rb1 molecules very weakly and did not react with other steroidal compounds like glycyrrhizin, digitoxin, tigogenin, tigonin, and solamargine. Anti-ginsenoside Rg1 MAb [7] and Re [18] have been prepared and set up their ELISA. Anti-ginsenoside Rg1 MAb was also high specific like anti-ginsenoside Rb1. On the other hand, anti-ginsenoside Re MAb showed wide cross-reactivity. Therefore, this MAb can be used for the analysis for the total ginsenoside concentration.

Figure 5 shows the H_2SO_4 staining and eastern blotting of ginsenoside standards and the traditional Chinese medicine (TCM) using anti-ginsenoside Rb1 MAb. It is impossible to determine the ginsenosides by TLC-staining by H_2SO_4 as indicated in Figure 5(a). On the other hand,

clear staining of ginsenoside Rb1 occurred by eastern blotting (Figure 5(b)). Furthermore, it became evident that kikyoto and daiokanzoto prescriptions related to TCM prescription that did not contain ginseng indicating no band of ginsenoside Rb1. The eastern blotting method was considerably more sensitive than that of H_2SO_4 staining. The H_2SO_4 staining detected all standard compounds. The Eastern blotting indicated only limited staining of ginsenoside Rb1, Rc, and Rd, of which cross-reactivities were under 0.02%. We suggest that an aglycon, protopanaxadiol, and a part of the sugars may be of importance to the immunization and may function as an epitopes for the structure of ginsenosides. In addition, it is suggested that the specific reactivity of sugar moiety in the ginsenoside molecule against anti-ginsenoside Rb1 MAb may be modified by the $NaIO_4$ treatment of ginsenosides on the PVDF membrane causing ginsenoside Rc and Rd to become detectable by eastern blotting.

When the mixture of anti-ginsenoside Rb1 and Rg1 MAbs and the pair of substrates were tested for staining of ginsenosides, all ginsenosides, ginsenoside-Rb1, -Rc,-Rd,-Re, and -Rg1 were stained blue although the purple color staining for ginsenoside Rg1 was expected because 3-amino-9-ethylcarbazole and 4-chloro-1-naphotol might be different. Therefore, we performed successive staining of the membrane using anti-ginsenoside Rg1 and then anti-ginsenoside Rb1. Finally, we succeeded: the double-staining of ginsenosides indicating that ginsenoside Rg1 and ginsenoside Re were stained purple and the other blue, as indicated in Figure 6. From this result, both antibodies can distinguish individual aglycon protopanaxatriol and protopanaxadiol. For this application, the crude extract of various *Panax* species were analyzed by the newly developed double-staining system. Major ginsenosides can be determined clearly by the double staining method indicated in Figure 6 as the fingerprint of ginsenosides. Therefore, it is possible to suggest that the staining color shows the pharmacological activity, such as the purple bands indicate ginsenosides having stimulation activity for the CNS. On the other hand, the blue color indicates ginsenosides possessing the depression affect for

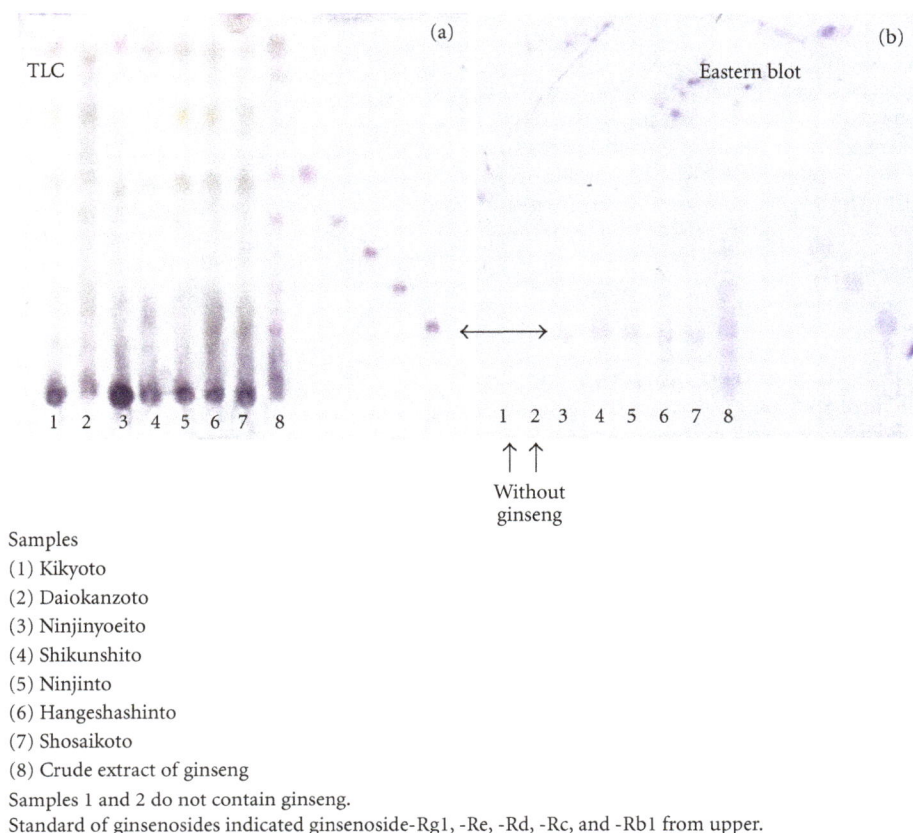

Samples
(1) Kikyoto
(2) Daiokanzoto
(3) Ninjinyoeito
(4) Shikunshito
(5) Ninjinto
(6) Hangeshashinto
(7) Shosaikoto
(8) Crude extract of ginseng
Samples 1 and 2 do not contain ginseng.
Standard of ginsenosides indicated ginsenoside-Rg1, -Re, -Rd, -Rc, and -Rb1 from upper.

FIGURE 5: TLC stained with sulfuric acid (a) and eastern blotting (b) of ginsenosides in traditional Chinese medicine (TCM) prescriptions by anti-ginsenoside Rb1 MAb.

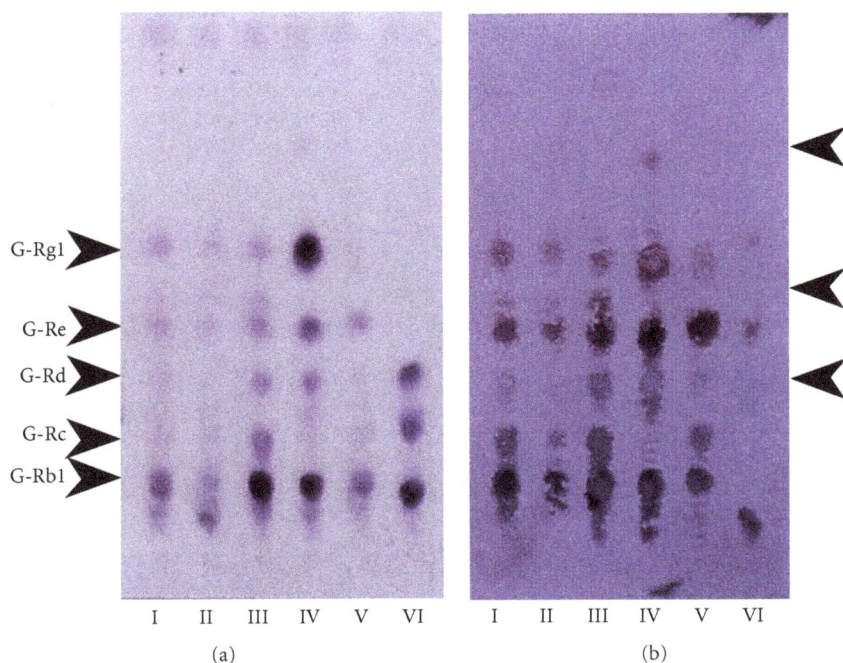

FIGURE 6: Double-staining of eastern blotting for ginsenosides contained in various ginseng using anti-ginseenoside-Rb1 and anti-ginsenoside-Rg1 monoclonal antibodies. (a) TLC profile stained by sulfuric acid, (b) Eastern blotting by anti-ginsenoside-Rb1 and anti-ginseside-Rg1 monoclonal antibodies. I, II, III, IV, V, and VI indicated white ginseng, red ginseng, fibrous ginseng (*Panax ginseng*), *Panax notoginseng*, *Panax quinquefolius,* and *Panax japonicus*, respectively. Upper purple color spots and lower blue color spots were stained by anti-ginsenoside-Rg1 and anti-ginsenoside-Rb1 monoclonal antibodies, respectively.

the CNS. Moreover, the Rf value of ginsenosides roughly suggests the number of sugars attached to the aglycon. Both analyses make it possible to identify which aglycon attaches and how many sugars combine, leading to the structure of ginsenosides. In fact, three kinds of ginsenosides possessing protopanaxadiol-ginsenoside Rh1, Rf, and 20-O-gluco-ginsenoside Rf in *P. ginseng* root were determined by coloring and Rf value by comparing them with the structures reported in the previous paper [19].

4. Conclusion

Fingerprint by eastern blotting method can open its application for the wide field of natural products, especially glycosides like solasodine glycosides and ginsenosides. An MAb having a wide cross-reactivity like anti-solamargine MAb can distinguish the total solasodine glycosides. On the other hand, a high specific MAb like anti-ginsenoside Rb1 MAb can be used for the detection of a single antigen compound.

When two kinds of MAbs can be used, the double-staining is possible as the staining system enhanced the separate staining of ginsenosides having protopanatriol or protopanaxadiol in a molecule. The staining color can monitor the pharmacological activity suggesting the stimulation activity and/or the depression affect for the CNS. Furthermore, the Rf value of solasodine glycosides and ginsenosides suggests the number of sugars attached to the aglycon. Both evidences make it possible to confirm which aglycon attaches and how many sugars combine to the aglycon resulting the confirmation of structure for glycosides. Therefore, the fingerprint stained by eastern blotting using MAb could provide a new strong methodology in the wide field of natural product investigation.

References

[1] R. Sakata, Y. Shoyama, and H. Murakami, "Production of monoclonal antibodies and enzyme immunoassay for typical adenylate cyclase activator, forskolin," *Cytotechnology*, vol. 16, no. 2, pp. 101–108, 1994.

[2] M. Ishiyama, Y. Shoyama, H. Murakami, and H. Shinohara, "Production of monoclonal antibodies and development of an ELISA for solamargine," *Cytotechnology*, vol. 18, no. 3, pp. 153–158, 1996.

[3] L. Xuan, H. Tanaka, Y. Xu, and Y. Shoyama, "Preparation of monoclonal antibody against crocin and its characterization," *Cytotechnology*, vol. 29, no. 1, pp. 65–70, 1999.

[4] H. Tanaka, Y. Goto, and Y. Shoyama, "Monoclonal antibody based enzyme immunoassay for marihuana (cannabinoid) compounds," *Journal of Immunoassay*, vol. 17, no. 4, pp. 321–342, 1996.

[5] Y. Shoyama, T. Fukada, and H. Murakami, "Production of monoclonal antibodies and ELISA for thebaine and codeine," *Cytotechnology*, vol. 19, no. 1, pp. 55–61, 1996.

[6] H. Tanaka, N. Fukuda, and Y. Shoyama, "Formation of monoclonal antibody against a major ginseng component, ginsenoside Rb1 and its characterization," *Cytotechnology*, vol. 29, no. 2, pp. 115–120, 1999.

[7] N. Fukuda, H. Tanaka, and Y. Shoyama, "Formation of monoclonal antibody against a major ginseng component, ginsenoside Rg1 and its characterization: monoclonal antibody for a ginseng saponin," *Cytotechnology*, vol. 34, no. 3, pp. 197–204, 2000.

[8] J. S. Kim, H. Tanaka, and Y. Shoyama, "Immunoquantitative analysis for berberine and its related compounds using monoclonal antibodies in herbal medicines," *Analyst*, vol. 129, no. 1, pp. 87–91, 2004.

[9] O. Morinaga, S. Nakajima, H. Tanaka, and Y. Shoyama, "Production of monoclonal antibodies against a major purgative component, sennoside B, their characterization and use in ELISA," *Analyst*, vol. 126, no. 8, pp. 1372–1376, 2001.

[10] Z. Lu, O. Morinaga, H. Tanaka, and Y. Shoyama, "A quantitative ELISA using monoclonal antibody to survey paeoniflorin and albiflorin in crude drugs and traditional Chinese herbal medicines," *Biological and Pharmaceutical Bulletin*, vol. 26, no. 6, pp. 862–866, 2003.

[11] H. Tanaka and Y. Shoyama, "Formation of a monoclonal antibody against glycyrrhizin and development of an ELISA," *Biological and Pharmaceutical Bulletin*, vol. 21, no. 12, pp. 1391–1393, 1998.

[12] S. J. Shan, H. Tanaka, and Y. Shoyama, "Enzyme-linked immunosorbent assay for glycyrrhizin using anti-glycyrrhizin monoclonal antibody and an eastern blotting technique for glucuronides of glycyrrhetic acid," *Analytical Chemistry*, vol. 73, no. 24, pp. 5784–5790, 2001.

[13] P. Loungratana, H. Tanaka, and Y. Shoyama, "Production of monoclonal antibody against ginkgolic acids in *Ginkgo biloba* Linn," *The American Journal of Chinese Medicine*, vol. 32, no. 1, pp. 33–48, 2004.

[14] K. Kido, K. Edakuni, O. Morinaga, H. Tanaka, and Y. Shoyama, "An enzyme-linked immunosorbent assay for aconitine-type alkaloids using an anti-aconitine monoclonal antibody," *Analytica Chimica Acta*, vol. 616, no. 1, pp. 109–114, 2008.

[15] K. Kido, O. Morinaga, Y. Shoyama, and H. Tanaka, "Quick analysis of baicalin in Scutellariae Radix by enzyme-linked immunosorbent assay using a monoclonal antibody," *Talanta*, vol. 77, no. 1, pp. 346–350, 2008.

[16] H. Tanaka, W. Putalun, C. Tsuzaki, and Y. Shoyama, "A simple determination of steroidal alkaloid glycosides by thin-layer chromatography immunostaining using monoclonal antibody against solamargine," *FEBS Letters*, vol. 404, no. 2-3, pp. 279–282, 1997.

[17] H. Tanaka, N. Fukuda, and Y. Shoyama, "Eastern blotting and immunoaffinity concentration using monoclonal antibody for ginseng saponins in the field of traditional Chinese medicines," *Journal of Agricultural and Food Chemistry*, vol. 55, no. 10, pp. 3783–3787, 2007.

[18] O. Morinaga, H. Tanaka, and Y. Shoyama, "Detection and quantification of ginsenoside Re in ginseng samples by a chromatographic immunostaining method using monoclonal antibody against ginsenoside Re," *Journal of Chromatography B*, vol. 830, no. 1, pp. 100–104, 2006.

[19] N. Fukuda, S. Shan, H. Tanaka, and Y. Shoyama, "New staining methodology: eastern blotting for glycosides in the field of Kampo medicines," *Journal of Natural Medicines*, vol. 60, no. 1, pp. 21–27, 2006.

[20] O. Morinaga, S. Zhu, H. Tanaka, and Y. Shoyama, "Visual detection of saikosaponins by on-membrane immunoassay and estimation of traditional Chinese medicines containing Bupleuri radix," *Biochemical and Biophysical Research Communications*, vol. 346, no. 3, pp. 687–692, 2006.

[21] B. B. Shashi, P. S. Niranjan, N. G. Amar, R. Kasai, and O. Tanaka, "Steroidal alkaloids from Solanum khasianum: application of ^{13}C NMR spectroscopy to their structural elucidation," *Phytochemistry*, vol. 19, no. 9, pp. 2017–2020, 1980.

[22] W. Putalun, F. Taura, W. Qing, H. Matsushita, H. Tanaka, and Y. Shoyama, "Anti-solasodine glycoside single-chain Fv antibody stimulates biosynthesis of solasodine glycoside in plants," *Plant Cell Reports*, vol. 22, no. 5, pp. 344–349, 2003.

[23] The Society for Korean Ginseng, *In Understanding of Korean Ginseng*, Seoul, Korea, 1995.

2

Ion Suppression Study for Tetracyclines in Feed

Joaquim Chico,[1] Frederique van Holthoon,[2] and Tina Zuidema[2]

[1] *Departament de Química Analítica, Facultat de Química, Universitat de Barcelona, c/Martí i Franquès 1, 08028 Barcelona, Spain*
[2] *RIKILT-Institute of Food Safety, Wageningen University and Research Center, Akkermaalsbos 2, 6708 WB, P.O. Box 230, 6700AE Wageningen, The Netherlands*

Correspondence should be addressed to Joaquim Chico, jchicorovira@gmail.com

Academic Editor: Andrew Cannavan

Ion suppression in analysis of tetracyclines in feed was studied. The conventional analysis consists of a liquid extraction followed by a clean-up step using solid phase extraction (SPE) technique and analysis of the tetracyclines by liquid chromatography and mass spectrometric detection. Various strategies for extraction and cleanup were tested in the present work, and the effectiveness to decrease the ion suppression on the MS/MS signals was evaluated. Four sample treatment methods were tested with five different feed samples. Extraction solvents tested were McIlvaine buffer and a mixture of McIlvaine buffer dichloromethane (3 : 1). SPE cartridges for cleanup were Oasis HLB, Oasis MCX, and Oasis MAX. The effectiveness of the methods was evaluated in terms of decreasing the ion suppression effect but also of decreasing the variability of ion suppression between samples. The method that provided the most satisfactory results involved a clean-up step based on SPE using mixed-mode cation exchange cartridges (Oasis MCX).

1. Introduction

Tetracyclines are a family of drugs belonging to the group of antibiotics. They are widely used in animal husbandry for therapeutic and prophylactic purposes. Oxytetracycline, tetracycline, chlortetracycline, and doxycycline are by far the most used antibiotics from this family. Their main chemical properties are their amphoteric behaviour due to their several acid-base equilibria and the tendency to act as chelating agents in presence of multivalent ions [1, 2]. Figure 1 shows structures and pKa values of the tetracyclines studied. At pH values below 3 they are positively charged. At pH between pKa_1 and pKa_2 they are neutral (zwitterionic state), and above pH 8 they are negatively charged.

The use of antibiotics in animal husbandry is strictly regulated to protect consumers, as the presence of antimicrobials residues in food products of animal origin can lead to resistance of bacteria to antibiotics. Therefore, the European Union has developed regulation concerning this issue [3, 4]. Analysis and control of antibiotics in feedstuffs for animals has become an important issue as only authorized feedstuffs can be medicated under specific conditions as stated in Council Directive 90/167/EEC [5]. Use of tetracyclines as feed additives is forbidden in the EU since 2006, as stated in annex II of Commission Recommendation 2005/925/EC [6].

Feed contamination can occur depending on a large number of factors such as human error or handling procedures, but production practices have been identified as the main source [7].

Nowadays, liquid chromatography-tandem mass spectrometry (LC-MS/MS) is the technique of choice for the analysis of veterinary residues in food. The analysis of antibiotics in animal feed, though, has proved to be quite a challenge because of the high complexity and variability of the composition of the matrix. Numerous raw materials and additives are added into the feeds, including grains, seeds, beans, rice, and soy, and thus many interfering components, such as oils, fats, proteins, and salts can occur at very high levels. This complexity causes a strong effect of ion suppression. Ion suppression can be defined as a change in the efficiency of droplet formation or evaporation in the ion source of mass spectrometer, caused mainly by interfering matrix compounds. This affects the amount of charged analyte that reaches the detector and so the signal obtained for it. During the last years a growing concern on this issue has been reported [8–17]. Some factors such

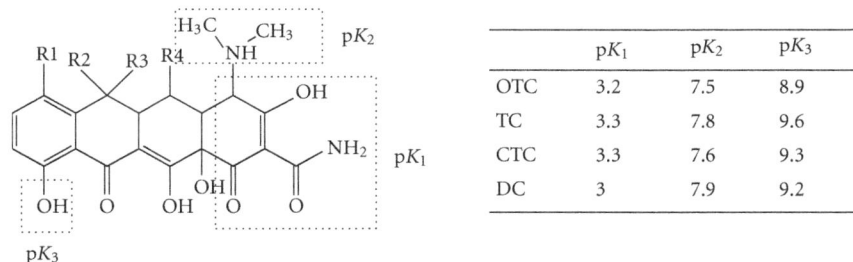

FIGURE 1: Structures and pKa values of tetracyclines.

	pK_1	pK_2	pK_3
OTC	3.2	7.5	8.9
TC	3.3	7.8	9.6
CTC	3.3	7.6	9.3
DC	3	7.9	9.2

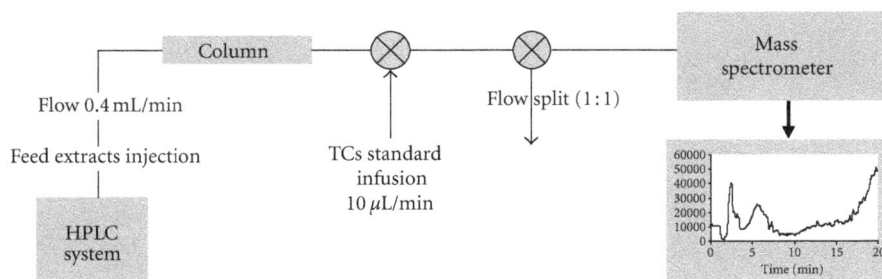

FIGURE 2: Schematic setup for ion suppression recording.

as mobile phase composition [9, 15] or the type of ion source and its geometry [17] have been reported to play a role in ion suppression, but matrix components reaching ion source are the most commonly reported of them. More knowledge on the removal of matrix interferences is needed to overcome ion suppression problems. Many authors have studied this phenomenon by improving sample treatment in residue analysis by HPLC-MS/MS in biological matrices such as whole blood, plasma, serum, or urine [12–14, 16, 18]. In the case of feed samples, it has been proved recently that they are an extreme case, regarding ion suppression, compared to other kind of matrices [19]. Moreover, the changeable composition of each individual feed leads to the obtention of sample extracts with high variation in matrix components, and that leads then to very different extents of ion suppression for each single feed sample. Therefore, not only is it much harder than with other kind of samples (like food) to avoid ion suppression effect, but it is also difficult to obtain at least a homogeneous sample-independent effect. This factor does not allow accurate quantification even when matrix-matched calibration approach is performed. A solution to overcome this effect has been found in the emergence of more isotopically labelled internal standards. The labelled internal standard coelutes with the analyte in question and has similar physicochemical properties. These internal standards, though, still do not ensure correct quantification in all cases [20]. Moreover, their commercial availability is still scarce, and they represent a high cost option. Dilution of the final extract to reduce matrix concentration is also a common option. However, when analyzing samples that may have been contaminated by error or by cross-contamination during production, levels can be very low (in the range of the few parts per billion) and so no great dilution factors are

recommended. Presently, the only way to make completely sure that HPLC-MS/MS quantification is fully reliable is to apply standard addition calibration. This ensures a correct quantification of each individual feed. Unfortunately this quantification tool is very time consuming and cumbersome, resulting in only a few feeds being analysed per day, which is hardly affordable for laboratories which have to handle a high number of samples.

Development of analytical methodologies for analysis of veterinary drugs in feed by HPLC-MS/MS has started to increase in number for the last recent years. Although almost all authors are aware about ion suppression/enhancement phenomena in feed analysis, only a few of them have developed their methods including standard addition calibration [21, 22]. Some others decided to perform this calibration technique by building calibration curve with spiked aliquots of the processed sample prior to HPLC-MS/MS analysis [23, 24], assuming that extraction recovery is a factor of much less impact in the final results than matrix effects. Others assume that their extraction and clean-up techniques are good enough to compensate this effect [25–27] or do not even mention it [28–30].

The aim of this work was to investigate some clean-up methodologies and to evaluate their effectiveness to reduce ion suppression in the analysis of antibiotic residues in feed by LC-MS. For this evaluation, some strategies mentioned in the literature [9, 11–13, 18] were used. Tetracyclines were chosen as a model group to perform the experiments. No papers have been found reporting ion suppression concern in LC-MS/MS tetracycline analysis, and only one in LC-MS analysis (single quadrupole and ion trap) in soils [31]. Several sample treatment procedures were tested and compared using different kinds of feed samples.

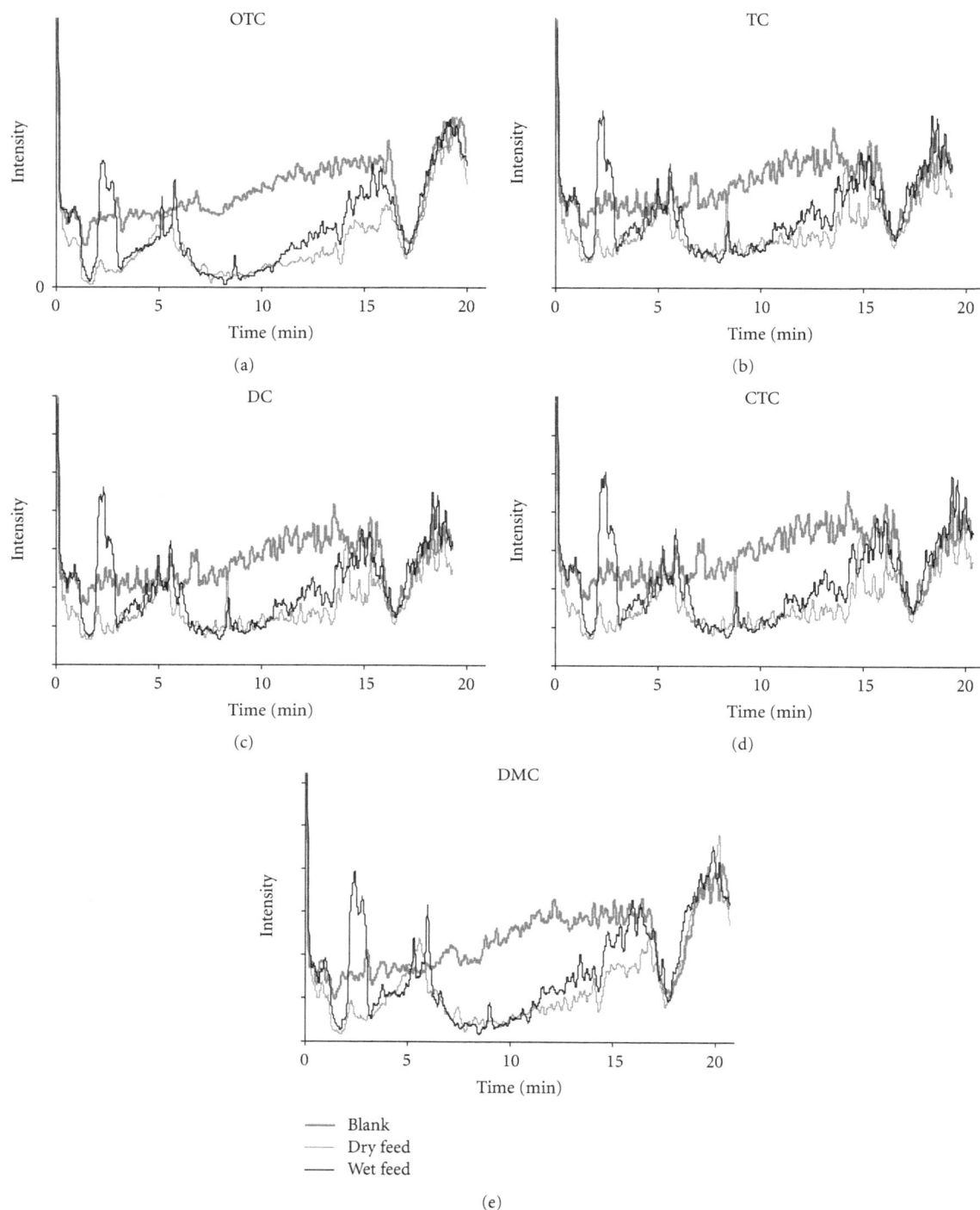

FIGURE 3: Ion suppression profiles of a blank, a dry, and a wet feed sample obtained with sample treatment method 1.

2. Experimental

2.1. Chemicals and Reagents. Oxytetracycline (OTC), tetracycline (TC), chlortetracycline (CTC), doxycycline (DC) and demeclocycline (DMC) were purchased from Sigma (St. Louis, MO, USA). Methanol (MeOH) and acetonitrile (ACN) were obtained from Biosolve (Valkenswaard, The

Netherlands) and acetic acid (AA), ammonia solution 25% v/v (NH₃), formic acid (FA), and dichloromethane (DCM) from Merck (Darmstadt, Germany). Solid reagents were purchased from Merck and included citric acid, potassium dihydrogen phosphate, ethylenediaminetetraacetic acid disodium salt (EDTA), and sodium hydroxide (NaOH). All reagents were analytical-reagent grade.

TABLE 1: Sample pretreatment methods tested for the cleanup of tetracyclines from animal feeds.

	Method			
	1	2	3	4
Extraction	40 mL McIlvaine buffer-EDTA 0,1 M	40 mL DCM McIlvaine buffer-EDTA 0,1 M (1 : 3)	40 mL McIlvaine buffer-EDTA 0,1 M	40 mL McIlvaine buffer-EDTA 0,1 M
SPE Loading pH	4.2	4.2	2.5	10
Cartridge	Oasis HLB (60 mg)	Oasis HLB (60 mg)	Oasis MCX (60 mg)	Oasis MAX (60 mg)
Wash (2 mL)	H_2O	H_2O	(1) FA 2% (v/v) (2) MeOH	(1) NH_3 (2) MeOH
Elution (2 mL)	MeOH	MeOH	MeOH : NH_3 (95 : 5)	MeOH : FA (95 : 5)
Final step	Evaporation with N_2 and reconstitution with mobile phase	Evaporation with N_2 and reconstitution with mobile phase	Dilution with 8 mL acetic acid 10% (v/v)	Dilution with 8 mL H_2O (v/v)

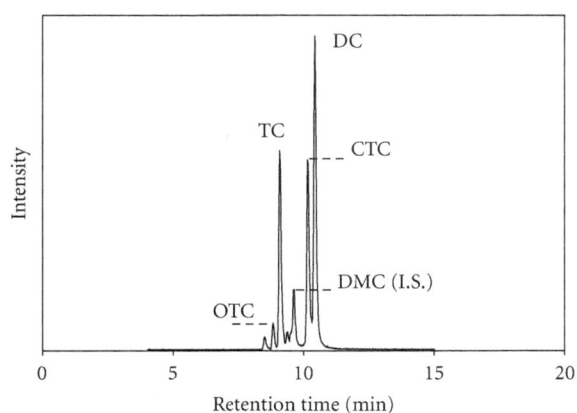

FIGURE 4: Chromatogram froms a standard injection of TC, CTC, OTC, DC, and DMC (1 mg L^{-1}).

TABLE 2: LC-MS/MS precursor/product ion combinations (quantifier bold) monitored in MRM ESI positive mode.

Tetracycline	Retention Time (min)	Precursor ion (m/z)	Product ions (m/z)
TC	9.2	445.2	**410.1**, 154.1
CTC	10.2	479.1	444.1, **154.1**
DC	10.4	445.2	**428.1**, 154.1
OTC	8.9	461.2	**337.1**, 201.1
DMC	9.7	465.1	154.1

The pH of the extract was adjusted when necessary, and filtration through glass fiber filters was performed before SPE step. Final extracts obtained were filtered through 0.45 μm nylon syringe filters before injection into the LC-MS/MS system.

Standard solutions (1000 mg L^{-1}) were prepared in methanol monthly and stored at 4°C. Mixtures of OTC, TC, CTC, and DC (10 and 100 mg L^{-1}) were prepared by dilution of the concentrated solutions and stored at 4°C for a week. An internal standard (IS) stock solution(DMC, 100 mg L^{-1}) in MeOH was prepared monthly. Working standard solutions were prepared daily by mobile phase dilution of the 10 and 100 mg L^{-1} mixtures.

SPE materials were obtained from Waters (Micromass/Waters, Manchester, UK).

2.2. Sample Treatment Procedures. Five porcine feed samples were used for the study: one premix sample, two dry feed samples, and two slurry feed samples. Feeds were chosen that were representative for the range of different feeds available. Four different sample treatment procedures, which are summarized in Table 1, were tested. Sample weight was 2 grams, and extraction solution volume was 40 mL. Extraction was carried out by means of a head-over-head shaker for 20 minutes in all cases, and samples were subsequently centrifuged at 3000 rpm for 15 minutes.

2.3. LC-MS/MS Conditions. A Waters 2690 separations module HPLC system (Waters Corporation, USA) coupled to a Quattro Ultima tandem mass detector (Micromass/Waters, Manchester, UK), both operating under MassLynx software, was used for sample analysis. The mass spectrometer was operated in electrospray positive mode, and data acquisition was in multiple reaction monitoring mode (MRM). The precursor/product ions monitored are listed in Table 2. The source settings were as follows: capillary voltage 2.7 kV, source temperature 120°C, desolvation temperature 300°C, cone nitrogen gas flow 180 Lh^{-1}, and desolvation gas flow 580 Lh^{-1}. Argon (3.2 × 10–3 mbar) was used as the collision gas, and the multiplier was operated at 750 V. The cone voltage was set at 20 V, and collision energy changed during analysis depending on the analyte (25 eV for TC and DC, 26 eV for OTC, and 30 eV for CTC and DMC). The HPLC system was equipped with a Symmetry C$_{18}$ (5 μm, 3.0 × 150 mm column, Waters) at 10°C. A binary gradient mobile phase was used at a flow rate of 0.4 mL min^{-1} with solvent A (ammonium acetate 1 mM, pH 2.6) and solvent B (ammonium acetate 10 mM : ACN, 10 : 90). The gradient started isocratic for 1 min at 0% B, followed by a linear

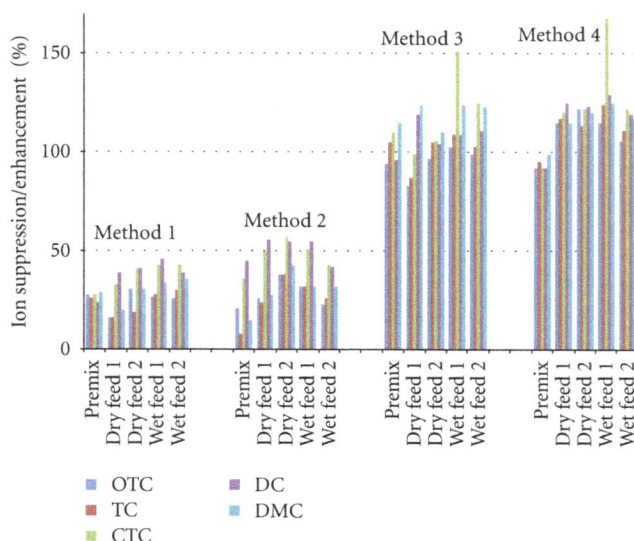

FIGURE 5: Quantitative results of ion suppression/enhancement for all analytes in all matrices and extraction methods tested.

increase to 50% B in 9 min. The gradient remained isocratic at 50% B for 3 min. Subsequently the gradient linearly increased to 100% B in 1 min. The gradient remained at this % B for a further 3 min. Afterwards the gradient returned to 0% B for equilibration of the column. Sample injection volume was $10\,\mu$L.

2.4. Qualitative Assessment of Ion Suppression. The experiments for qualitative assessment of ion suppression were carried out using a postcolumn infusion setup coupled to the chromatographic system described in Section 2.3 through a T piece. The setup is shown in Figure 2. The infusion pump flushes a constant flow at $10\,\mu$L min^{-1} of a $5\,$mg L^{-1} standard solution of all tetracyclines in mobile phase. The quantification transition for each tetracycline is monitored in the MS/MS system. When mobile phase is injected into the system, a reference baseline for each transition is obtained due to the constant infusion of the standard solution of analytes. When feed extracts free from tetracyclines are injected, ion suppression profiles for each transition are obtained, and the influence of ion suppression in the tetracyclines infusion baseline due to the eluted matrix components can be evaluated. In fact, these profiles show the effect of compounds eluting from the chromatographic system on the analytes MS/MS signals. The signal intensity of the baseline decreases when matrix components causing ion suppression elute, and it increases when substances enhancing ionization elute.

Observing the signal variation at the time window where every analyte elutes, a good qualitative prediction can be made, whether suppression or enhancement are expected for that analyte.

2.5. Quantitative Assessment of Ion Suppression. $1\,$mg L^{-1} tetracyclines standard solutions and matrix-matched recovery standards (MMRSs) at the same concentration were injected into the LC-MS/MS system. MMRSs are extracts from blank feed samples that have been spiked with the analytes at the end of the sample treatment process. Ion suppression or enhancement percentages were determined for each tetracycline as the peak area ratio of the MMRS to the standard in solution multiplied by 100. Values lower than 100% were an indicative of ion suppression whereas values higher than 100% indicated ion enhancement.

For each analyte, the response factors (Area$_{analyte}$/Area$_{IS}$) in the 5 studied feed samples were determined in MMRS for the set of samples. Each MMRS was analysed by triplicate, and the average was calculated. The RSD (%) between the averages of the five tested samples ($n = 5$) was used to quantify the variation in ion suppression due to differences between feed samples for each sample treatment method.

3. Results and Discussion

Four different sample treatments, summarized in Table 1, were tested in this study. All of them are based in an extraction step using McIlvaine buffer-EDTA $0.1\,$M (pH 4.2) and a further clean-up step of solid phase extraction. McIlvaine buffer has been extensively reported to be efficient for tetracycline extraction in a large number of matrices, as stated in some reviews [1, 2].

Method 1 is currently in use at this laboratory for routine analysis. In this method, the cleanup of the extracts is performed with the reversed phase Oasis HLB cartridges. The loading of the extract into the cartridge does not require any pH adjustment since the maximum interaction of TCs with the sorbent occurs when the neutral form of the analytes is prevalent, like at pH 4.2 (Figure 1). Finally the elution of TCs is achieved with methanol.

The effect of addition of dichloromethane (DCM) in the extraction step was investigated in method 2, as DCM might assist to the removal of some nonpolar matrix compounds

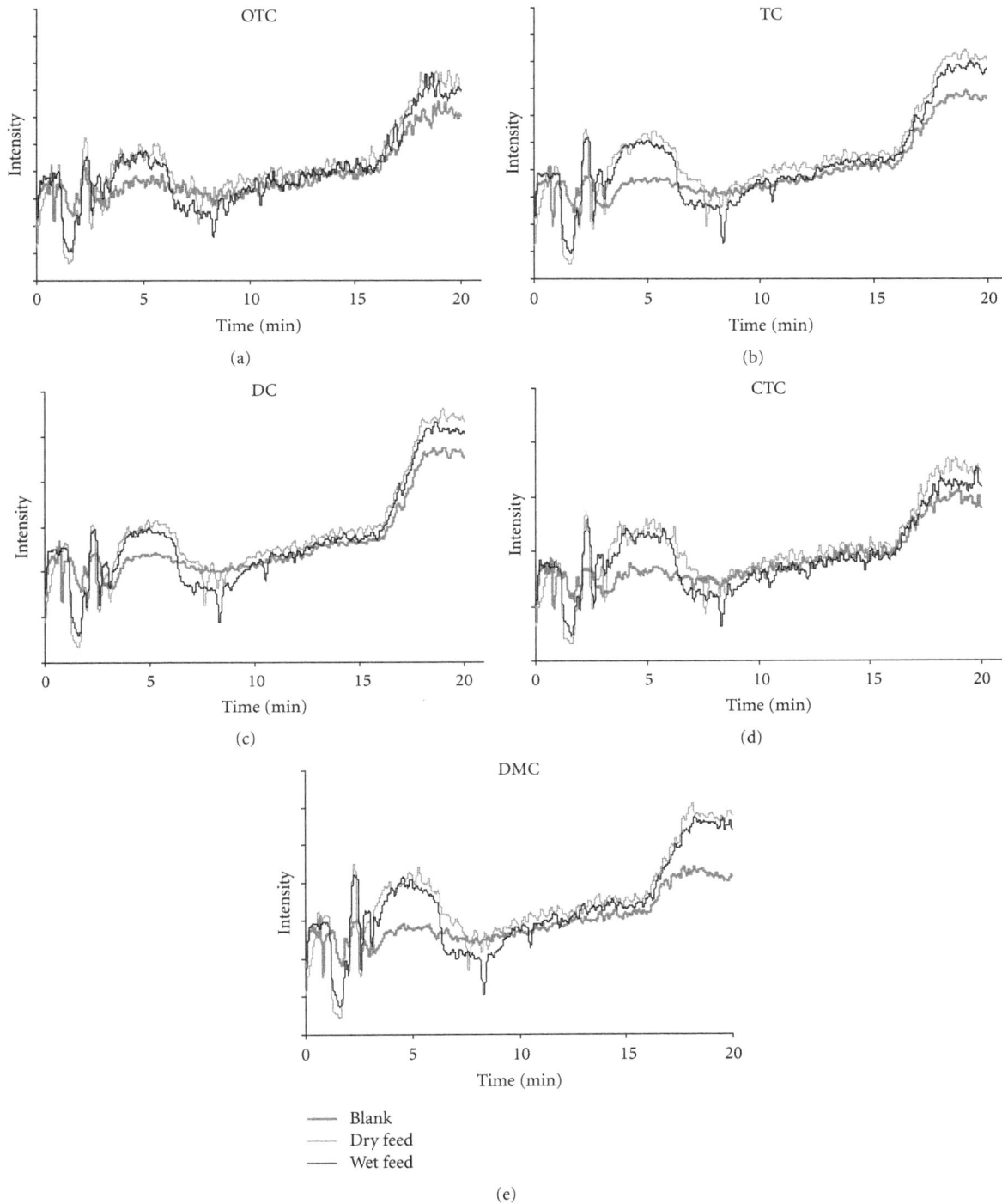

FIGURE 6: Ion suppression profiles of a blank, a dry, and a wet feed sample obtained with sample treatment method 3.

and thus provide cleaner extracts. After extraction, the aqueous layer was processed throughout an Oasis HLB cartridge as in method 1.

The pH of the McIlvaine buffer extract was modified after extraction in methods 3 and 4 to reach a suitable pH for SPE. Mixed-mode cation exchange (Oasis MCX, method 3) or mixed-mode anion exchange (Oasis MAX, method 4) cartridges were used. These cartridges base their performance in a combination of ion exchange and reverse phase mechanisms. Therefore, they are expected to be more selective for targeted analytes and so to provide a more efficient cleanup [18, 32–34].

For method 3, before loading into the cartridge, pH of the extract was decreased to 2.5 in order to have the analytes positively charged. Theoretically SPE cartridge performance may be compromised, as the pH is only slightly lower than pKa_1 (Table 1). However decreasing pH more is not recommended, as it has been reported to induce epimerization of tetracyclines [1, 2]. At these conditions good SPE recoveries for all analytes were obtained for the complete optimized method (74–100%).

For method 4, pH of the extract was brought to 10 to ensure that all analytes were in anionic form. At these conditions, SPE recoveries were good for the analytes (97–100%) except for CTC (ca. 30%). This is possibly due to partial degradation of this analyte to iso-CTC, as this analyte is particularly prone to form this derivative at high pH values [2].

Ion suppression profiles of feed extracts obtained for the four sample pretreatment methods were recorded. These profiles were studied separately for each analyte in every feed extract and compared. Focusing in the behaviour of the profile at the retention time at which every tetracycline elutes, and comparing it with the reference "blank" signal (which corresponds to the injection of mobile phase), a good qualitative assessment can be made, whether suppression or enhancement are expected in a significant extent for each tetracycline in each studied feed extract.

Ion suppression profiles of a blank, a dry and a slurry (wet) feed sample extract obtained with sample treatment method 1 are shown in Figure 3. Both feeds exhibit ion suppression in the chromatogram time window where tetracyclines elute (8–12 min, Figure 4).

A similar trend was observed when using method 2. The addition of DCM to the extraction solution was therefore not an improvement. Numeric results from the ion suppression quantification experiment agreed with these results, and the suppression factors (see Section 2.5) for all the analytes in the set of feed samples were clearly below 100% for both methods (Figure 5).

Some improvements were clearly observed in the ion suppression profiles of the extracts obtained according to methods 3 and 4. No ion suppression or enhancement was observed in the profiles between 8 and 12 min As a matter of example Figure 6 shows the ion suppression profiles of a blank, a dry and a slurry (wet) feed sample extract obtained with method 3. Similar profiles were obtained when using method 4. The results of the ion suppression quantification experiment are consistent with these qualitative results. The percentage of signal obtained in $1\,mg L^{-1}$ MMRS in the five feeds studied compared to a standard solution was close to 100% (Figure 5). Only in the case of CTC in one of the two slurry feeds, significant enhancement (>150%) was found, what seems to show that this sample contains some particular substances that enhance the ionization of this particular analyte under the mentioned sample treatment conditions. The overall clear improvement provided by methods 3 and 4 has been achieved by the combination of the use of ion exchange cartridges (more selective) and the dilution of the extracts instead of evaporation and reconstitution (less introduction of contaminants and human error).

TABLE 3: Variation (RSD %) between the five different studied feed samples processed with the four different sample pretreatment methods at $1\,mg L^{-1}$.

	Method			
	1	2	3	4
	RSD (%)	RSD (%)	RSD (%)	RSD (%)
TC	12.3	19.4	11.1	2.6
CTC	17.9	25.8	12.3	21.1
DC	30.8	39.7	4.9	6.5
OTC	14.7	24.3	9.4	8.6

As expected, a high variation between the different feed samples was found for all tetracyclines (Table 3). Methods 3 and 4 provide the best results regarding the variability on the response factors (Area$_{analyte}$/Area$_{IS}$) in MMS due to the feed sample. CTC is an exception (method 4) due to degradation at pH 10. That indicates that the extracts obtained for the five feeds with these two methods are more uniform than the ones obtained by methods 1 and 2.

4. Conclusions

Ion suppression in LC-MS/MS analysis of tetracyclines in feed was studied. The study of this phenomenon in feed samples has proven to be of a high level of complexity. Four sample pretreatment methods were tested with five different feed samples in terms of ion suppression profiles, ion suppression quantification, and variation. The method that seemed to provide less ion suppression and more uniform extracts without significant degradation of any analyte was method 3, which involved SPE with Oasis MCX cartridges. Although these results are still not sufficient to replace the current protocol (method 1) which relies heavily on standard addition, they provided valuable information for future research, which should include studies with larger numbers of different feed samples, different concentration levels, and different levels of extract dilution.

References

[1] H. Oka, Y. Ito, and H. Matsumoto, "Chromatographic analysis of tetracycline antibiotics in foods," *Journal of Chromatography A*, vol. 882, no. 1-2, pp. 109–133, 2000.

[2] C. R. Anderson, H. S. Rupp, and W. H. Wu, "Complexities in tetracycline analysis—chemistry, matrix extraction, cleanup, and liquid chromatography," *Journal of Chromatography A*, vol. 1075, no. 1-2, pp. 23–32, 2005.

[3] Commission Decision. Regulation (EC) No 882/2004 of the European Parliament and of the Council of 29 April 2004 on official controls performed to ensure the verification of compliance with feed and food law, animal health and animal welfare rules, 2004.

[4] Regulation. Regulation (EC) No 1831/2003 of the European Parliament and of the Council of 22 September 2003 on additives for use in animal nutrition, 2003.

[5] Council Directive 90/167/EEC. Council Directive 90/167/EEC of 26 March 1990 laying down the conditions governing the

preparation, placing on the market and use of medicated feedingstuffs in the Community, 1990.

[6] Commision Recommendation 2005/925/EC of 14 December 2005 on the coordinated inspection programme in the field of animal nutrition for the year 2006 in accordance with Council Directive 95/53/EC.

[7] J. D. G. McEvoy, "Contamination of animal feedingstuffs as a cause of residues in food: a review of regulatory aspects, incidence and control," *Analytica Chimica Acta*, vol. 473, no. 1-2, pp. 3–26, 2002.

[8] L. L. Jessome and D. A. Volmer, "Ion suppression: a major concern in mass spectrometry," *LC-GC North America*, vol. 24, no. 5, pp. 498–510, 2006.

[9] T. M. Annesley, "Ion suppression in mass spectrometry," *Clinical Chemistry*, vol. 49, no. 7, pp. 1041–1044, 2003.

[10] J. P. Antignac, K. De Wasch, F. Monteau, H. De Brabander, F. Andre, and B. Le Bizec, "The ion suppression phenomenon in liquid chromatography-mass spectrometry and its consequences in the field of residue analysis," *Analytica Chimica Acta*, vol. 529, no. 1-2, pp. 129–136, 2005.

[11] Y. Hsieh, M. Chintala, H. Mei et al., "Quantitative screening and matrix effect studies of drug discovery compounds in monkey plasma using fastgradient liquid chromatography/tandem mass spectrometry," *Rapid Communications in Mass Spectrometry*, vol. 15, no. 24, pp. 2481–2487, 2001.

[12] B. K. Matuszewski, M. L. Constanzer, and C. M. Chavez-Eng, "Strategies for the assessment of matrix effect in quantitative bioanalytical methods based on HPLC-MS/MS," *Analytical Chemistry*, vol. 75, no. 13, pp. 3019–3030, 2003.

[13] C. Müller, P. Schäfer, M. Störtzel, S. Vogt, and W. Weinmann, "Ion suppression effects in liquid chromatography-electrospray-ionisation transport-region collision induced dissociation mass spectrometry with different serum extraction methods for systematic toxicological analysis with mass spectra libraries," *Journal of Chromatography B*, vol. 773, no. 1, pp. 47–52, 2002.

[14] R. Bonfiglio, R. C. King, T. V. Olah, and K. Merkle, "The effects of sample preparation methods on the variability of the electrospray ionization response for model drug compounds," *Rapid Communications in Mass Spectrometry*, vol. 13, no. 12, pp. 1175–1185, 1999.

[15] C. R. Mallet, Z. Lu, and J. R. Mazzeo, "A study of ion suppression effects in electrospray ionization from mobile phase additives and solid-phase extracts," *Rapid Communications in Mass Spectrometry*, vol. 18, no. 1, pp. 49–58, 2004.

[16] J. X. Shen, R. J. Motyka, J. P. Roach, and R. N. Hayes, "Minimization of ion suppression in LC-MS/MS analysis through the application of strong cation exchange solid-phase extraction (SCX-SPE)," *Journal of Pharmaceutical and Biomedical Analysis*, vol. 37, no. 2, pp. 359–367, 2005.

[17] M. Holčapek, K. Volná, P. Jandera et al., "Effects of ion-pairing reagents on the electrospray signal suppression of sulphonated dyes and intermediates," *Journal of Mass Spectrometry*, vol. 39, no. 1, pp. 43–50, 2004.

[18] E. Chambers, D. M. Wagrowski-Diehl, Z. Lu, and J. R. Mazzeo, "Systematic and comprehensive strategy for reducing matrix effects in LC/MS/MS analyses," *Journal of Chromatography B*, vol. 852, no. 1-2, pp. 22–34, 2007.

[19] H. G. J. Mol, P. Plaza-Bolaños, P. Zomer, T. C. De Rijk, A. A. M. Stolker, and P. P. J. Mulder, "Toward a generic extraction method for simultaneous determination of pesticides, mycotoxins, plant toxins, and veterinary drugs in feed and food matrixes," *Analytical Chemistry*, vol. 80, no. 24, pp. 9450–9459, 2008.

[20] N. Lindegardh, A. Annerberg, N. J. White, and N. P. J. Day, "Development and validation of a liquid chromatographic-tandem mass spectrometric method for determination of piperaquine in plasma. Stable isotope labeled internal standard does not always compensate for matrix effects," *Journal of Chromatography B*, vol. 862, no. 1-2, pp. 227–236, 2008.

[21] L. Kantiani, M. Farré, J. M. Grases I Freixiedas, and D. Barceló, "Determination of antibacterials in animal feed by pressurized liquid extraction followed by online purification and liquid chromatography- electrospray tandem mass spectrometry," *Analytical and Bioanalytical Chemistry*, vol. 398, no. 3, pp. 1195–1205, 2010.

[22] F. Van Holthoon, P. P. J. Mulder, E. O. Van Bennekom, H. Heskamp, T. Zuidema, and H. J. A. Van Rhijn, "Quantitative analysis of penicillins in porcine tissues, milk and animal feed using derivatisation with piperidine and stable isotope dilution liquid chromatography tandem mass spectrometry," *Analytical and Bioanalytical Chemistry*, vol. 396, no. 8, pp. 3027–3040, 2010.

[23] M. J. G. de la Huebra, U. Vincent, and C. von Holst, "Determination of semduramicin in poultry feed at authorized level by liquid chromatography single quadrupole mass spectrometry," *Journal of Pharmaceutical and Biomedical Analysis*, vol. 53, no. 4, pp. 860–868, 2010.

[24] U. Vincent, Z. Ezerskis, M. Chedin, and C. von Holst, "Determination of ionophore coccidiostats in feeding stuffs by liquid chromatography-tandem mass spectrometry. Part II. Application to cross-contamination levels and non-targeted feed," *Journal of Pharmaceutical and Biomedical Analysis*, vol. 54, no. 3, pp. 526–534, 2011.

[25] M. Cronly, P. Behan, B. Foley, E. Malone, P. Shearan, and L. Regan, "Determination of eleven coccidiostats in animal feed by liquid chromatography-tandem mass spectrometry at cross contamination levels," *Analytica Chimica Acta*, vol. 700, no. 1-2, pp. 26–33, 2011.

[26] R. Liu, W. Hei, P. He, and Z. Li, "Simultaneous determination of fifteen illegal dyes in animal feeds and poultry products by ultra-high performance liquid chromatography tandem mass spectrometry," *Journal of Chromatography B*, vol. 879, no. 24, pp. 2416–2422, 2011.

[27] W. Li, T. J. Herrman, and S. Y. Dai, "Determination of aflatoxins in animal feeds by liquid chromatography/tandem mass spectrometry with isotope dilution," *Rapid Communications in Mass Spectrometry*, vol. 25, no. 9, pp. 1222–1230, 2011.

[28] P. Delahaut, G. Pierret, N. Ralet, M. Dubois, and N. Gillard, "Multi-residue method for detecting coccidiostats at carry-over level in feed by HPLC-MS/MS," *Food Additives and Contaminants—Part A*, vol. 27, no. 6, pp. 801–809, 2010.

[29] C. Van Poucke, F. Dumoulin, and C. Van Peteghem, "Detection of banned antibacterial growth promoters in animal feed by liquid chromatography-tandem mass spectrometry: optimisation of the extraction solvent by experimental design," *Analytica Chimica Acta*, vol. 529, no. 1-2, pp. 211–220, 2005.

[30] C. Van Poucke, K. De Keyser, A. Baltusnikiene, J. D. G. McEvoy, and C. Van Peteghem, "Liquid chromatographic-tandem mass spectrometric detection of banned antibacterial growth promoters in animal feed," *Analytica Chimica Acta*, vol. 483, no. 1-2, pp. 99–109, 2003.

[31] S. O'Connor, J. Locke, and D. S. Aga, "Addressing the challenges of tetracycline analysis in soil: extraction, clean-up, and matrix effects in LC-MS," *Journal of Environmental Monitoring*, vol. 9, no. 11, pp. 1254–1262, 2007.

[32] M. Lavén, T. Alsberg, Y. Yu, M. Adolfsson-Erici, and H. Sun, "Serial mixed-mode cation- and anion-exchange solid-phase

extraction for separation of basic, neutral and acidic pharmaceuticals in wastewater and analysis by high-performance liquid chromatography-quadrupole time-of-flight mass spectrometry," *Journal of Chromatography A*, vol. 1216, no. 1, pp. 49–62, 2009.

[33] A. Tölgyesi, L. Tölgyesi, V. K. Sharma, M. Sohn, and J. Fekete, "Quantitative determination of corticosteroids in bovine milk using mixed-mode polymeric strong cation exchange solid-phase extraction and liquid chromatography-tandem mass spectrometry," *Journal of Pharmaceutical and Biomedical Analysis*, vol. 53, no. 4, pp. 919–928, 2010.

[34] D. R. Baker and B. Kasprzyk-Hordern, "Multi-residue analysis of drugs of abuse in wastewater and surface water by solid-phase extraction and liquid chromatography-positive electrospray ionisation tandem mass spectrometry," *Journal of Chromatography A*, vol. 1218, no. 12, pp. 1620–1631, 2011.

Development and Validation of Selective High-Performance Liquid Chromatographic Method Using Photodiode Array Detection for Estimation of Aconitine in Polyherbal Ayurvedic Taila Preparations

Nitin Dubey,[1] Nidhi Dubey,[2] and Rajendra Mehta[3]

[1] College of Pharmacy, IPS Academy, Indore, India
[2] School of Pharmacy, Devi Ahilya Vishwavidyalaya, Indore, India
[3] A. R .College of Pharmacy, Vallabh Vidyanagar, India

Correspondence should be addressed to Nidhi Dubey, nidhidubeympharm@yahoo.com

Academic Editor: Sibel A. Ozkan

A simple, sensitive, and selective high-performance liquid chromatographic (HPLC) method has been developed and validated for the analysis of aconitine in marketed ayurvedic *taila* (oil) formulations containing roots of *Aconitum chasmanthum*. Chromatography of methanolic extracts of these formulations was performed on C_{18} (5 µm × 25 cm × 4.6 mm i.d.) column using isocratic mobile phase consisting of (65 : 35% v/v) acetonitrile and buffer solution (aqueous 0.01 M ammonium bicarbonate buffer, adjusted to pH 9.6 using 30% ammonia solution) at a flow rate of 1 mL/min and SPD-10 A_{VP} photodiode array (PDA) UV-Visible detector. The analytical reference, aconitine, was quantified at 238 nm. The retention time of aconitine was about 42.54 min. The linear regression analysis data for the calibration plot showed a good linear relationship with correlation coefficient of 0.9989 in the concentration range of 15 to 90 µg/mL for aconitine with respect to peak area. The limit of detection and limit of quantitation values were found to be 0.03 µg/mL and 0.1 µg/mL respectively. Repeatability of the method was found to be 0.551–1.689 RSD. Recovery values from 97.75 to 99.91% indicate excellent accuracy of the method. The developed HPLC method is accurate and precise and it can be successfully applied for the determination of aconitine in marketed ayurvedic oil formulations containing *Aconitum chasmanthum*.

1. Introduction

Aconitum chasmanthum (Family, Ranunculaceae) is a most valuable medicinal plant, widely used in the traditional and folk medicines of a number of countries of south east Asia. The chief chemical constituents of *Aconitum chasmanthum* root are aconitine, mesaconitine, and hypaconitine, and their respective hydrolyzed analogs are called monoester alkaloids, that is, benzoylaconine, benzoylmesaconine, and benzoylhypaconine [1–3]. Aconitine being the major diterpenoid responsible for the biopotency of *Aconitum chasmanthum* is recognized as the reference compound [3, 4]. Many polyherbal oil formulations in Indian and Chinese traditional systems of medicine used for control of skin diseases contain aconitum root as major active ingredient [5, 6]. Polyherbal oil formulations are made with the main objective of incorporating the fat-soluble fraction of the component of herbal drugs to a suitable oil base. Standardization of these formulations in terms of composition is important to ensure quality and safety. Aconitine can be used as analytical reference in the quality control of polyherbal oil formulations containing *Aconitum chasmanthum* [7–10].

There are reports on the application of various analytical methods for isolation and quantitation of aconitine present in *Aconitum chasmanthum*, biological fluids and other botanical sources [10–14]. Its estimation in polyherbal traditional medicines especially oil formulation is challenging. But no reported method deals with estimation of aconitine in complex matrix of ayurvedic *taila* formulations. High-performance liquid chromatography has emerged as an efficient tool for the phytochemical evaluation of herbal drugs because of its simplicity, sensitivity, accuracy, suitability for high-throughput screening, and so forth [15–21]. Hence it was thought worthwhile to develop a simple chromatographic method for determination of aconitine in ayurvedic *taila* formulations. The method was validated and found to be sensitive and reproducible [22, 23].

2. Material and Methods

2.1. Reference Compound and Reagents. Reference aconitine (95% w/v purity) was purchased from Sigma-Aldrich (Germany). All the chemicals used in analysis were of AR grade except those used for HPLC analyses which were of HPLC grade. All A.R. grade chemicals were procured from S.D. fine chemicals, Mumbai and all HPLC grade solvents were procured from Merck, Mumbai.

2.2. Polyherbal Oil Formulations. Commercial marketed ayurvedic polyherbal oil formulations "*Varnaraksasa taila*" and "*Vipritmalla taila*" which contain *Aconitum chasmanthum* as one of the components were selected for studies. *Vranaraksasa taila* is a polyherbomineral oil containing mercury, sulphur, cinnabar (HgS), realgar (As_4S_4), fine copper powder, orpiment (As_2S_3), *Allium sativum* bulb (family, Alliaceae), and *Aconitum chasmanthum* root (family, Ranunculaceae) digested in mustard seed oil (*Brassica nigra*. Family, Brassicaceae) as per classical Ayurvedic text [5, 6]. *Vipritmalla taila* contains cinnabar (HgO), *Saussurea lappa* root (family, Asteraceae), *Aconitum chasmanthum* root (family, Ranunculaceae), *Ferula asafetida* resin (family, Apiaceae), *Allium sativum* bulb (family, Alliaceae), *Plumbago zylenicum* root (family, Plumbaginaceae), *Valeriana wallichii* root (family, Valerianaceae), and *Gloriosa superba* root (family, Liliaceae) digested in mustard seed oil as per classical Ayurvedic text [5, 6].

Samples of the same formulations in triplicate, manufactured by three different reputed ayurvedic drug manufacturers were collected from retail pharmacies in Indore, Madhya Pradesh, India.

2.3. Preparation of Standard Solutions. The stock solution of 1 mg/mL in methanol was prepared after keeping the purity of reference aconitine into consideration. Solution was filtered through Whatman filter paper (no. 1). Aliquots of stock solution were diluted to 5 mL using methanol to obtain working standards in concentration range from 15 to 90 μg/mL.

2.4. Chromatographic Conditions. The mobile phase consisted of acetonitrile: aqueous 0.01 M ammonium bicarbonate buffer, adjusted to pH 9.6 using ammonia solution (65 : 35% v/v) at a flow rate of 1 mL/min. Before use, the mobile phase was degassed by an ultrasonic bath and filtered using 0.4 μm membrane filter. Separation was performed at room temperature on HPLC system having a pump (Shimadzu LC 10AT$_{VP}$) with 20 μL Rheodyne injector, Phenomenex Luna C$_{18}$ (5 μm × 25 cm × 4.6 mm i.d) column,

and SPD-10 A$_{VP}$ photodiode array (PDA) UV-Visible detector set at 238 nm and equipped with CLASS-VP software (Shimadzu. Kyoto, Japan).

2.5. HPLC Analysis of Ayurvedic Taila Preparations. Oil formulations (10 gm) were homogenated using homogenizer (Scientific instruments ltd, Indore) using methanol (50 mL) in proportion of 1 : 5, w/v at 50°C for 20 min. The mixture was centrifuged at 2000 rpm for 20 min at 4°C and the supernatant was collected. The residue was resuspended in methanol and the extraction was repeated five more times similarly. The supernatants were pooled and concentrated under vacuum at room temperature and made up to a volume of 20 mL using methanol. The extracts were filtered through 0.45 μm filter and HPLC was performed under the conditions optimized for the reference compound. The amount of aconitine was quantified using calibration curves plotted with the reference compound.

2.6. Validation of Method

(a) Calibration Graph (Linearity of the HPLC Method). The calibration curve was obtained at 6 concentration levels of aconitine standard solutions (15–90 μg/mL). The solutions (20 μL) were injected into the HPLC system ($n = 6$) with the chromatographic conditions previously given. The linearity was evaluated by the least-squares regression method.

(b) Limits of Detection and Quantification. For determination of the limit of detection (LOD) and the limit of quantification (LOQ) different dilutions of the standard solution of aconitine were analyzed using mobile phase as the blank. The LOD and LOQ were determined on the basis of signal-to-noise ratio until the average responses were approximately 3 and 10 times the responses of the blank, respectively.

(c) Accuracy (Recovery). The accuracy of the methods was determined by calculating recovery of aconitine by the standard addition method. Known amounts of standard solution of aconitine (at three levels 50%, 100%, 150%) were added to prequantitated sample solutions. The amount of aconitine was estimated by applying values of peak area to the regression equations of the calibration graph. Five replicate samples of each concentration level were prepared.

(d) Method Precision (Repeatability). The precision of the instruments was checked by repeatedly injecting and analyzing ($n = 6$) standard solutions of aconitine (45 μg/mL). The results are reported in terms of relative standard deviation (RSD).

(e) Intermediate Precision (Reproducibility). The intraday and interday precision of the proposed method were determined by analyzing standard solution of aconitine at 3 different concentrations (15, 45, and 90 μg/mL) three times on the same day and on three different days. The results are reported in terms of RSD.

Development and Validation of Selective High-Performance Liquid Chromatographic Method Using Photodiode Array Detection for Estimation of Aconitine in Polyherbal Ayurvedic Taila Preparations

19

(f) Solution Stability and Mobile Phase Stability. Solution stability in the assay method was evaluated by leaving test solutions of sample and reference standard in tightly capped volumetric flasks at room temperature in the dark for 24 h. The same sample solutions were assayed every 6 h interval in the study period. Mobile phase stability was studied by assaying the freshly prepared sample solutions against freshly prepared reference standard solutions at 6 h intervals up to 24 h. Mobile phase was prepared and kept constant during the study period. The relative standard deviation (RSD) of the assay of aconitine was calculated for the study period during mobile phase and solution stability experiments.

2.7. Statistical Analysis. The statistical analysis was performed using Microsoft Excel 2003.

3. Result and Discussion

The literature revealed that methanol is preferred for extraction of aconitine from *Aconitum chasmanthum* [10–14]. The same was used for extraction of aconitine from oil formulations. It is advantageous as the base oil in all the selected ayurvedic oil formulations was immiscible in this solvent. The immiscibility of oil in solvent will help in reducing number of interfering components in further chromatographic development. Multiple extractions were carried out to ensure complete extraction.

3.1. Development of the HPLC Method. The method development and selection of a suitable mobile phase involved several trials because of the complexity of the chemical composition of the herbals and the affinities of the components towards various solvents. The proportions of the organic and aqueous phases were adjusted to obtain a simple assay method with a reasonable run time and suitable retention time. Further optimization of mobile phase was performed based on resolution, asymmetric factor, and theoretical plates obtained for aconitine. Different mobile phases were tried like methanol: acetonitrile: water (45 : 45 : 10) which gave broad peak for aconitine. Combination of acetonitrile: methanol (10 : 90) was tried which gave broad and tailed peak. A mixture of methanol: water (60 : 40) was tried, which gave unresolved peak at retention time 48.9 min. Under optimized conditions HPLC with C18 column and UV detector at 238 nm using mobile phase (acetonitrile: aqueous 0.01 M ammonium bicarbonate buffer, adjusted to pH 9.6 using 30% ammonia solution (65 : 35%v/v)) gave well-resolved symmetric band for aconitine from its oil formulation (Figure 1). The resolution was found to be 1.7. Retention time was found to be around 42 minutes and aconitine appeared on chromatogram at 42.54 minutes. Retention time of aconitine was found to be of 42.54 minutes.

The method consumes less volume of HPLC solvents. When the same drug solution was injected 6 times, the retention time of the drug was found to be the same (Figure 2).

3.2. Validation of Method. The calibration curve was prepared by plotting the peak area against aconitine concentration; it was found linear in the range of 15–90 μg/mL. The

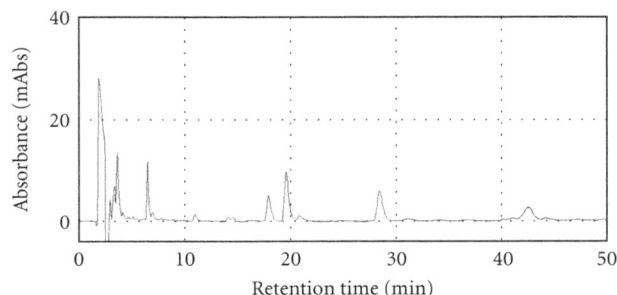

FIGURE 1: HPLC chromatogram of a 20 μL injection of 40 μg/mL reference aconitine at 238 nm.

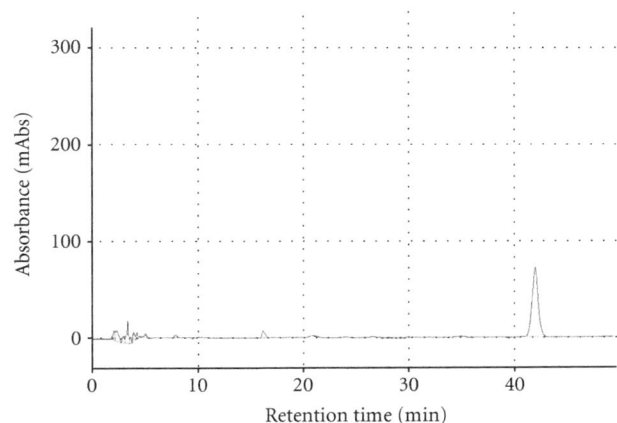

FIGURE 2: HPLC chromatogram of a 20 μL injection of the sample of taila preparation at 238 nm.

TABLE 1: Summary of validation parameters and system suitability parameters.

Parameters	Observations ± % RSD ($n = 06$)
Retention time (min)	42.54 ± 0.33
Tailing factor	1.1 ± 0.1
Resolution	1.7 ± 0.23
Theoretical plates	3213 ± 0.42
Linearity range (μg/mL)	15–90
Correlation coefficient (r^2)	0.9989
Regression equation	$y = 71.5x - 24.5$
Limit of detection (μg/mL)	0.03
Limit of quantification (μg/mL)	0.1
Repeatability (% RSD, $n = 6$)	0.55–1.68

regression equation was found as $y = 71.5x - 24.5$ ($r^2 = 0.9989$), showing excellent linearity (Figure 3). The method was validated in terms of precision, repeatability, accuracy, and other validation parameters (Table 1). The repeatability of the HPLC method and the intermediate precisions for intraday and interday variations are given in Table 2. The LOD value was found to be 0.03 μg/mL, which is the concentration that yields a signal-to-noise (S/N) ratio of 3/1. The LOQ value under the described conditions was 0.1 μg/mL

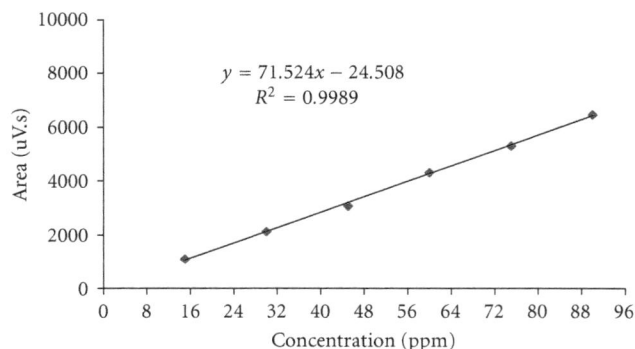

$$y = 71.524x - 24.508$$
$$R^2 = 0.9989$$

FIGURE 3: Calibration curve of reference aconitine in blank oil sample matrix at 238 nm.

TABLE 2: Precision studies data for aconitine.

Concentration (μg/mL)	Intraday		Interday*	
	Peak area ± SD	% RSD*	Peak area ± SD	% RSD*
15	1101.6 ± 16.7	1.52	1112.0 ± 6.3	0.57
45	3059.9 ± 29.5	0.96	3063.0 ± 29.5	0.96
75	5282.5 ± 42.3	0.80	5292.6 ± 47.8	0.90

*Relative standard deviation (% RSD, $n = 3$).

with an S/N ratio of $10:1$. This confirmed the sensitivity for quantitation of aconitine in *taila* preparations. Recovery values from 97.7 to 99.9% indicate excellent accuracy of the method (Table 3). The RSD values of assay of aconitine during solution stability, mobile phase stability and robustness studies (Table 4) were within 2.0%. The data obtained in both experiments proves that the sample solutions and mobile phase used during assay were stable up to 24 h.

3.3. HPLC Analysis of Ayurvedic Taila Preparations. Quantitative estimation of aconitine in polyherbal oil formulations given in Table 5 revealed variation in its content in different brands, which indicates the need of standardization of raw material used and uniformity in method of manufacturing to be followed by different ayurvedic manufacturers. The method developed here does not require separation of unsaponifiable matter for quantification as reported for some active ingredients in oil formulation. Oil extract can be directly used for analysis. Avoidance of long and tedious step therefore makes this method more amendable to the high-throughput screening.

4. Conclusion

A method for analysis of *Aconitum chasmanthum* using aconitine as analytical reference in polyherbal oil formulation was developed. Proposed method does not require tedious steps of saponification for separation of fatty acid which are the major interfering component in analysis of oils. Further the method does not require any chemical transformation of active moiety aconitine and it is analyzed as such. The method was found to be simple, precise, specific,

TABLE 3: Accuracy study of the proposed method for aconitine.

Amount detected in sample (μg/mL)	Amount of standard drug added (μg/mL)	Amount of drug recovered (μg/mL)*	% Recovery*
Varnaraksasa taila			
28.3	0	27.6 ± 0.4	97.7 ± 1.6
28.3	14	41.7 ± 0.2	98.6 ± 0.5
28.3	28	56.2 ± 0.3	99.9 ± 0.6
28.3	42	69.2 ± 0.4	98.5 ± 0.5
Vipritmalla taila			
32.2	0	31.8 ± 0.2	98.7 ± 0.6
32.2	16	47.9 ± 0.4	99.4 ± 0.2
32.2	32	65.4 ± 0.3	101.8 ± 0.5
32.2	48	80.4 ± 0.5	100.2 ± 0.7

*Mean value ± relative standard deviation (% RSD, $n = 3$).

TABLE 4: Results of robustness studies.

Parameter	% RSD peak area
Mobile phase composition	1.22
Amount of mobile phase	0.83
Temperature	0.92
Chamber saturation time	0.71
Chamber dimensions	0.43

TABLE 5: Aconitine content found in various *taila* preparations.

Sample	Brand	Aconitine content in oil (mg/100 gm)
Varnaraksasa taila	1	0.27 ± 0.003
	2	0.28 ± 0.005
	3	0.26 ± 0.004
Vipritmalla taila	1	0.31 ± 0.05
	2	0.32 ± 0.07
	3	0.32 ± 0.08

*Mean ± relative standard deviation (RSD, $n = 3$).

sensitive, and accurate. It can be used for routine quality control of polyherbal oil formulations containing *Aconitum chasmanthum*.

Acknowledgments

The authors are thankful to School of Pharmacy, DAVV University, Indore (Madhya Pradesh) and Director, SICART, V. V. Nagar (Gujarat) for providing analytical facilities.

References

[1] *The Ayurvedic Pharmacopoeia of India-part I*, Ministry of Health and Family Welfare, Government of India, New Delhi, India, 1st edition, 2000.

[2] H. M. Chang and P. P. H. But, *Pharmacology and Applications of Chinese Materia Medica*, vol. 1-2, World scientific press, Singapore, Singapore, 1987.

Development and Validation of Selective High-Performance Liquid Chromatographic Method Using Photodiode Array Detection for Estimation of Aconitine in Polyherbal Ayurvedic Taila Preparations

21

[3] G.E. Trease and W.C. Evans, *Pharmacognosy*, Saunders/Elsevier Science, Amsterdam, The Netherlands, 15th edition, 2002.

[4] P.K. Mukherjee, *Quality Control of Herbal Drugs*, Business Horizons, New Delhi, India, 1st edition, 2002.

[5] *The Ayurvedic formulary of India-part III*, Ministry of Health and Family Welfare, Government of India, New Delhi,India, 1st edition, 2000.

[6] R. N. Sharma, *Ayurveda-sarsangrha*, Shri Baidhyanath Ayurveda Bhavan Ltd., Varnasi, India, 13th edition, 1985.

[7] Anonymous, "Natural health product directorate," 2006, Canada, version 2, http://www.hc-sc.gc.ca/dhp-mps/prodnatur/ legislation/docs/compendium_mono_table_8-eng.php.

[8] EMEA/HMPC/253629/2007, Reflection paper on references used for quantitative and qualitative analysis of herbal medicinal products and traditional herbal medicinal products.

[9] EMEA/HMPC/CHMP/CVMP/214869/2006, Guideline on quality of combination herbal medicinal products / traditional herbal medicinal products.

[10] N. Dubey, N. Dubey, R. Mehta, and A. Saluja, "Selective determination of aconitine in polyherbal oils containing aconitum chasmanthum using HPTLC," *Journal of AOAC International*, vol. 92, no. 6, pp. 1617–1621, 2009.

[11] H. Hikino, O. Ishikawa, C. Konno, and H. Watanabe, "Determination of aconitine alkaloids by high-performance liquid chromatography," *Journal of Chromatography*, vol. 211, no. 1, pp. 123–128, 1981.

[12] H. Hikino, M. Murakami, C. Konno, and H. Watanabe, "Determination of aconitine alkaloids in aconitum roots," *Planta Medica*, vol. 48, no. 2, pp. 67–71, 1983.

[13] Y. Xie, Z.H. Jiang, H. Zhou, H.X. Xu, and L. Liu, "Simultaneous determination of six Aconitum alkaloids in proprietary Chinese medicines by high-performance liquid chromatography," *Journal of Chromatography A*, vol. 1093, no. 1-2, pp. 195–203, 2005.

[14] Z. Wang, J. Wen, J. Xing, and Y. He, "Quantitative determination of diterpenoid alkaloids in four species of Aconitum by HPLC," *Journal of Pharmaceutical and Biomedical Analysis*, vol. 40, no. 4, pp. 1031–1034, 2006.

[15] E. Frérot and E. Decorzant, "Quantification of total furocoumarins in citrus oils by HPLC coupled with UV, fluorescence, and mass detection," *Journal of Agricultural and Food Chemistry*, vol. 52, no. 23, pp. 6879–6886, 2004.

[16] Y. J. Hsieh, L. C. Lin, and T. H. Tsai, "Determination and identification of plumbagin from the roots of Plumbago zeylanica L. by liquid chromatography with tandem mass spectrometry," *Journal of Chromatography A*, vol. 1083, no. 1-2, pp. 141–145, 2005.

[17] Y. Novikova and A. A. Tulaganov, "Using HPLC for the quality control of psoralen and related preparations," *Pharmaceutical Chemistry Journal*, vol. 38, no. 5, pp. 50–52, 2004.

[18] A. P. Neilson, R. J. Green, K. V. Wood, and M. G. Ferruzzi, "High-throughput analysis of catechins and theaflavins by high performance liquid chromatography with diode array detection," *Journal of Chromatography A*, vol. 1132, no. 1-2, pp. 132–140, 2006.

[19] X. R. Yang, C. X. Ye, J. K. Xu, and Y. M. Jiang, "Simultaneous analysis of purine alkaloids and catechins in Camellia sinensis, Camellia ptilophylla and Camellia assamica var. kucha by HPLC," *Food Chemistry*, vol. 100, no. 3, pp. 1132–1136, 2007.

[20] F. Q. Yang, Y. T. Wang, and S. P. Li, "Simultaneous determination of 11 characteristic components in three species of *Curcuma* rhizomes using pressurized liquid extraction and high-performance liquid chromatography," *Journal of Chromatography A*, vol. 1134, no. 1-2, pp. 226–231, 2006.

[21] T. Wu, S. W. Annie Bligh, L. Gu et al., "Simultaneous determination of six isoflavonoids in commercial Radix Astragali by HPLC-UV," *Fitoterapia*, vol. 76, no. 2, pp. 157–165, 2005.

[22] D.M. Bliesner, *Validating Chromatographic Methods: A Practical Guide*, John Wiley & Sons, Toronto, Canada, 2006.

[23] ICH, "Validation of analytical procedures: methodology Q2 (R1)," in *Proceedings of The International Conference on Harmonisation of Technical Requirements for Registration of Pharmaceuticals for Human Use*, IFPMA, November 1996.

4

HPTLC Fingerprint Profile and Isolation of Marker Compound of *Ruellia tuberosa*

Daya L. Chothani,[1] M. B. Patel,[2] and S. H. Mishra[2]

[1] *Pharmacognosy Department, Pioneer Pharmacy Degree College, Ajwa-Nimeta Road, Sayajipura, Vadodara 390019, India*
[2] *Herbal Drug Technology Lab, Pharmacy Department, Faculty of Technology and Engineering, The M. S. University of Baroda, Vadodara 390 001, India*

Correspondence should be addressed to Daya L. Chothani, daya.herb@gmail.com

Academic Editor: Monika Waksmundzka-Hajnos

The present study was aimed to identification, isolation, and quantification of marker in *R. tuberosa* (Acanthaceae). HPTLC fingerprinting was carried out for various extract of root, stem, and leaf of *R. tuberosa*. From the HPTLC fingerprint the florescent band (under 366 nm) at R_f: 0.56 (mobile phase chloroform : toluene : ethyl acetate (6 : 3 : 1, v/v)) was found in leaf, root, and stem of *R. tuberosa*. So, the florescent band (under 366 nm) at R_f: 0.56 was isolated as marker compound RT-F2 from root of *R. tuberosa*. The marker compound RT-F2 was quantified by using HPTLC technique. The percentage (W/W) amount of RT-F2 was found to 40.0% and 44.6% in petroleum ether and ethyl acetate extract of *R. tuberosa* roots, respectively. Further study is suggested to characterization and biological nature of marker compound.

1. Introduction

Marker compound means chemical constituents within a medicinal that can be used to verify its potency or identity. For sometimes, the marker compounds may be described as active ingredients or chemicals that confirm the correct botanical identity of the starting material. It is very difficult to identify correct marker compounds for all traditional medicinals, because some medicinals have unknown active constituents and others have multiple active constituents. A chromatographic fingerprint of a herbal medicine is a chromatographic pattern of the extract of some common chemical components of pharmacologically active and/or chemical characteristics. By using chromatographic fingerprints, the authentication and identification of herbal medicines can be accurately conducted even if the amount and/or concentration of the chemically characteristic constituents is not exactly the same for different samples of drug. Hence it is very important to obtain reliable chromatographic fingerprints that represent pharmacologically active and chemically characteristic component of the herbal drug [1–5].

Ruellia tuberosa is an erect, suberect, or diffuse perennial herb up to 60–70 cm tall herb and belongs to family Acanthaceae, a native of Central America, introduced into Indian garden as ornament. It is used medicinally in West Indies, Central America, Guiana, and Peru. *R. tuberosa* is commonly known as *"Cracker plant"* [6–8]. In Siddha system of medicine, leaves are given with liquid copal as remedy for gonorrhea and ear diseases [9], used in stomach cancer [10]. Dried and ground roots in dose of two ounces cause abortion and also used in sore eyes [11]. The herb also exhibits emetic activity and employed substitute of ipecac, also used in bladder stones and decoction of leaves used in treatment of Bronchitis [12]. In Suriname's traditional medicine system, it is used as anthelmintic and also in management of joint pain and strained muscles. In folk medicine, it has been used as diuretic, antipyretic, antidiabetic, antidotal, thirst-quenching agent and analgesic and anti-hypertensive activity [13, 14]. *Ruellia tuberosa* is used as cooling in urinary problem, uterine fibroids [15, 16]. It has recently been incorporated as a component in a herbal drink in Taiwan [17]. It has been experimentally proved to possess antioxidant [18], antimicrobial

[19], anticancer [20], gastroprotective activity [21], antinociceptive, and anti-inflammatory activity [22]. It is reported that it contains flavonoids, steroids, and triterpenoids and alkaloid [23–26]. But there is no any identified marker reported; so the present study is aimed to identification, isolation, and quantification of marker in *R. tuberose*.

2. Materials and Methods

2.1. Plant Material.
Fresh plant of *Ruellia tuberosa* was collected from the campus of The M. S. University of Baroda in the month of August 2008. Plant was authenticated at Botany Department of The M. S. University. Voucher specimen (PHR/HDT/DC-RT-08) was stored in herbarium of our laboratory. Roots were separated and sun dried separately. Dried plant material was powdered.

2.2. Chemicals.
All other reagents were analytical grade, purchased from Merck (Darmstadt, Germany). All UV-Vis measurements were recorded on a Shimadzu UV-1800.

2.3. Preparation of Extracts [27].
Powdered air dried drug, weighing about 50 g, was extracted successively in soxhlet apparatus with the series of solvents of increasing polarity as follows: petroleum ether, toluene, chloroform, ethyl acetate, and methanol. Each time before extracting with the next solvent, the material was dried. All the extracts were filtered through Whatman filter paper and concentrated. Concentrated extracts were applied on the TLC plate as sample solution.

2.4. HPTLC Finger Print Profiles for Various Extracts [28, 29]

2.4.1. TLC Conditions.
TLC plate consists of 20×10 cm, precoated with silica gel 60 F254 TLC plates (E. Merck) (0.2 mm thickness) with aluminum sheet support. The spotting device was a CAMAG Linomat V Automatic Sample Spotter (Camag Muttenz, Switzerland); the syringe, $100 \mu L$ (from Hamilton); the developing chamber was a CAMAG glass twin trough chamber (20×10 cm); the densitometer consisted of a CAMAG TLC scanner 3 linked to WINCATS software. Mobile phase was chloroform : toluene : ethyl acetate ($6:3:1$, v/v). Saturation time for mobile phase was 2 hours.

2.4.2. Procedure.
Various extracts of roots, leaf, and stem of *R. tuberosa* were applied on TLC plate and the plate was developed in chloroform : toluene : ethyl acetate ($6:3:1$, v/v) solvent system to a distance of 8 cm. The plates were dried at room temperature in air. The plate was scanned at 254 nm (Figure 1) and 366 nm (Figure 2) before spraying and at 600 nm (Figure 3) after spraying with detection reagent (Anisaldehyde sulfuric acid reagent and plate was heated at $110°C$ for 5 minutes). The R_f values and color of the resolved bands were noted.

2.5. Isolation and Characterization of Chemical Marker

2.5.1. Isolation of Compound RT-F2 from Petroleum Ether and Ethyl Acetate Extract of Root.
The dried powder of root

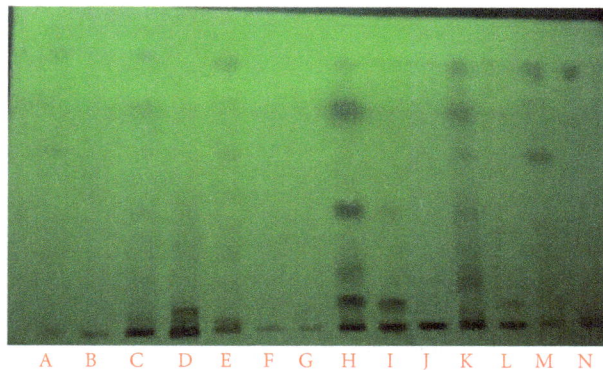

FIGURE 1: HPTLC fingerprint of various extracts of *R. tuberosa* at 254 nm.

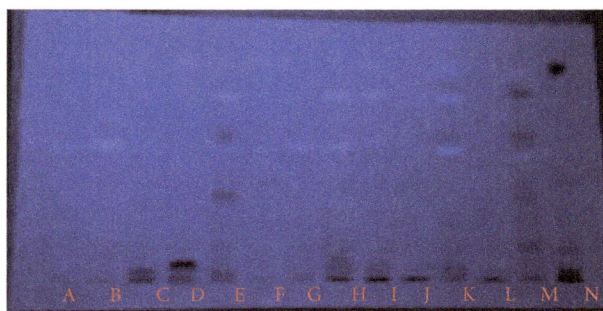

FIGURE 2: HPTLC fingerprint of various extracts *R. tuberosa* at 366 nm.

(200 g) was extracted with petroleum ether and ethyl acetate (500 mL) separately in soxhlet apparatus for 2 days. Then the extracts were concentrated by distilling the solvent and concentrated extracts were subjected to repetitive preparative thin layer chromatography using Silica Gel G as stationary phase (20×20 cm glass plates) and chloroform : toluene : ethyl acetate ($6:3:1$ v/v/v) as mobile phase. Fluorescents bands under 366 nm at R_f value 0.56 were identified RT-F2 compound. RT-F2 bands were scraped. RT-F2 was separated from silica Gel G by treating with methanol and Chloroform mixture ($1:1$), filtered through Whatman filter paper, and filtrates were combined, concentrated, and dried. Isolated compounds were subjected to TLC and HPTLC, UV spectroscopy (Figure 4), IR spectroscopy (Figure 5), and Mass spectroscopy (Figure 6).

2.5.2. Quantification of RT-F2 in Petroleum Ether and Ethyl Acetate Extract Using HPTLC Method

Standard Stock Solution. A solution of F2 compound ($500 \mu g/mL$) was prepared in chloroform.

Sample Preparation

Ethyl Acetate Extract. Stock solution of sample 2 mg/mL of extract was prepared in chloroform.

FIGURE 3: HPTLC fingerprint of various extracts of *R. tuberosa* at 600 nm. (A) Petroleum Ether extract (R). (B) Toluene extract (R). (C) Chloroform extract (R). (D) EtOH extract (R). (E) Petroleum Ether Fraction (R). (F) Hexane Fraction (R). (G) Toluene Fract (R). (H) Chloroform Fract (R). (I) EtOH Fract (R). (J) Methanol Fract (R). (K) EtOH extract (R). (L) methanol extract (R). (M) Petroleum Ether extract (S). (N) petroleum ether (L). R: root; Fract: hydroalcoholic fraction; L: Leaf; S: stem.

FIGURE 6: Mass spectra of RT-F2 compound.

FIGURE 7: Densitogram of RT-F2 compound at 600 nm, where T1, T2, T3, T6, and T7 are 1.25, 2.5, 3.75, 5.0, and 6.25 concentration of standard RT-F2 (μg/mL), respectively. T4 and T5 are petroleum ether extract and ethyl acetate extract (μg/mL), respectively.

FIGURE 4: UV spectra of RT-F2 compound (λ_{max} 208 nm, 272 nm).

FIGURE 8: Chromatogram of RT-F2 standard compound at 600 nm.

FIGURE 5: IR spectra of RT-F2 compound.

FIGURE 9: Chromatogram of petroleum ether extract at 600 nm.

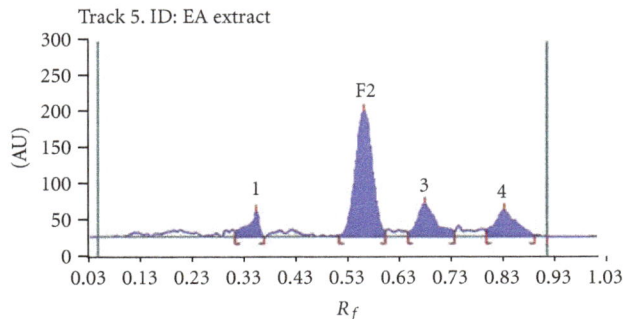

FIGURE 10: Chromatogram of ethyl acetate extract at 600 nm.

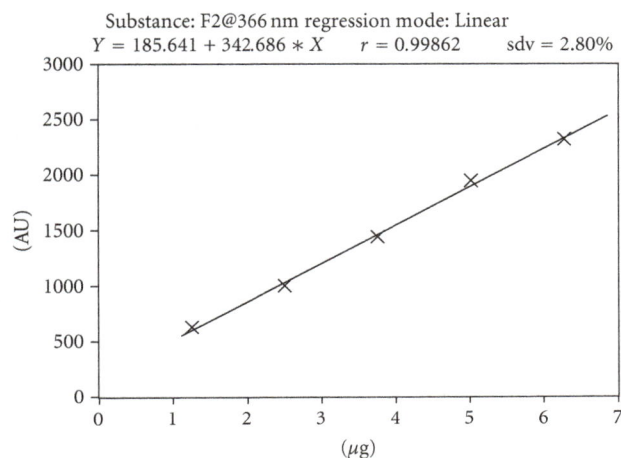

FIGURE 11: Calibration curve of RT-F2 compound.

Petroleum Ether Extract. Stock solution of sample 1 mg/mL of extract was prepared in chloroform.

Calibration Curve. From the standard stock solution 2.5–12.5 μL solutions were applied on precoated plate of Silica Gel G, to produce the range of 1.25–6.25 μg of RT-F2 per spot, respectively (Figure 7). Calibration curve is given in Figure 11.

Sample. A 10 μL of each extract was applied.

Mobile Phase. The mobile phase was chloroform : toluene : ethyl acetate (6 : 3 : 1).

Stationary Phase. The stationary phase was Precoated plate, Silica Gel G 60 F254.

Applicator. The applicator phase was CAMAG LINOMAT 5.

Development. Plate was developed in a twin trough chamber.

Detection. We spray with Anisaldehyde sulfuric acid reagent and heat at 110°C for 5 minutes.

The plate was scanned at 366 nm under fluorescent mode before spraying and at 600 nm (Figure 7) after spraying. The Chromatograph of RT-F2 standard compound (Figure 8),

petroleum ether extracts (Figure 9), and ethyl acetate extract (Figure 10) were reported.

3. Results and Discussion

3.1. HPTLC Fingerprint Profile. HPTLC fingerprint showed that purple colored band (after derivatisation) (Figure 3) at R_f: 0.56 was found in leaf, root, and stem of *R. tuberosa*. The florescent band (under 366 nm) at R_f: 0.56 was selected as marker compound and identified as RT-F2.

3.2. Isolation and Characterization of Marker RT-F2 Compound. Data from fingerprinting results provide information about presence of major terpenoid in petroleum ether extract and ethyl acetate extract of root, targeted for isolation.

3.3. Compound RT-F2. Isolated compound F2 has sticky type of nature. It gives violet purple color with Anisaldehyde sulfuric acid reagent and Liebermann-Burchard reagent.

Analysis

TLC. R_f: 0.56, Solvent system-toluene : chloroform : ethyl acetate (3 : 6 : 1).

Detection. Anisaldehyde sulfuric acid reagent (heat at 105°C for 5 minutes.

IR (KBr, cm^{-1}). 3409, 1622.

MS. (m/z) 279, 167, 149, 113, 83, 55.

3.4. Quantification of RT-F2 in Petroleum Ether and Ethyl Acetate Extract Using HPTLC. Figure 7 shows the HPTLC chromatogram of standard RT-F2 compound, petroleum ether extract and ethyl acetate extract.

The percentage (W/W) amount of RT-F2 was found to 40.0% and 44.6% in petroleum ether and ethyl acetate extract of *R. tuberosa* roots, respectively (Tables 1 and 2).

4. Conclusion

Herbal medicines are composed of many constituents and are therefore very capable of variation. Hence it is very important to obtain reliable chromatographic fingerprints that represent pharmacologically active and chemically characteristic components of the herbal medicine. HPTLC fingerprinting profile is very important parameter of herbal drug standardization for the proper identification of medicinal plants. A TLC densitometric method for the quantification of isolated marker compound RT-F2 was established in petroleum ether and ethyl acetate extract of roots of *R. tuberosa*. The present HPTLC fingerprinting profile can be used as a diagnostic tool to identity and to determine the quality and purity of the *R. tuberosa* in future studies.

TABLE 1: Calibration curve data for RT- F2 compound of HPTLC method.

Track	R_f	Concentration of RT-F2	Height of peak	Calculated RT-F2	Area of peak	Calculated RT-F2
1	0.57	1.250 μg	51.72		641.96	
2	0.56	2.500 μg	48.37		1005.73	
3	0.56	3.750 μg	71.26		1446.38	
4	0.56	Unknown*	73.75	3.758 μg	1558.45	4.006 μg
5	0.57	Unknown**	158.38	>6.875 μg	3246.91	>6.875 μg
6	0.57	5.000 μg	86.79		1945.75	
7	0.58	6.250 μg	110.10		2313.74	

*Petroleum ether extract, **Ethyl acetate extract
Regression equation (Height) $Y = 27.094 + 12.415X$, $r = 0.9574$
Regression equation (Area) $Y = 185.641 + 342.686X$, $r = 0.9986$.

TABLE 2: Quantification of RT-F2 compound.

Stationary phase	Precoated Silica Gel 60 GF$_{254}$
Mobile phase	Chloroform : toluene : ethyl acetate (6 : 3 : 1)
Calibration range of F2	2.5–12.5 μg/spot
Detection	Anisaldehyde sulphuric acid reagent heated at 110°C for 5 min and detected at 600 nm.
Regression equation	$Y = 185.6 + 342.686X$ (area wise)
R value	0.99862 (area wise)

References

[1] http://www.bluepoppy.com/blog/blogs/blog1.php/what-are-marker-compounds.

[2] P. S. Patil and R. Shettigar, "An advancement of analytical techniques in herbal research," *Journal of Advanced Scientific Research*, vol. 1, no. 1, pp. 8–14, 2010.

[3] Y. Z. Liang, P. Xie, and K. Chan, "Quality control of herbal medicines," *Journal of Chromatography B*, vol. 812, no. 1-2, pp. 53–70, 2004.

[4] E. S. Ong, "Chemical assay of glycyrrhizin in medicinal plants by pressurized liquid extraction (PLE) with capillary zone electrophoresis (CZE)," *Journal of Separation Science*, vol. 25, no. 13, pp. 825–831, 2002.

[5] P. S. Xie, "A feasible strategy for applying chromatography fingerprint to assess quality of Chinese herbal medicine," *Traditional Chinese Drug Research & Clinical Pharmacology*, vol. 12, no. 3, pp. 141–169, 2001.

[6] C. N. Pandey, *Medicinal Plants of Gujarat*, Gujarat Ecological Education and Research Foundation, Gujarat, India, 2005.

[7] Medicinal Plants of the Guiana's (Guyana, Surinam, French Guiana).

[8] D. L. Chothani, M. B. Patel, H. U. Vaghasiya, and S. H. Mishira, "Review on *Ruellia tuberosa* (cracker plant)," *Pharmacognosy Journal*, vol. 2, no. 12, pp. 506–512, 2010.

[9] L. Suseela and S. Prema., "Pharmacognostic study on *Ruellia tuberosa*," *Journal of Medicinal and Aromatic Plant Sciences*, vol. 29, pp. 117–122, 2007.

[10] M. B. Reddy, K. R. Reddy, and M. N. Reddy, "Ethnobotany of Cuddapah district, Andhra Pradesh, India," *International Journal of Pharmacognosy*, vol. 29, no. 4, pp. 273–280, 1991.

[11] B. D. Kirtikar and B. D. Basu, *Indian Medicinal Plants*, vol. 3, International Book Distributors, Deheradun, India, 1935.

[12] The Wealth Of India, *A Dictionary Of Indian, Raw Material and Industrial Product*, Publication and Information Directorate, Council of Scientific and Industrial Research, New Delhi, India, 1972.

[13] N. Y. Chiu and K. H. Chang, "The illustrated medicinal plants of Taiwan," *Mingtong Medical Journal*, vol. 226, no. 1, 1995.

[14] F. A. Chen, A. B. Wu, P. Shieh, D. H. Kuo, and C. Y. Hsieh, "Evaluation of the antioxidant activity of *Ruellia tuberosa*," *Food Chemistry*, vol. 94, no. 1, pp. 14–18, 2006.

[15] C. A. Lans, *Creole remedies. Case studies of ethnoveterinary medicine in Trinidad and Tobago*, Ph.D.Dissertation, Wageningen University, Wageningen, The Netherlands, 2001. no. 2992.

[16] C. A. Lans, "Ethnomedicines used in Trinidad and Tobago for urinary problems and diabetes mellitus," *Journal of Ethnobiology and Ethnomedicine*, vol. 2, article 45, pp. 1–11, 2006.

[17] M. J. Balick, F. Kronenberg, A. L. Ososki et al., "Medicinal plants used by latino healers for women's health conditions in New York City," *Economic Botany*, vol. 54, no. 3, pp. 344–357, 2000.

[18] F. A. Chen, A. B. Wu, P. Shieh, D. H. Kuo, and C. Y. Hsieh, "Evaluation of the antioxidant activity of *Ruellia tuberosa*," *Food Chemistry*, vol. 94, no. 1, pp. 14–18, 2006.

[19] C. Wiart, M. Hannah, M. Yassim, H. Hamimah, and M. Sulaiman, "Anti-microbial activity of *Ruellia tuberosa* L," *American Journal of Chinese Medicine*, vol. 33, no. 4, pp. 683–685, 2005.

[20] S. Arun, P. Giridharan, A. Suthar et al., *Isolation of Tylocrebrine from Ruellia tuberosa through Bioassay Directed Column Chromatography and Elucidating its Anti-Cancer and Anti-Inflammatory Potential*, 7th Joint Meeting of GA, AFERP, ASP, PSI & SIF, Athens, Greece, 2008.

[21] L. S. R. Arambewela, R. Thambugala, and W. D. Ratnasooriya, "Gastroprotective activity of *Ruellia tuberosa* root extract in rats," *Journal of Tropical Medicinal Plants*, vol. 4, no. 2, pp. 191–194, 2003.

[22] M. A. Alam, N. Subhan, M. A. Awal et al., "Antinociceptive and anti-inflammatory properties of *Ruellia tuberosa*," *Pharmaceutical Biology*, vol. 47, no. 3, pp. 209–214, 2009.

[23] C. F. Lin, Y. L. Huang, L. Y. Cheng et al., "Bioactive flavonoid from *Ruellia tuberosa*," *The Journal of Chinese Medicine*, vol. 17, no. 3, pp. 103–109, 2006.

[24] S. S. Subramanian and A. G. R. Nair, "Apigenin glycoside from thunbergia fragrans and *Ruellia tuberosa*," *Current Science*, p. 480, 1974.

[25] R. S. Singh, H. S. Pandey, B. K. Singh, and R. P. Pandey, "A new triterpenoid from *Ruellia tuberosa* linn," *Indian Journal of Chemistry B*, vol. 41, no. 8, pp. 1754–1756, 2002.

[26] C. K. Andhiwal and R. P. Chandra Haas Varshney, "Phytochemical investigation of *Ruellia tuberosa* L," *Indian Drugs*, vol. 23, no. 49, 1985.

[27] C. K. Kokate, *Practical Pharmacognosy*, Vallabh Prakashan, New Delhi, India, 4th edition, 2005.

[28] H. Wagner, S. Blade, and G. M. Zgainsky, *Plant Drug Analysis*, Springer, Great Britain, UK, 1984.

[29] E. Stahl, "Thin layer chromatography," in *A Laboratory Hand Book*, Springer, Berlin, Germany, 2nd edition, 1965.

5

The HPLC/DAD Fingerprints and Chemometric Analysis of Flavonoid Extracts from the Selected Sage (*Salvia*) Species

Mieczysław Sajewicz,[1] Dorota Staszek,[1] Michał S. Wróbel,[1] Monika Waksmundzka-Hajnos,[2] and Teresa Kowalska[1]

[1] *Institute of Chemistry, University of Silesia, 9 Szkolna Street, 40-006 Katowice, Poland*
[2] *Department of Inorganic Chemistry, Faculty of Pharmacy, Collegium Pharmaceuticum, Medical University of Lublin, 4a Chodźki Street, 20-093 Lublin, Poland*

Correspondence should be addressed to Teresa Kowalska, teresa.kowalska@us.edu.pl

Academic Editor: Yvan Vander Heyden

The results of spectrophotometric and HPLC/DAD analysis are discussed, and a comparison is made of selectively extracted flavonoid fractions derived from twenty six sage species belonging to the *Salvia* genus. The sage samples were harvested in the vegetation seasons 2007, 2008, and 2009. It was a goal of this study to find out which species contain the highest yields of flavonoids (recognized for their free-radical-scavenging activity), as those with the highest yields could be applied in official medicine. It was spectrophotometrically established that the four sage species can be recognized for their highest flavonoid levels, while the HPLC/DAD analysis pointed out to the four other species. The source of the discrepancy between the two evaluation approaches was discussed. Moreover, the HPLC/DAD fingerprints of the flavonoid fraction underwent a chemometric pre-treatment, and then the purified fingerprints were analyzed by means of Principal Component Analysis (PCA) for the differences in the harvesting period. A difference was revealed between the herbs harvested in the 2007 season, and those harvested in 2008 and 2009. The main source of this difference could be the seasonal weather variation and the relatively longest storage period with the plants harvested in 2007.

1. Introduction

In spite of a vast number of the different species and a wide popularity of the plants belonging to the *Salvia* genus, relatively little attention has been paid to phytochemical analysis of these plants, prior to our own systematic research (e.g., [1–4]). In the course of the centuries, different sage species have gained the repute for their outstanding therapeutic, culinary, and decorative valor. However, the official European medicine recognizes just one sage species for its curative properties, which is *Salvia officinalis*. It is a well-known fact that the curative properties of many plants are due to the high contents of phenolics, which act as the free-radical scavengers. Thus, it became an objective of this study to analyze a selection of the different sage species popular in Central and South Europe (where they grow both in a natural habitat and as cultivars) and to fingerprint the flavonoid fraction present therein by means of HPLC/DAD. It was our intent to ultimately point out to these *Salvia* species, which might compete with *S. officinalis* in terms of the high levels of flavonoids.

To this effect, we spectrophotometrically determined an overall content of flavonoids in the sage extracts which were selectively obtained following the pharmacopeial procedure [5]. This preliminary assessment allowed selecting four out of twenty six sage species with the highest overall contents of flavonoids (which were *S. glutinosa*[a], *S. pratensis ssp. Haematodes*, *S. staminea*, and *S. triloba*). For these four sage species and additionally for *S. officinalis*, a comparison was carried out of their respective HPLC/DAD fingerprints. In our earlier study [6], an analogous comparison was performed of the total sage extracts derived with the use of methanol and, hence, containing a wider spectrum of the polar components. It was expected that the fingerprints registered for the selectively derived flavonoid fraction could better characterize the sage species than the nonselectively derived methanol extracts.

However, the flavonoid contents in the sage species, as monitored through the prism of the sums of all the separated chromatographic peak areas had to be arranged in a different order and, in this case, the four out of twenty-six sage species with the highest overall contents of flavonoids were *S. nemorosa*, *S. forskahlei*, *S. azurea*, and *S. amplexicaulis*.

The source of the discrepancy between the two assessment approaches was discussed and, additionally, a chemometric comparison was performed by means of principal component analysis (PCA) of all the sage species considered in this study, which based on the HPLC/DAD chromatograms of the flavonoid fractions. Upon the obtained results, certain conclusions were drawn regarding seasonal differences in flavonoid composition among the individual plant species.

2. Experimental

2.1. Herbal Material and Reagents. Samples of the twenty-six different sage species (which are listed in Table 1 as species 1–5, 7–26, and 28) investigated in this study were collected in the Pharmacognosy Garden of the Medical University, Lublin, Poland, in three harvesting seasons (2007, 2008, and 2009). Botany specialists identified each investigated species, and the voucher specimens were deposited in the herbarium of the Department of Pharmacognosy, Medical University, Lublin, Poland. This plant material was dried for 40 h in an oven with a forced air flow at 35 to 40°C. The obtained dry material was stored in a refrigerator until commencement of the analysis. Species 6 was *S. glutinosa*[a], which originated from the natural habitat in the Ostrowsko region of south Poland and species 27 was *S. officinalis*, which originated from the natural habitat in the Zlatibor region of central Serbia. Species 6 was harvested in all three vegetation seasons (2007, 2008, and 2009), and species 27 was harvested in the summer of 2008 only and purchased dried. Summarized information about all the investigated herbal material is given in Table 1.

Methenamine was purchased from Pharma Cosmetics (Cracow, Poland), and methanol, ethyl acetate, acetone, glacial acetic acid, aluminium chloride, and hydrochloric acid used for the experiments were of analytical purity grade and purchased from POCh (Gliwice, Poland). Water was double distilled and deionized in the laboratory conditions by means of Elix Advantage model Millipore system (Molsheim, France).

2.2. Selective Extraction of Flavonoids from Herbal Material. Stock extract solution of each investigated sage species was prepared from 1 g medium powdered crude plant material. To this plant material, 20 mL acetone, 2 mL HCL (281 g L^{-1}), and 1 mL methenamine (5 g L^{-1}) were added. The entity was kept boiling on the water bath under the reflux for 30 min. Hydrolysate was filtered to the volumetric flask (100 mL). The separated plant material was extracted for the second and the third time with the 20 mL portions of

TABLE 1: Basic characteristics of the investigated plant material.

Sample no.	Sage species	Harvesting year
1	*S. amplexicaulis*	2007, 2008, 2009
2	*S. azurea*	2007, 2008, 2009
3	*S. cadmica*	2007, 2008, 2009
4	*S. deserta*	2007, 2008, 2009
5	*S. forskahlei*	2007, 2008, 2009
6	*S. glutinosa*[a]	2007, 2008, 2009
7	*S. hians*	2007, 2008, 2009
8	*S. jurisicii*	2007, 2008, 2009
9	*S. nemorosa*	2007, 2008, 2009
10	*S. pratensis ssp. Haematodes*	2007, 2008, 2009
11	*S. sclarea*	2007, 2008, 2009
12	*S. staminea*	2007, 2008, 2009
13	*S. stepposa*	2007, 2008, 2009
14	*S. tesquicola*	2007, 2008, 2009
15	*S. triloba*	2007, 2008, 2009
16	*S. verticillata*	2007, 2008, 2009
17	*S. officinalis*	2007, 2008, 2009
18	*S. lavandulifolia*	2007, 2009
19	*S. atropatana*	2007
20	*S. canariensis*	2007
21	*S. argentea*	2008, 2009
22	*S. austriaca*	2008, 2009
23	*S. nutans*	2008, 2009
24	*S. regeliana*	2008, 2009
25	*S. superba*	2008, 2009
26	*S. glutinosa*	2008
27	*S. officinalis*[b]	2008
28	*S. pratensis*	2009

[a]Sample originating from the natural habitat in the Ostrowsko region of south Poland.
[b]Sample originating from the natural habitat in the Zlatibor region of central Serbia.

acetone kept boiling for 10 min. All extracts were filtered to the same volumetric flask, and acetone was added to make up to 100 mL. Then, 20 mL of the obtained solution was transferred to the separation funnel, 20 mL water was added, and the entity was extracted with ethyl acetate (firstly with the 15 mL, and then three times with the 10 mL portions of ethyl acetate). The separated organic layers were collected jointly and twice washed with the 40 mL portions of water. The organic layer was filtered to the 50 mL volumetric flask and filled up to the volume with ethyl acetate. Selective extraction of flavonoids from the investigated plant material was carried out in triplicate from the three different plant samples belonging to each individual batch of the sage species. The procedure of selective extraction is described in [5].

2.3. Spectrophotometric Determination of the Overall Content of Flavonoids. Flavonoids were determined spectrophotochemically according to the procedure described in [5], using the selectively obtained flavonoid fraction extracts. Each result presented in this study is a mean value from the three independent spectrophotometric measurements obtained for each individual extract. An overall (%) content of flavonoids was recalculated for hyperoside, using the recalculation factor (k).

For each spectrophotometric measurement, two solutions were prepared. Solution 1 was prepared in the following way: to the 10 mL stock extract solution, 2 mL aluminium chloride (20 g L^{-1}) solution was added and filled up to 25 mL with the 1 : 19 mixture of acetic acid and methanol. Solution 2 was the reference sample, and it was prepared as follows: 10 mL stock solution was filled up to 25 mL with the 1 : 19 mixture of acetic acid and methanol. After 45 min from the preparation of these two solutions, absorbance of sample 1 was measured at the wavelength $\lambda = 425$ nm, using sample 2 as a blank. The percentage (%) content of total flavonoids (X) was calculated, using the following formula [5]:

$$X = \frac{A \cdot k}{m}, \qquad (1)$$

where A is the absorbance of the examined solution, k is the recalculation factor for hyperoside ($k = 1.25$), and m is the weight of the crude plant material (in grams).

The obtained quantitative results (i.e., the overall contents of flavonoids in the percentage scale) for all the investigated sage species are given in form of the bar diagram in Figure 1.

2.4. The HPLC/DAD Fingerprinting.

The HPLC analysis was carried out with a Varian 920-LC model liquid chromatograph (Harbor City, Calif, USA) equipped with a 900-LC model autosampler, gradient pump, 330 model DAD, and the Galaxie software for data acquisition and processing. The analyses were carried out in the gradient mode using a Pursuit C18 (5 μm particle size) column (250 × 4.6 mm id; Varian; Cat. no. 1215-9307). Methanol (A) and water (with a 1% (v/v) amount of glacial acetic acid; (B)) were used in the following A + B gradient program (v/v): from 0 to 17 min, 50 + 50; from 18 to 26 min, 80 + 20; from 27 to 46 min, 90 + 10; from 47 to 58 min, 100 + 0; from 59 to 70 min, 50 + 50. The flow rate was 0.6 mL min^{-1}. The chromatograms were registered at the wavelength $\lambda = 254$ nm.

2.5. Chemometric Baseline and Noise Correction of Chromatograms.

Dealing with the chromatograms of natural samples is not an easy task, although the fingerprinting approach has long been used for rapid screening of complex analytical signals. It has also been used in our earlier phytochemical study of the sage species [7]. In this section, a short description of the applied chemometric techniques is given, aiming to remove the background and noise from the HPLC/DAD chromatograms, and in that way to prepare the input data for the further exploration and visualization thereof with the use of the principal component analysis (PCA).

The first step was elimination of the background component from the chromatograms. One of numerous baseline elimination techniques is the penalized asymmetric least squares approach (PALS) [8]. This method applies the least squares approach to fit a baseline to the signal. Each point of a signal gets a different weight which locates it above or below the original signal. The weights are modified according to an iterative procedure such that the points above the original signal have very small weights and the points below this signal have the weights close to 1. There are two adjustable parameters, that is, the order of the differences and the penalty parameter, which are to be optimized. Usually, the order of the differences is set as equal to 2. The larger the penalty parameter is, the smoother baseline is obtained.

Chromatographic signals were smoothed to suppress the white (Gaussian) noise. In this study, the Savitzky-Golay differentiation filter [9] was used. This filter helps to reduce the peak overlapping and the linear baseline drift by constructing the first and the second derivative spectra. The Savitzky-Golay filter technique resembles the local polynomial regression with a window of at least $f + 1$ points, where f is the polynomial degree.

Also, the other undesired effects that could be present in the raw data were eliminated using the standard normal variate transformation (SNV) [10]. The pretreated data were used as the input data for principal component analysis (PCA) and discriminant partial least squares (DPLS).

2.6. Principal Component Analysis (PCA).

Principal component analysis (PCA) is the data exploration and visualization technique [11]. It allows to construct a set of new variables called principal components (PCs). The principal components are the orthogonal vectors that are linear combinations of the original variables and represent the data structure by maximizing the description of data variance. The PCA model consists of k principal components, where k is selected by the user. The original data matrix $\mathbf{X}(m \times n)$ is decomposed according to the following formula:

$$\mathbf{X} = \mathbf{T}\mathbf{P}^{\mathrm{T}} + \mathbf{E}, \qquad (2)$$

where $\mathbf{T}(m \times k)$ is the matrix of scores, $\mathbf{P}(n \times k)$ is the matrix of loadings, $\mathbf{E}(m \times n)$ is the residual matrix, and the superscript T denotes transposition of the matrix.

As a projection method, PCA enables projection of the objects or variables on the planes which they define [12]. Projection of the samples on the plane defined by the selected pairs of PCs allows studying the similarities among the samples (in form of the score plots). The loading plots are the projections of the variables on the planes of the selected principal components and allow tracing correlations among the data variables.

2.7. Discriminant Partial Least Squares (DPLS).

The discriminant partial least squares approach (DPLS) [13] is widely applied in chemistry, because of the multivariate character of the data studied and the high correlation usually observed among the explanatory variables. With the DPLS model, a linear relationship between a property of interest and a set of the explanatory variables is described. The property of interest is usually a binary or a bipolar coded vector. The

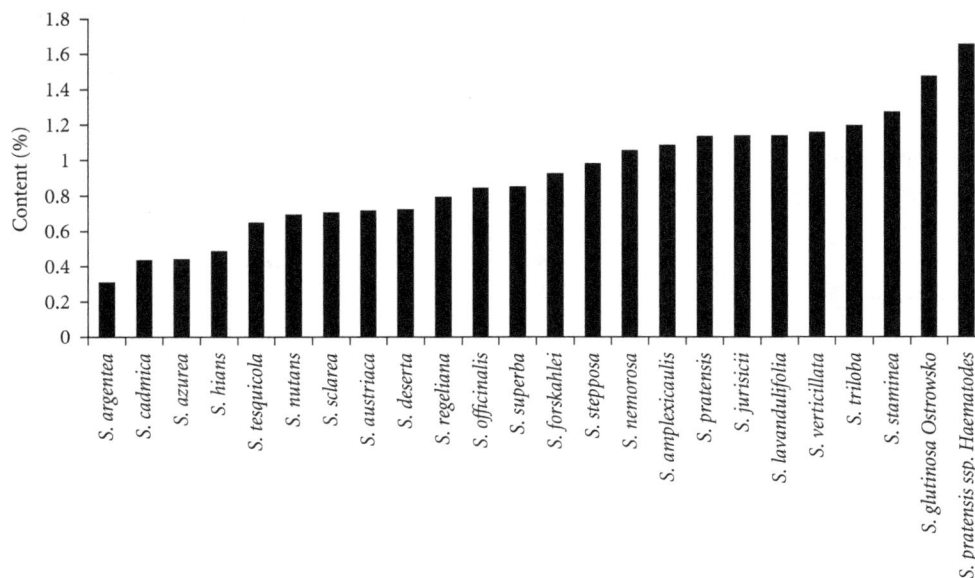

FIGURE 1: A bar diagram comparison of the overall contents of flavonoids in all the sage species harvested in 2009.

explanatory variables can be the sets of instrumental signals, for example, the chromatograms. Via the DPLS model, a set of a few orthogonal factors is constructed, aiming to maximize the covariance of the explanatory variables with the property of interest. When constructing the model, a number of orthogonal factors need to be estimated, which is usually done through the cross-validation mode. The final model is delivered in the form of the regression coefficients vector.

3. Results and Discussion

3.1. The HPLC/DAD Fingerprinting and Spectrophotometric Results. From a comparison of the HPLC/DAD fingerprints obtained for twenty-four selective extracts of flavonoids derived from the different sage species harvested in 2009, it was observed that the chromatogram of *S. nemorosa* shows both the highest number of the fourteen separated peaks and the highest sum of the separated peak areas (1244 mAV × min). The same species showed one of the highest numerical values of the sum of the separated peak heights (1893.5 mAV).

To the contrary, the chromatogram of the flavonoid extract from *S. cadmica* is characterized with the lowest number of the four separated peaks. Accordingly, the sums of the separated peak heights and the sums of the separated peak areas for this particular sage species were obtained among the lowest numerical levels (630.4 mAV × min and 1507.7 mAV, resp.).

On the basis of the chromatographic results, the following sage species are those with the highest sums of the separated peak areas and/or the highest sums of the separated peak heights: *S. amplexicaulis*, *S. azurea*, *S. forskahlei*, *S. hians*, and *S. nemorosa*. Sage species showing the lowest sums of the separated peak areas and/or the lowest sums of the separated peak areas are the following ones: *S. cadmica*,

S. nutans, *S. officinalis*, *S. regeliana*, and *S. triloba*. Numbers of the separated chromatographic peaks, sums of the separated peak areas, and sums of the separated peak heights for all the investigated sage species are given in Table 2. For the sake of graphical illustration, in Figure 2(a), we presented selected chromatographic fingerprints of the four sage species with the highest sums of the separated peak areas and also the fingerprint of *S. officinalis*. *Salvia officinalis* was compared with these four species, due to its unique position in the traditional European medicine, in spite of the lowest overall percentage content of flavonoids among the five compared species. The chromatogram of the *S. officinalis* extract fully confirmed the spectrophotometric findings. Its fingerprint is characterized with a relatively low number of the six separated peaks, and with the relatively low sums of the separated peak areas and the peak heights (676.1 mAV × min and 1377.4 mAV, resp.).

According to the spectrophotometric results (Figure 1), the following sage species: *S. glutinosa*[a], *S. pratensis ssp. Haematodes*, *S. staminea*, and *S. triloba* were characterized with the highest overall percentage contents of flavonoids, as recalculated to hyperoside.

Salvia triloba is one of the four sage species showing the higher overall percentage contents of flavonoids, as established spectrophotometrically. However, on the chromatogram of this species, we found seven separated peaks only, and the sums of their areas and heights are not very impressive either (although higher than with *S. officinalis*).

The second out of the four is *S. staminea*, which according to the spectrophotometric result is richer in flavonoids than *S. triloba* and *S. officinalis*. The chromatographic results confirmed the spectrophotometric ones. On the chromatogram of the flavonoid fraction derived from *S. staminea*, we found nine separated peaks, and the sums of their areas and heights are the highest ones among the selected four.

TABLE 2: A comparison of the numbers of the separated chromatographic peaks, and of the sums of the separated peak heights and peak areas with twenty-four different sage species harvested in 2009. Numbering of the sage samples is in conformity with Table 1.

Sample no.	Sage species	No. of separated peaks	Sum of separated peak heights (mAV)	Sum of separated peak areas (mAV × min)
1	S. amplexicaulis	7	1906.3	1038.2
2	S. azurea	6	1856.3	1168.3
3	S. cadmica	4	1507.7	630.4
4	S. deserta	6	1666.3	740.8
5	S. forskahlei	8	2041.6	1238.3
6	S. glutinosa[a]	8	1630.3	895.8
7	S. hians	11	1951.7	1009.0
8	S. jurisicii	13	1643.7	847.0
9	S. nemorosa	14	1893.5	1244.0
10	S. pratensis ssp. Haematodes	12	1635.5	903.4
11	S. sclarea	5	1610.7	852.3
12	S. staminea	9	1980.8	948.0
13	S. stepposa	9	1646.1	632.3
14	S. tesquicola	9	1665.8	667.4
15	S. triloba	7	1586.4	669.4
16	S. verticillata	7	1711.7	878.8
17	S. officinalis	6	1377.4	676.1
18	S. lavandulifolia	9	1614.3	837.4
21	S. argentea	9	1634.1	981.5
22	S. austriaca	8	1561.0	930.0
23	S. nutans	7	1464.7	609.4
24	S. regeliana	7	1487.9	806.6
25	S. superba	6	1993.5	886.4
28	S. pratensis	9	1600.7	927.6

The third species (S. glutinosa[a]) showed the highest overall content of flavonoids among the four species. On its fingerprint chromatogram, eight separated peaks can be found, and the sums of their areas and heights are 895.8 mAV × min and 1630.3 mAV, respectively. These values are lower than those chromatographically obtained for S. staminea, yet higher than with S. officinalis and S. triloba.

The fourth spectrophotometrically selected species is S. pratensis ssp. Haematodes as that with the relatively highest percentage content of flavonoids. On its chromatogram, the highest number of the twelve separated peaks was observed among the four compared species (and S. officinalis). Sums of their areas and heights were, respectively, 903.4 mAV × min and 1635.5 mAV. For the sake of graphical illustration, in Figure 2(b), we presented selected chromatographic fingerprints of the four sage species with the highest overall percentage contents of flavonoids, as spectrophotometrically assessed and recalculated to hyperoside, and also the fingerprint of S. officinalis.

The perceptible discrepancy between the spectrophotometric and the chromatographic results (which can anyway be considered as semiquantitative only) is due to the different principles and also different sensitivities of the two analytical approaches. Spectrophotometric analysis assumes a very simplifying recalculation of the overall flavonoid contents to hyperoside. On the other hand, the chromatographic fingerprinting is certainly more sensitive, although in spite of the selective and flavonoids-oriented extraction, one cannot exclude the presence of the compounds other than flavonoids in the chromatographed extracts, which might result in a different source of the estimation error. This is the reason why these two approaches have been presented and compared in this study.

3.2. Chemometric Evaluation of Chromatographic Fingerprints. In this study, a set of herbal fingerprints obtained from HPLC/DAD for the Salvia species was analyzed with the use of the chemometric techniques. Firstly, we enhanced the signal-to-noise ratio. To this effect, the background removal was carried out by application of the PALS method. Also, the noise influence was reduced with the use of the Savitzky-Golay smoothing filter, which delivered smoothed signals. For all the assessed fingerprints, the penalty parameter used in the PALS method was set to 10^7 and the Savitzky-Golay standard Matlab command of 51 frame size was applied. In Figure 3, we showed the baselines of the chromatograms and

(a)

(b)

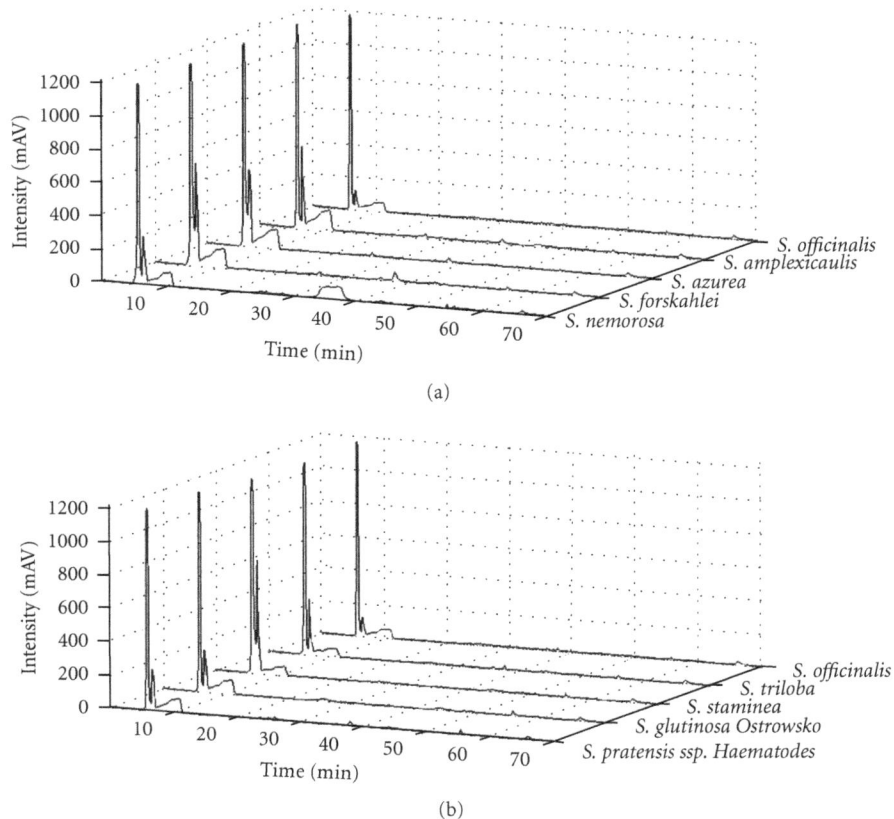

FIGURE 2: (a) A comparison of the HPLC/DAD fingerprints for the five different sage species (*S. nemorosa, S. forskahlei, S. azurea, S. amplexicaulis,* and *S. officinalis*). The chromatograms were registered at the wavelength λ = 254 nm. All fingerprints except for *S. officinalis* are those with the highest sums of the separated chromatographic peak areas. (b) A comparison of the HPLC/DAD fingerprints for the five different sage species (*S. glutinosa. S. pratensis ssp. Haematodes. S. staminea. S. triloba.,* and *S. officinalis*). The chromatograms were registered at the wavelength λ = 254 nm. All fingerprints except for *S. officinalis* are those with the highest overall sums of flavonoids, as spectrophotometrically established.

also the signals after performance of the preprocessing step. Finally, the SNV transformation was applied to the signals which were analyzed further in that form.

On the score plots shown in Figure 4, three groups of the *Salvia* samples can be distinguished. One can easily notice that the sage samples collected in the 2007 vegetation season markedly differ from the remaining ones in the space of PC1, which describes nearly 70% of data variance (Figure 5). Real cause of this difference remains unknown, yet it can be due to the local weather changes and/or due to the relatively longest storage period with the plant samples collected in 2007.

We tried to distinguish samples originating from the 2007 vegetation season from the remaining ones in another way also, which was achieved by application of the discriminant partial least squares model (DPLS) [13]. All samples were split into the two sets using the Kennard and Stone approach [14], namely into the training and the test set. With this algorithm, all kinds of samples are included in the training set, what provides the representativeness of the training set. The training set consisted of 17 samples from each of the two classes (class 1 was valid for the 2007 vegetation season samples and class 2 for the 2008 and 2009 samples), and the

test set contained all the remaining samples. Due to a rather limited number of the available samples, the cross-validation leave-3-out method was used to estimate the complexity of the DPLS model and the eight latent factors were chosen (Figure 6). The correct classification rate (CCR) was used as the model characterization parameter, and the sensitivity and specificity parameters were calculated (Table 3). The CCR for the model set characterizes the fitting of the model to the data, and for the test set it describes predictive power of the model. Finally, we obtained the 84.38% correctly classified samples from the independent test set, which was a reasonably satisfying result (additionally confirming correctness of distinguishing the 2007 sage samples from the remaining specimens studied). All calculations and chemometric treatment were applied via the R2010a Matlab by the MathWorks and its toolboxes.

4. Conclusions

Upon the spectrophotometric results, we compared twenty-four different sage species harvested in 2009 in terms of the overall percentage contents of flavonoids (recalculated

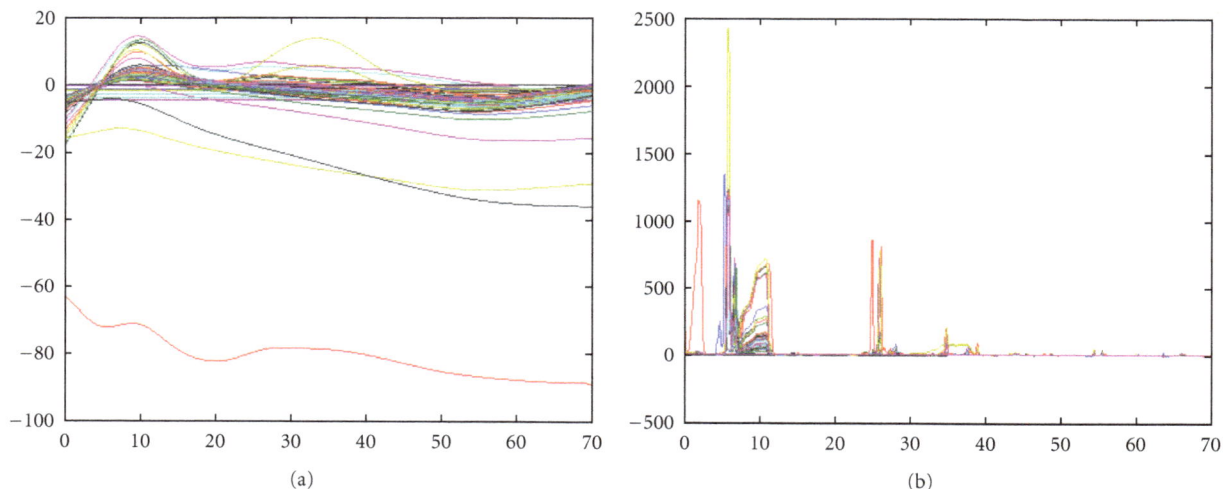

FIGURE 3: The HPLC/DAD profiles of the different *Salvia* species extracts: (a) with the baseline and (b) after the baseline and noise removal.

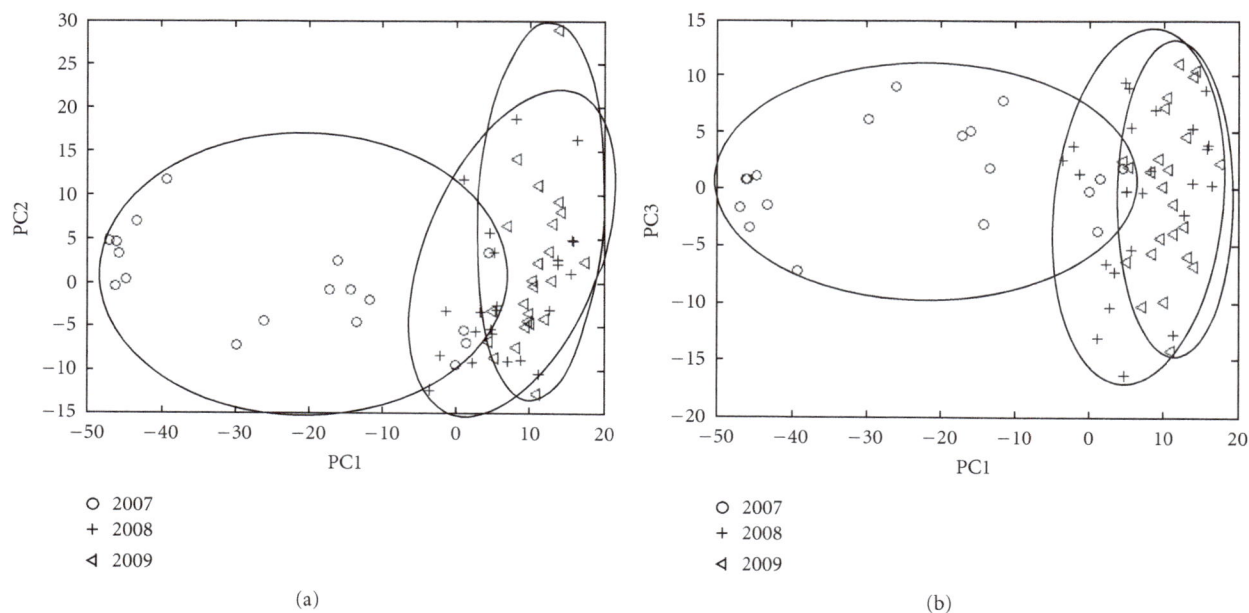

FIGURE 4: Plots of the *Salvia* samples on the plane determined by (a) the first and the second principal component and (b) the first and the third principal component.

to the contents of hyperoside) and, on this basis, we selected those showing the highest overall percentage contents (i.e., *S. glutinosa*[a], *S. pratensis ssp. Haematodes*, *S. staminea*, and *S. triloba*).

From the HPLC/DAD comparison of the fingerprints valid for the same twenty-four different sage species and from the comparison of the chromatograms with the highest sums of the chromatographic peak areas, the following sage species: *S. nemorosa*, *S. forskahlei*, *S. azurea*, and *S. amplexicaulis* could be selected as those with the highest overall contents of flavonoids.

A comparison of the spectrophotometric data with the results of the chromatographic fingerprinting allowed for

TABLE 3: Model parameters. where SE is sensitivity, SP is specificity, and CCR is the correct classification rate for the training and the test set, respectively.

	Training set	Test set
SE	94.18	88.00
SP	100	71.43
CCR	97.06	84.38

a conclusion as to the perceptible discrepancy between the results of these two approaches. This discrepancy apparently is due to the different principles and also different

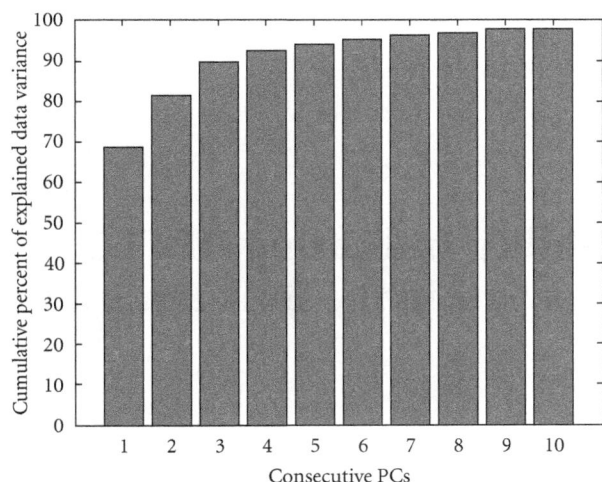

FIGURE 5: The cumulative percent of the explained data variance by the consecutive principal components.

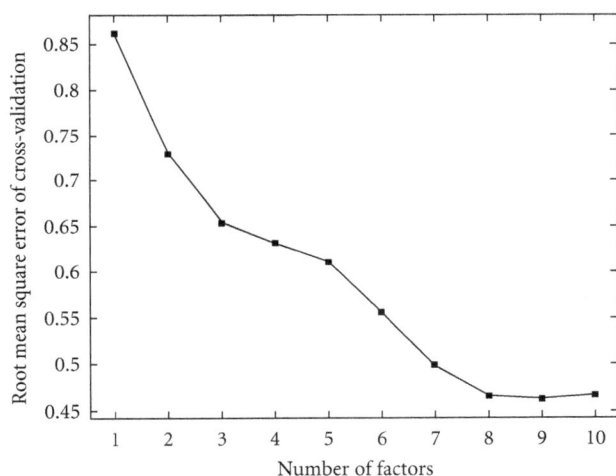

FIGURE 6: The cross-validation error for estimation of the model complexity.

sensitivities of each analytical technique applied. Due to completely different error sources in each approach and also to an unknown chemical composition of the fingerprinted extracts, for the time being, it seems noteworthy to pay roughly equal attention to the two series of the obtained results.

In spite of the differences in the spectrophotometric and chromatographic results, it can be concluded that, in terms of the flavonoid fraction contents, many individual sage species outperform *S. officinalis*.

The chromatographic fingerprints of the selectively derived flavonoid extracts proved useful for the construction of the chemometric models. In this study, they proved helpful in differentiating among the harvesting years with the investigated *Sage* species.

Acknowledgments

The work of two authors (D. Staszek and M. S. Wróbel) was partially supported by the Ph.D. scholarship granted to then in 2010 within the framework of the University as a Partner of the Economy Based on Science project subsidized by the European Social Fund of the European Union.

References

[1] J. Rzepa, Ł. Wojtal, D. Staszek et al., "Fingerprint of selected *Salvia* species by HS-GC-MS analysis of their volatile fraction," *Journal of Chromatographic Science*, vol. 47, no. 7, pp. 575–580, 2009.

[2] Ł. Ciesla, M. Hajnos, D. Staszek, Ł. Wojtal, T. Kowalska, and M. Waksmundzka-Hajnos, "Validated binary high-performance thin-layer chromatographic fingerprints of polyphenolics for distinguishing different *Salvia* species," *Journal of Chromatographic Science*, vol. 48, no. 6, pp. 421–427, 2010.

[3] Ł. Cieśla, D. Staszek, M. Hajnos, T. Kowalska, and M. Waksmundzka-Hajnos, "Development of chromatographic and free radical scavenging activity fingerprints by thin-layer chromatography for selected *Salvia* species," *Phytochemical Analysis*, vol. 22, no. 1, pp. 59–65, 2011.

[4] M. Sajewicz, D. Staszek, M. Natić, Ł. Wojtal, M. Waksmundzka-Hajnos, and T. Kowalska, "TLC-MS versus TLC-LC-MS fingerprints of herbal extracts. Part II. Phenolic acids and flavonoids," *Journal of Liquid Chromatography and Related Technologies*, vol. 34, no. 10-11, pp. 864–887, 2011.

[5] Polish Pharmaceutical Society, *Polish Pharmacopoeia VI*, Polish Pharmaceutical Society, Warsaw, Poland, 2002.

[6] M. Sajewicz, D. Staszek, Ł. Wojtal, T. Kowalska, M. Hajnos, and M. Waksmundzka-Hajnos, "Binary HPLC-Diode Array Detector and HPLC-Evaporative Light Scattering Detector fingerprints of methanol extracts from the selected sage (*Salvia*) species," *Journal of AOAC International*, vol. 94, pp. 71–76, 2011.

[7] M. Daszykowski, M. Sajewicz, J. Rzepa et al., "Comparative analysis of the chromatographic fingerprints of twenty different sage (*Salvia* L.) species," *Acta Chromatographica*, vol. 21, no. 4, pp. 513–530, 2009.

[8] P. H. C. Eilers, "A perfect smoother," *Analytical Chemistry*, vol. 75, no. 14, pp. 3631–3636, 2003.

[9] A. Savitzky and M. J. E. Golay, "Smoothing and differentiation of data by simplified least squares procedures," *Analytical Chemistry*, vol. 36, no. 8, pp. 1627–1639, 1964.

[10] R. J. Barnes, M. S. Dhanoa, and S. J. Lister, "Standard normal variate transformation and de-trending of near-infrared diffuse reflectance spectra," *Applied Spectroscopy*, vol. 43, no. 5, pp. 772–777, 1989.

[11] S. Wold, K. Esbensen, and P. Geladi, "Principal component analysis," *Chemometrics and Intelligent Laboratory Systems*, vol. 2, no. 1–3, pp. 37–52, 1987.

[12] M. Daszykowski, B. Walczak, and D. L. Massart, "Projection methods in chemistry," *Chemometrics and Intelligent Laboratory Systems*, vol. 65, no. 1, pp. 97–112, 2003.

[13] D. L. Massart, B. G. M. Vandeginste, L. M. C. Buydens, S. de Jong, P. J. Lewi, and J. Smeyers-Verbeke, *Handbook of Chemometrics and Qualimetrics: Part A*, Elsevier, Amsterdam, The Netherlands, 1997.

[14] R. W. Kennard and L. A. Stone, "Computer aided design of experiments," *Technometrics*, vol. 11, pp. 137–148, 1969.

6

Development and Validation of a Stability-Indicating HPTLC Method for Analysis of Rasagiline Mesylate in the Bulk Drug and Tablet Dosage Form

Singaram Kathirvel,[1] Suggala Venkata Satyanarayana,[2] and Garikapati Devalarao[3]

[1] Department of Pharmaceutical Analysis, Hindu College of Pharmacy, Amaravathi Road, Guntur 522002, India
[2] Department of Chemical Engineering, JNTU College of Engineering, Anantapur 515002, India
[3] Department of Pharmaceutical Analysis, KVSR Siddhartha College of Pharmaceutical Sciences, Vijayawada 520008, India

Correspondence should be addressed to Garikapati Devalarao, devalarao2007@gmail.com

Academic Editor: Wenkui Li

A simple and sensitive thin-layer chromatographic method has been established for analysis of rasagiline mesylate in pharmaceutical dosage form. Chromatography on silica gel 60 F_{254} plates with $6:1:2(v/v/v)$ butanol-methanol water as mobile phase furnished compact spots at R_f 0.76 ± 0.01. Densitometric analysis was performed at 254 nm. To show the specificity of the method, rasagiline mesylate was subjected to acid, base, neutral hydrolysis, oxidation, photolysis, and thermal decomposition, and the peaks of degradation products were well resolved from that of the pure drug. Linear regression analysis revealed a good linear relationship between peak area and amount of rasagiline mesylate in the range of 100–350 ng/band. The minimum amount of rasagiline mesylate that could be authentically detected and quantified was 11.12 and 37.21 ng/band, respectively. The method was validated, in accordance with ICH guidelines for precision, accuracy, and robustness. Since the method could effectively separate the drug from its degradation products, it can be regarded as stability indicating.

1. Introduction

Rasagiline mesylate (Figure 1) is a chemical inhibitor of the enzyme monoamine oxidase type-B which has a major role in the inactivation of biogenic and diet-derived amines in the central nervous system. Rasagiline is a propargylamine-based drug indicated for the treatment of idiopathic Parkinson's disease. It is designated chemically as (R)-N-(prop-2-ynyl)-2,3-dihydro-1H-inden-1-amine. Rasagiline is freely soluble in water and ethanol and sparingly soluble in isopropyl alcohol. It is a chiral compound with one asymmetric carbon atom in a five-member ring with an absolute R-configuration which is produced as single enantiomer [1].

There are many methods reported in the literature for analysis of rasagiline mesylate, for example, GC-MS [2], HPLC methods [3–5], LC-MS/MS in human plasma [6], and spectrophotometric methods [7, 8]. However, no available densitometric method for the simultaneous

separation of rasagiline mesylate from degradation products has been reported. Hence, the objective of the present study is to develop and validate a new HPTLC method for the estimation of rasagiline mesylate in bulk drug and its dosage form.

Today, HPTLC is rapidly becoming a routine analytical technique due to its advantages of low operating costs high sample throughput, and need for minimum sample preparation. The major advantage of HPTLC is that several samples can be run simultaneously using a small quantity of mobile phase unlike HPLC, thus reducing the analysis time and cost per analysis [9]. Accordingly, the aim of the present study was to establish the inherent stability of rasagiline mesylate through stress studies under a variety of ICH recommended test conditions [10, 11] and to develop a stability-indicating HPTLC assay [12]. The proposed method was validated according to the ICH guidelines [13].

Development and Validation of a Stability-Indicating HPTLC Method for Analysis of Rasagiline Mesylate in the Bulk Drug and Tablet Dosage Form

37

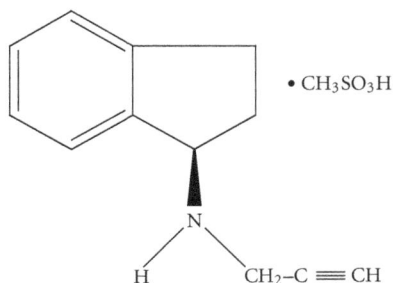

FIGURE 1: Structure of rasagiline mesylate.

2. Experimental

2.1. Materials and Reagents.
Rasagiline mesylate was obtained from Orchid Pharmaceuticals Ltd (Chennai, India). It was used without further purification and certified to contain 99.67% (w/w) on dry weight basis. Rasagiline Mesylate Pharmaceutical preparation, Rasipron was purchased from a local drug store; it contained 1 mg active material. The other chemicals and reagents used were of AR grade and procured from S.D. Fine-Chem (New Delhi, India).

2.2. HPTLC Instrumentation.
Chromatography was performed on 10 cm × 10 cm aluminium foil plates precoated with 0.2 mm layers of silica gel 60 F_{254} (E. Merck, Germany). Before use, the plates were prewashed by development with methanol then dried in the current of dry air and activated at 60°C for 5 min. Samples were applied as bands 6 mm wide, 15 mm apart, by use of a Camag (Switzerland) Linomat 5 equipped with a microlitre syringe. A constant application rate of 150 nL s^{-1} was used. Butanol methanol water 6 : 1 : 2 (v/v/v) was used as the mobile phase. Linear ascending development was performed in a twin-trough glass chamber previously saturated with mobile phase vapour for 30 min at room temperature (RT, 25 ± 2°C) and relative humidity 60 ± 5%. The development distance was approximately 80 mm. After development, the plates were dried in current of air by use of an air dryer. Densitometric scanning, at 254 nm, was performed with a Camag TLC scanner 3 in absorbance mode. The source of radiation was a deuterium lamp emitting a continuous UV spectrum in the range 190–400 nm.

2.3. Calibration Plots of Rasagiline Mesylate.
A stock solution containing 100 μg/mL of rasagiline mesylate was prepared by dissolving an accurately weighed 10 mg portion of the drug in methanol in 100 mL volumetric flask. Different volumes of stock solution (1, 1.5, 2, 2.5, 3, 3.5 μL) were spotted on an HPTLC plate in triplicate to obtain concentrations of 100, 150, 200, 250, 300, and 350 ng/band of rasagiline mesylate, respectively. The data of peak area versus drug concentration were treated by linear least-squares regression.

2.4. Method Validation

2.4.1. Precision.
The precision of the method was verified by repeatability and intermediate precision studies. Repeatability studies were performed by analyzing three different concentrations (100, 200, and 300 ng/spot) of the drug by six times on the same day. The intermediate precision of the method was checked by repeating on three different days.

2.4.2. Robustness.
The analytical conditions were deliberately changed, by introducing small changes in mobile phase composition (±2%), mobile phase volume (±2%), chamber saturation period (±10%), and development distance (±10%), and time from application to development (0, 10, 15, 20 min) and time from development to scanning (0, 30, 60, 90 min) were carried out.

2.4.3. LOD and LOQ.
The method was used to determine the LOD and LOQ. Blank methanol was spotted six times, and the SD (S_b) of the peak area of the blanks was calculated. The limits were determined from the slope (S) of the calibration plot and the SD of the response for the blank sample (S_b) by use of the formula:

$$LOD = 3.3 \times \frac{S_b}{S}, \qquad LOQ = 10 \times \frac{S_b}{S}. \qquad (1)$$

2.4.4. Accuracy.
To check the degree of accuracy of the method, recovery studies were performed in triplicate by standard addition method at 50, 100, and 150%. Known amount of standard rasagiline mesylate was added to pre-analysed samples and was subjected to the proposed HPTLC method.

2.4.5. Specificity.
The specificity of the method was determined by comparing the results for the standard drug and the sample. The peak purity of the sample was assessed by comparing the spectra at peak start, peak apex, and peak end positions of the band.

2.4.6. Forced Degradation Studies of Rasagiline Mesylate.
A stock solution containing 10 mg rasagiline mesylate in 100 mL methanol (100 μg/mL) was used for forced degradation to provide an indication of the stability-indicating ability and specificity of the proposed method. Studies of acid-induced decomposition were performed by exposing the solution of the drug to 5 M hydrochloric acid by heating the solution under reflux at 80°C for 5 h. Studies of base-induced decomposition were performed in 5 M sodium hydroxide, and the solution was heated under reflux for 3 h at 80°C. The resulting solutions were applied on TLC plate in triplicate (2.5 μL each, i.e. 250 ng/band). The plate was chromatographed as described above. Oxidative degradation was studied by refluxing the drug solution to 30% hydrogen peroxide for 4 h at 80°C. The resulting solution was applied to TLC plates such that 250 ng per band was applied to the plates. Neutral hydrolysis was carried out by accurately weighing 25 mg of the drug in 25 mL HPLC grade water, and the solution was refluxed for 12 hrs at 50°C. Appropriate aliquot was taken from the above solution and diluted with methanol to obtain a final concentration of 100 ng/μL. The resultant solution (2.5 μL) was applied to TLC plate (250 ng/spot), and the chromatogram was run to evaluate the degradation effect. Similarly, dry heat degradation also

carried out by placing the standard drug in solid form in an oven at 100°C for 10 hrs. Rasagiline mesylate 1 mg was accurately weighed and separately dissolved in methanol in 10 m L volumetric flask. The resultant solution was applied to TLC plate in such a way that final concentration achieved was 250 ng/spot. The photochemical stability of the drug was also studied by exposing the stock solution to direct sunlight (60,000–70,000 lux) for 24 h on a wooden plank and kept on terrace. The solution (2.5 μL, equivalent to 250 ng/band) was then applied to TLC plates, and densitograms were obtained as described above. The densitogram was run in triplicate in all the conditions to evaluate the degradative effect.

2.5. Analysis of Rasagiline Mesylate in Tablet Dosage Form. To determine the rasagiline mesylate content of conventional tablets, twenty tablets were weighed and powdered in a glass mortar. An amount of powder equivalent to 10 mg rasagiline mesylate was transferred to a 100 mL volumetric flask, extracted with methanol, sonicated for 20 min, and diluted to volume with same solvent. The resulting solution was filtered through a 0.45 μm filter (Millifilter; Milford, MA; USA). The solution (2.5 μL, 250 ng rasagiline mesylate) was applied in triplicate on an HPTLC plate for quantification using the proposed method.

3. Results and Discussion

3.1. Optimization of the Mobile Phase. Several solvent mixtures in different ratios were tested to obtain a compact band of rasagiline mesylate. Butanol : methanol : water 6 : 1 : 2 v/v/v) was found to give a compact band for rasagiline mesylate with an R_f value of 0.76 ± 0.01 (Figure 2). This mobile phase gave good resolution for the separation of rasagiline mesylate, all degradation products, and was selected for the proposed stability-indicating method. Thirty minutes was found to be sufficient for saturation of the development chamber with the mobile phase vapor in order to obtain separation of the compounds. A 20 mL aliquot of mobile phase was used for a 20 min development over a distance of 80 mm.

3.2. Calibration Plots of Rasagiline Mesylate. The linear regression analysis data for the calibration plots showed a good linear relationship ($r^2 = 0.9993 ± 0.02$) with respect to peak area in the concentration range of 100–350 ng/band (Figure 3). The mean values of the slope and intercept were 5.40 ± 0.257 and 23.15 ± 1.791, respectively, for densitometric analysis at 254 nm (Table 1).

3.3. Method Validation

3.3.1. Precision. The results of the repeatability and intermediate precision experiments are shown in Table 2. Inter precision studies were checked by both interday and intraday analysis. Interday analysis was carried out by repeating the experiments on three different days, whereas intraday analysis was done for 3 times on the same day. The developed method was found to be precise as the RSD values for

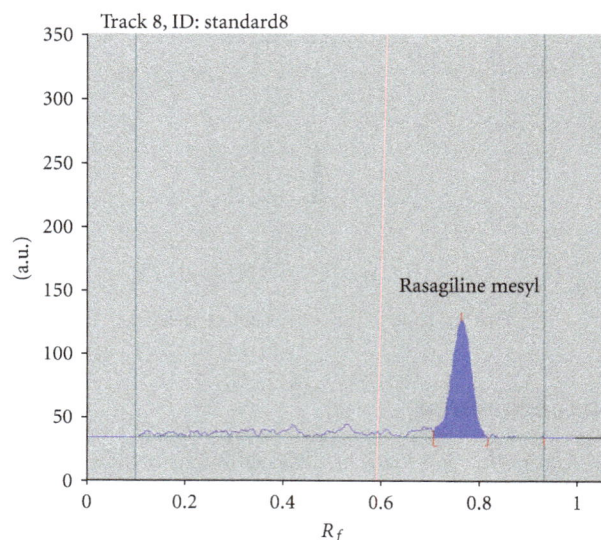

FIGURE 2: A typical HPTLC densitogram of rasagiline mesylate at 254 nm ($R_f - 0.76 ± 0.01$).

TABLE 1: Linear regression data for the calibration curves ($n = 6$).

Parameter	Rasagiline mesylate
Linearity	100–350 ng/band
Correlation coefficient, r±SD	0.9993 ± 0.02
Slope ±SD	5.40 ± 0.257
Intercept ±SD	23.15 ± 1.791
LOD, ng/band	11.12
LOQ, ng/band	37.21

repeatability and intermediate precision studies were <2%, respectively, as recommended by ICH guideline.

3.3.2. Robustness. The low values of RSD obtained after introducing small, deliberate changes in the mobile phase composition, mobile phase volume, chamber saturation time, time from application to development, and time from development to scanning in the developed HPTLC method indicated the robustness of the method. The % RSD values of all the above-mentioned parameters are shown in Table 3.

3.3.3. LOD and LOQ. LOD and LOQ were determined by the SD method and were found to be 11.12 and 37.21 ng/band, respectively (Table 1).

3.3.4. Accuracy. Accuracy of the method was obtained by recovery after spiking with 50, 100, and 150% of additional drug. The study was carried out in triplicate by standard addition method and found to be in the range of 99.10–101.0% (Table 4).

3.3.5. Specificity. The R_f value (0.76 ± 0.01) of the sample and standard was almost identical, and spectra of the sample and the standard were superimposable. These results indicated the specificity of the method.

Development and Validation of a Stability-Indicating HPTLC Method for Analysis of Rasagiline Mesylate in the Bulk Drug and Tablet Dosage Form

39

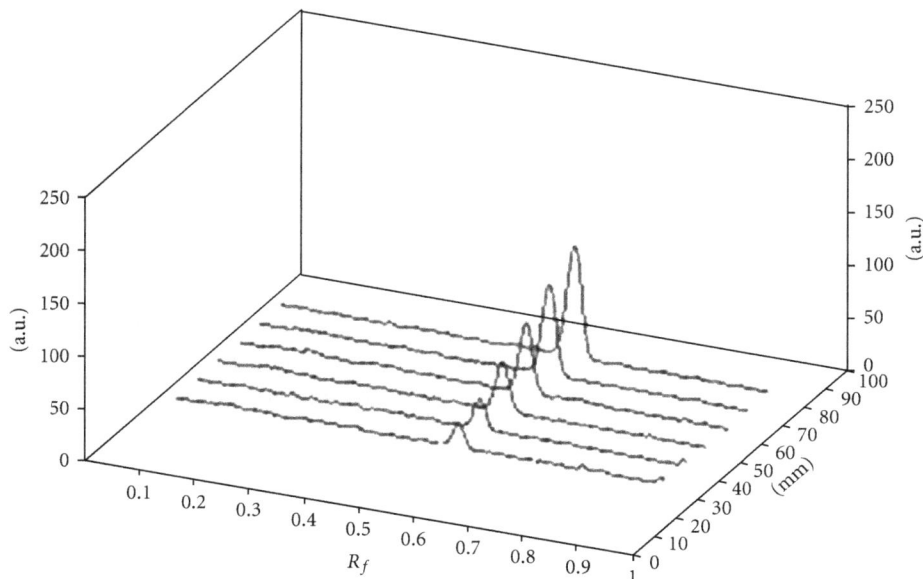

FIGURE 3: 3D representation of rasagiline mesylate sample. That is, track 1 is 100 ng/band, track 2 is 150 ng/band, track 3 is 200 ng/band, track 4 is 250 ng/band, track 5 is 300 ng/band and track 6 is 350 ng/band.

TABLE 2: Precision studies ($n = 6$).

Concentration (ng/spot)	Repeatability		Intermediate precision	
	Mean $\pm$ SD (ng/spot)	RSD (%)	Mean $\pm$ SD (ng/spot)	RSD (%)
100	98.20 $\pm$ 1.95	1.98	99.88 $\pm$ 1.76	1.76
200	197.41 $\pm$ 1.67	0.84	196.50 $\pm$ 1.50	0.76
300	298.77 $\pm$ 1.08	0.36	297.31 $\pm$ 1.80	0.60

TABLE 3: Robustness testing ($n = 6$).

Parameters	% RSD
% change in mobile phase	0.72 $\pm$ 0.01
Chamber saturation time	0.74 $\pm$ 0.02
Development distance	0.76 $\pm$ 0.01
Time from application to development	0.75 $\pm$ 0.02
Time from development to scanning	0.74 $\pm$ 0.01

TABLE 4: Recovery studies[a].

Excess drug added to the analyte (%)	Amount of drug found (ng)	% Recovery	% RSD
0	100.0	99.10	0.72
50	151.6	101.06	0.49
100	198.4	99.21	0.31
150	252.2	100.88	0.44

Matrix containing 100 ng of drug. [a]$n = 6$.

TABLE 5: Stability of rasagiline mesylate in sample solution[a].

Actual (ng)	Area mean	Area range	% RSD
	813.1943	809.0121–819.5817	0.58
200	1120.8268	1116.3661–1129.7529	1.58

[a]$n = 6$.

the same TLC plate after development, the chromatogram was evaluated as listed in (Table 5) for additional spots if any. The % RSD for the samples analyzed at different elapsed assay times was found to be <2%. Thus, the drug was stable in solution state. There was no indication of compound instability in the sample solution.

3.3.7. Forced Degradation Studies of Rasagiline Mesylate.
The bands of degradation products were well resolved from the drug bands. The peak of rasagiline mesylate was not significantly shifted in the presence of degradation peaks, which indicated the stability-indicating property of the proposed method. The number of degradation products with their R_f values under different stress conditions is shown in Table 6.

The chromatograms, of the samples treated with acid, base, hydrogen peroxide, and sunlight showed well-separated bands of pure rasagiline mesylate as well as some additional

3.3.6. Stability in Sample Solution.
Solutions of two different concentrations (150 and 200 ng/spot) were prepared from sample solution and stored at room temperature for 6.0, 12.0, 24.0, 48.0, and 72.0 h, respectively. They were then applied on

TABLE 6: Degradation of rasagiline mesylate.

Condition	Time (h)	Recovery (%)	(%) value of degraded products
5 M HCl refluxed at 80°C	5	47.50	7.53, 44.96
5 M NaOH refluxed at 80°C	3	46.92	22.92, 11.96, 18.75
H_2O_2 30% v/v refluxed at 80°C	4	63.30	5.45, 9.89, 21.37
Daylight	24	100	None detected
Dry heat degradation	10	100	None detected
Neutral hydrolysis (Reflux at 50° C)	12	100	None detected

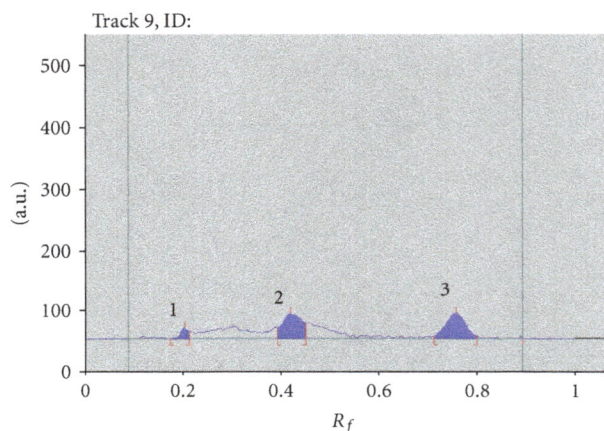

FIGURE 4: Densitogram of acid (5NHCl, reflux for 5 h, temp 80°C) treated rasagiline mesylate; peak 1 (degraded—R_f = 0.20), peak 2 (degraded—R_f = 0.42), and peak 3 (rasagiline mesylate—R_f = 0.76).

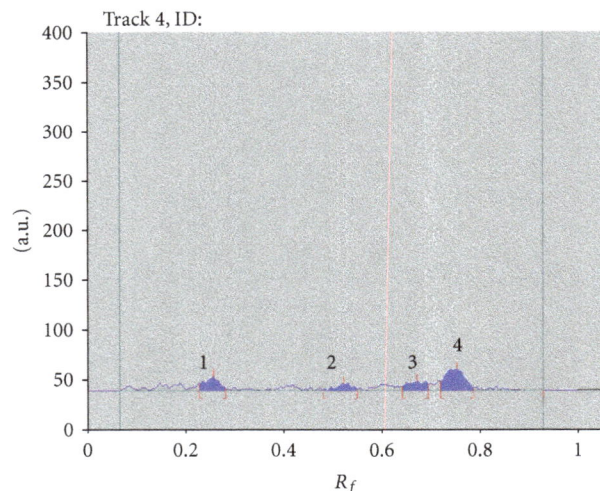

FIGURE 5: Densitogram of base (5 N NaOH, reflux for 3 h, temp 80°C) treated rasagiline mesylate; peak 1 (degraded—R_f = 0.26), peak 2 (degraded—R_f = 0.52), peak 3, (degraded—R_f = 0.67), and peak 4 (rasagiline mesylate—R_f = 0.76).

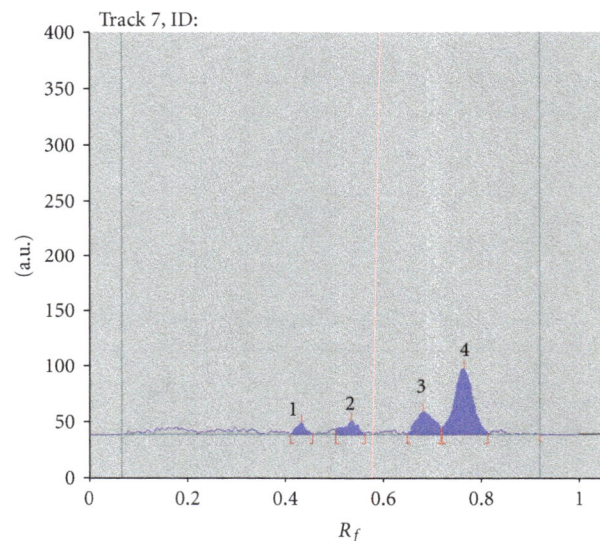

FIGURE 6: Densitogram of hydrogen peroxide (30% v/v, reflux for 4 h, temp 80°C) treated rasagiline mesylate; peak 1 (degraded—R_f = 0.43), peak 2 (degraded—R_f = 0.53), peak 3 (degraded—R_f = 0.68), and peak 4 (rasagiline mesylate—R_f = 0.76).

bands of the degradation products at different R_f values (Table 6).

Drug recovery from the acid-stressed sample was found to be 47.50%. Base-stressed samples showed recovery at the level of 46.92% (Table 6). The chromatogram of the acid-degraded sample showed two additional bands at R_f values of 0.20 and 0.42 (Figure 4). The chromatogram of the base-degraded sample showed three additional bands at R_f values of 0.26, 0.52, and 0.67 (Figure 5).

The chromatogram of the sample of rasagiline mesylate treated with 30% (v/v) H_2O_2 had three additional bands at R_f 0.43, 0.53, and 0.68 showing that rasagiline mesylate was susceptible to oxidation-induced degradation (Figure 6). Rasagiline mesylate remained stable in neutral, thermal, and photolytic conditions. All the above conditioned parameters showed recovery at the level of 100% suggesting that the drug does not undergo degradation in neutral, thermal, and photolytic conditions.

3.3.8. Analysis of Rasagiline Mesylate in Tablet Dosage Form.

A single spot at R_f − 0.76 was observed in the chromatogram of the drug samples extracted from tablets. There was no interference from the excipients commonly present in the tablets. The drug content was found to be 99.98% with a % RSD of 0.57 (n = 6). The low % RSD value indicated the suitability of this method for routine analysis of rasagiline mesylate in pharmaceutical dosage form.

4. Conclusion

The introduction of HPTLC to pharmaceutical analysis is a major step in quality assurance. This HPTLC technique is precise, specific, accurate, and stability indicating. Statistical analysis proved the method is suitable for analysis of rasagiline mesylate as the bulk drug and in a pharmaceutical formulation without interference from excipients. This is a typical stability-indicating assay, established in accordance with the recommendations of the ICH guidelines. The method can be used to determine the purity of a drug, and it is also proposed for analysis of drug and degradation products in stability samples obtained during industrial production.

References

[1] J. J. Chen, D. M. Swope, and K. Dashtipour, "Comprehensive review of rasagiline, a second-generation monoamine oxidase inhibitor, for the treatment of Parkinson's Disease," *Clinical Therapeutics*, vol. 29, no. 9, pp. 1825–1849, 2007.

[2] J. J. Thébault, M. Guillaume, and R. Levy, "Tolerability, safety, pharmacodynamics, and pharmacokinetics of rasagiline: a potent, selective, and irreversible monoamine oxidase type B inhibitor," *Pharmacotherapy*, vol. 24, no. 10, pp. 1295–1305, 2004.

[3] M. Fernández, E. Barcia, and S. Negro, "Development and validation of a reverse phase liquid chromatography method for the quantification of rasagiline mesylate in biodegradable PLGA microspheres," *Journal of Pharmaceutical and Biomedical Analysis*, vol. 49, no. 5, pp. 1185–1191, 2009.

[4] R. Narendrakumar, G. Nageswara Rao, and P. Y. Naidu, "Stability indicating RP-LC method for determination of rasagiline mesylate in bulk and pharmaceutical Dosage forms," *International Journal of Applied Biology and Pharmaceutical Technology*, vol. 1, no. 2, pp. 247–259, 2010.

[5] M. Vijayalakshmi, J. V. L. N. Seshagiri Rao, and A. Lakshmana Rao, "Development and validation of RP-HPLC method for the estimation of rasagiline tablet dosage forms," *Rasayan Journal of Chemistry*, vol. 3, no. 4, pp. 621–624, 2010.

[6] J. Chen, X. Duan, P. Deng, H. Wang, and D. Zhong, "Validated LC-MS/MS method for quantitative determination of rasagiline in human plasma and its application to pharmacokinetic studies," *Journal of Chromatography B*, vol. 873, no. 2, pp. 203–208, 2008.

[7] G. Devalarao, S. Kathirvel, and S. V. Satyanarayana, "Simple spectrophotometric methods for the determination of rasagiline mesylate in pharmaceutical dosage form," *Journal of Pharmacy Research*, vol. 4, no. 1, pp. 61–62, 2011.

[8] B. Rama and K. Preeti, "UV Spectrophotometric method for the determination of rasagiline mesylate in bulk and pharmaceutical formulations," *International Journal of Pharmaceutical Sciences Review and Research*, vol. 5, no. 1, pp. 5–7, 2010.

[9] G. Y. S. K. Swamy, K. Ravikumar, L. K. Wadhwa, R. Saxena, and S. Singh, "$(2R^*, 3R^*, 6S^*)$-N,6-Bis(4-fluorophenyl)-2-(4-hydroxyphenyl)-3,4,5,6-tetrahydro-2H-pyran-3-carboxamide," *Acta Crystallographica Section E*, vol. 61, no. 11, pp. 3608–3610, 2005.

[10] ICH, "Stability testing of new drug substances and products Q1A (R2)," in *Proceedings of the International Conference on Harmonization*, International Federation of Pharmaceutical Manufacturers and Association, Geneva, Switzerland, 2003.

[11] S. Singh and M. Bakshi, "Guidance on conduct of stress tests to determine inherent stability of drugs," *Pharmacy Technician Online*, vol. 24, pp. 1–14, 2000.

[12] M. Bakshi and S. Singh, "Development of validated stability-indicating assay methods—critical review," *Journal of Pharmaceutical and Biomedical Analysis*, vol. 28, pp. 1011–1040, 2002.

[13] ICH Topic Q2 (R1), Validation of analytical procedures: Methodology, Geneva, The European Agency for the Evaluation of Medicinal Products, September 2005.

7

Separation of Cyclic Dipeptides (Diketopiperazines) from Their Corresponding Linear Dipeptides by RP-HPLC and Method Validation

Mareike Perzborn, Christoph Syldatk, and Jens Rudat

Institute of Process Engineering in Life Sciences, Section II: Technical Biology, Karlsruhe Institute of Technology (KIT), 76131 Karlsruhe, Germany

Correspondence should be addressed to Mareike Perzborn; mareike.perzborn@kit.edu

Academic Editor: Toyohide Takeuchi

Simple, rapid, sensitive, precise, and accurate methods for detection and separation of seven diketopiperazines (DKPs), cyclo(Gly-Gly), cyclo(DL-Ala-DL-Ala), cyclo(L-Asp-L-Phe), cyclo(L-Asp-L-Asp), cyclo(Gly-L-Phe), cyclo(L-Pro-L-Tyr), and cyclo(L-Arg-L-Arg), from their corresponding linear dipeptides and related amino acids L-Phe and L-Tyr by reversed-phase high-performance liquid chromatography (RP-HPLC) were established. Moreover, for the racemic DKP cyclo(DL-Ala-DL-Ala) and dipeptide DL-Ala-DL-Ala, separation of the diastereomers was achieved. All methods can be performed within 15 min. For all DKPs, dipeptides, and amino acids, linear ranges with correlation coefficients R^2 greater than 0.998 were determined. Lowest limits of detection were found to be between 0.05 and 10 nmol per 10 μL injection, depending on the substance. For all tested substances intrarun and interrun precision ranged from 0.5 to 4.7% and 0.7 to 9.9% relative standard deviation, and accuracy was between −4.2 and 8.1% relative error. Short-term and freeze-thaw stabilities were 93% or greater for all substances. Recovery rate after heat treatment was determined to be at least 97%. These methods will be useful for quantitative determination of DKPs and their potential biodegradation products: dipeptides and amino acids.

1. Introduction

Diketopiperazines (DKPs) are the smallest possible cyclic peptides composed of two α-amino acids. They are abundant natural compounds produced by various bacteria like *Streptomyces* sp. [1], *Pseudomonas aeruginosa* [2], or *Lactobacillus plantarum* [3], fungi, e.g., *Aspergillus flavus* [4] or *Alternaria alternata* [5], and marine sponges like *Dysidea herbacea* [6]. Recently, the interest in this substance class has increased due to their immense bioactivities including antibacterial activity [7], antifungal function [3], cytotoxicity [4], phytotoxicity [5], and inhibition of plasminogen activator inhibitor-1 [8]. DKPs were shown to act as quorum sensing molecules; e.g., cyclo(L-Pro-L-Tyr), used in this study, was identified in culture supernatant of *Pseudomonas aeruginosa* and was identified as an activator of an N-acylhomoserine lactone biosensor [2]. Besides their widespread biosynthesis in nature, DKPs occur as chemical degradation products of, for

example, amoxicillin, an aminopenicillin antibiotic [9], neuropeptide substance P [10], angiotensin converting enzyme inhibitor enalapril [11], or the sweetener aspartame with cyclo(L-Asp-L-Phe) as degradation product [12–17]. Amoxicillin and especially its degradation products can be detected in aquatic environment and food of animal origin, thus they are under discussion as a health problem due to their allergenic potential [9]. Analysis of these compounds can be performed with HPLC-MS/MS [9] or UHPLC-MS/MS [18]. Enalapril and two degradation products are determined with RP-HPLC and detection at 215 nm [11]. Aspartame and its degradation products can be analyzed via RP-HPLC and detection of peptide bond at 195 nm [12], 200 nm and 220 nm [13] or 210 nm [14], capillary electrophoresis and detection at 214 nm [15], HPLC coupled with MS/MS [16] or analysis of trimethylsilyl derivatives with GC [17]. Cyclo(Gly-Gly) and Gly are separated by RP-HPLC and detected at 200 nm [19]. Thus, there are some analysis methods of selected

Separation of Cyclic Dipeptides (Diketopiperazines) from Their Corresponding Linear Dipeptides by RP-HPLC and
Method Validation

43

DKPs available, but to the best of our knowledge there is no publication of a comprehensive study for separation of DKPs and their corresponding linear dipeptides and amino acids and method validation. However, this would be of great interest for studies on microbial, enzymatical, and chemical degradation of this biologically active substance class. Until now, there are only few bacterial strains reported which can hydrolyze DKPs to the corresponding linear dipeptides. One strain is *Paenibacillus chibensis* (DSM 329) hydrolyzing the aspartame degradation product cyclo(L-Asp-L-Phe) [20]. This strain was chosen as biological matrix for method validation in this work.

The aim of this study was to establish rapid methods for detection and separation of seven selected DKPs from their corresponding dipeptides and in some cases the amino acids. Moreover, these methods were validated for sensitivity, linearity, precision, accuracy, stability, and recovery.

2. Experimental

2.1. Chemicals and Reagents. The standards cyclo(L-Ala-L-Ala), L-Asp-L-Phe, L-Phe-L-Asp, cyclo(L-Pro-L-Tyr), L-Pro-L-Tyr, L-Tyr-L-Pro, cyclo(L-Asp-L-Asp), L-Asp-L-Asp, cyclo(Gly-L-Phe), Gly-L-Phe, L-Phe-Gly, and L-Arg-L-Arg acetate salt were purchased from Bachem Holding (Bubendorf, Switzerland), cyclo(Gly-Gly), Gly-Gly, and DL-Ala-DL-Ala were obtained from TCI Europe N.V. (Zwyndrecht, Belgium), cyclo(DL-Ala-DL-Ala), L-Ala-L-Ala, cyclo(L-Asp-L-Phe) and L-Phe were received from Sigma-Aldrich (St. Louis, USA), L-Tyr was purchased from Carl Roth Corporation (Karlsruhe, Germany), and cyclo(L-Arg-L-Arg) acetate salt was provided by Taros Chemicals Corporation (Dortmund, Germany).

Methanol ROTISOLV HPLC Gradient Grade (MeOH), peptone from casein, K_2HPO_4, KH_2PO_4, $Na_2HPO_4 \times 2H_2O$, $NaH_2PO_4 \times 2H_2O$ were supplied by Carl Roth Corporation (Karlsruhe, Germany). Yeast extract was obtained from Becton, Dickinson and Company (Franklin Lakes, USA). $MgSO_4 \times 7H_2O$ was purchased from Sigma-Aldrich (St. Louis, USA).

2.2. Instrumentation and Chromatographic Conditions. The HPLC analysis was performed using an Agilent 1200 system (Agilent Technologies, Santa Clara, USA) equipped with a quaternary pump, a degasser, an autosampler, a thermostated column compartment, and a variable wavelength detector. Separations were carried out with isocratic elution on a reversed-phase column NUCLEODUR Sphinx RP (4.6 mm ID × 250 mm, 5 μm particle size, Macherey-Nagel Corporation, Düren, Germany) connected with a C_{18} security guard column (3.0 mm ID × 4 mm, Phenomenex, Torrance, USA). The mobile phase consisted of varying ratios of MeOH and 20 mM sodium phosphate buffer (pH 5.5). The buffer was filtered through a nitrocellulose membrane with 0.22 μm pore size (Merck Millipore, Darmstadt, Germany). Best separation conditions were investigated by determination of retention time (t_R), separation factor (α), and peak resolution (R_s) for each standard with 10%, 20%, and 30% MeOH. After

determination of best separation conditions, the following compositions of mobile phase were used. Separation of cyclo(Gly-Gly) from Gly-Gly, cyclo(DL-Ala-DL-Ala) from DL-Ala-DL-Ala, cyclo(L-Asp-L-Asp) from L-Asp-L-Asp, and cyclo(L-Arg-L-Arg) from L-Arg-L-Arg was done with 10% MeOH and 90% buffer. 20% MeOH and 80% buffer were used to separate cyclo(L-Asp-L-Phe), L-Asp-L-Phe, L-Phe-L-Asp, and L-Phe. With 30% MeOH and 70% buffer cyclo(L-Pro-L-Tyr) was separated from L-Pro-L-Tyr, L-Tyr-L-Pro, and L-Tyr, and cyclo(Gly-L-Phe) from Gly-L-Phe, L-Phe-Gly, and L-Phe. The injection volume was 10 μL and detection wavelength was 210 nm. The flow rate was 0.7 mL min^{-1}, and the column temperature was set to 20°C (for analysis with 10 and 20% MeOH) or to 30°C (for 30% MeOH). Differing from these standard conditions cyclo(L-Asp-L-Asp) and L-Asp-L-Asp were separated with 15°C column temperature and 0.3 mL min^{-1} flow rate; cyclo(L-Arg-L-Arg) was detected at a wavelength of 268 nm.

Validation of each substance was performed under best separation conditions and run time was 15 min for each sample.

2.3. Preparation of Standard Solutions. Stock solutions of DKPs, dipeptides, and amino acids were prepared in 50 mM sodium phosphate buffer (pH 7.5) and stored at −20°C. The concentration of each stock solution was adjusted to 50 mM, except for 25 mM cyclo(L-Asp-L-Phe), 10 mM cyclo(Gly-L-Phe), and 5 mM L-Tyr. These stock solutions were serially diluted and used directly for determination of the linear range. Calibration curves were established based on seven concentrations within the linear range. For method validation the stock solutions were diluted to low, medium, and high concentration and were mixed (1 : 1) with crude extract of DSM 329.

2.4. Crude Extract Preparation. Crude extract was prepared from the bacterial strain *Paenibacillus chibensis* (DSM 329) purchased from German Collection of Microorganisms and Cell Cultures (DSMZ, Braunschweig, Germany). The strain was cultivated in modified complex medium (10 g L^{-1} peptone, 10 g L^{-1} yeast extract, 3 g L^{-1} K_2HPO_4, 1 g L^{-1} KH_2PO_4, 0.5 g L^{-1} $MgSO_4 \times 7 H_2O$, adjusted to pH 7.2) [20] at 30°C and 100 rpm for 24 h until reaching an OD$_{600}$ of 2.1. Cells were washed three times by centrifugation (4,816 ×g, 30 min, 4°C), discarding the supernatant and resuspension in 50 mM sodium phosphate buffer (pH 7.5). Washed cell pellet was resuspended in the same buffer to an OD$_{600}$ of approximately 17. Cells were disrupted by sonication for 25 min using alternate intervals of 30 sec pulsation on and 30 sec pulsation off and 35% amplitude (Sonopuls HD 3100 with ultrasonic probe MS 72, Bandelin electronic Corporation, Berlin, Germany). After centrifugation (4,816 ×g, 30 min, 4°C), the crude extract was inactivated by heating at 90°C (20 min, 1,500 rpm) with Thermomixer comfort (Eppendorf Corporation, Hamburg, Germany) and centrifuged (24,725 ×g, 20 min, 4°C). The protein concentration was determined by Bio-Rad Protein Assay (Bio-Rad Laboratories, Hercules, USA) according to

TABLE 1: Retention factor (k) for DKPs, dipeptides, and amino acids and separation factor (α) and resolution (R_s) of two adjacent peaks at chosen separation conditions.

Separation conditions	Substance	Retention factor k	Separation factor α	Resolution R_s
10% MeOH, 20°C, 0.7 mL min^{-1}	Cyclo(Gly-Gly)	1.03	1.63	5.47
	Gly-Gly	0.63		
	Cyclo(L-Ala-D-Ala)	2.34	1.18	3.26
	Cyclo(L-Ala-L-Ala) and cyclo(D-Ala-D-Ala)	2.75		
	L-Ala-L-Ala and D-Ala-D-Ala	0.67	1.25	1.83
	L-Ala-D-Ala	0.84		
	Cyclo(L-Ala-D-Ala)	2.34	2.79	13.77
	L-Ala-D-Ala	0.84		
	Cyclo(L-Arg-L-Arg)	1.08	1.36	1.55
	L-Arg-L-Arg	0.79		
10% MeOH, 15°C, 0.3 mL min^{-1}	Cyclo(L-Asp-L-Asp)	0.61	1.15	1.04
	L-Asp-L-Asp	0.53		
20% MeOH, 20°C, 0.7 mL min^{-1}	Cyclo(L-Asp-L-Phe)	3.33	1.14	2.31
	L-Phe	2.93		
	L-Asp-L-Phe	2.10	1.40	5.45
	L-Phe	2.93		
	L-Asp-L-Phe	2.10	1.59	5.88
	L-Phe-L-Asp	1.32		
30% MeOH, 30°C, 0.7 mL min^{-1}	Cyclo(L-Pro-L-Tyr)	3.75	2.26	9.44
	L-Tyr-L-Pro	1.66		
	L-Tyr-L-Pro	1.66	1.76	4.23
	L-Tyr	0.95		
	L-Pro-L-Tyr	0.93	1.02	0.22
	L-Tyr	0.95		
	Cyclo(Gly-L-Phe)	4.36	2.31	16.12
	L-Phe-Gly	1.89		
	L-Phe-Gly	1.89	1.01	0.17
	L-Phe	1.87		
	Gly-L-Phe	1.84	1.02	0.27
	L-Phe	1.87		

Hold-up time = 2.167 min (manufacturer information).

the manufacturers' instructions. The obtained crude extract was diluted with the described buffer to a final protein concentration of 0.3 mg mL^{-1}.

2.5. Method Validation. The detection of DKPs, dipeptides, and amino acids was validated for sensitivity, linearity, precision, accuracy, stability, and recovery according to the bioanalytical method validation guidelines of the FDA [21]. Validation of each substance was done under best separation conditions, which were determined in this study.

Intrarun precision and accuracy were assessed using five replicates at low, medium, and high concentration, within consecutive runs. Interrun precision was determined using these experimental conditions on three different days. Precision was expressed as relative standard deviation, RSD (%) = 100∗ (standard deviation/mean). Accuracy was calculated as relative error, RE (%) = 100 ∗ (mean measured concentration − true concentration/true concentration). The acceptance limits were set at a maximum of ±15% at medium and high concentrations and ±20% at low concentration. For each substance the mean value of low, medium, and high concentrations was calculated for precision and accuracy.

Short-term temperature stability and freeze-thaw stability were determined with freshly spiked samples at low and high concentration in triplicate. Stability was calculated by comparing the peak areas obtained by direct injection, after storage at 21°C for 24 h, and after three freeze and thaw cycles. For determination of freeze and thaw stability samples were frozen at −20°C for 24 h, completely thawed and refrozen for 21 h. The freeze-thaw cycle was repeated two times and samples were analyzed on the third cycle.

Recovery was determined by using four replicates at low, medium, and high concentrations. Evaluation was done by comparing the peak areas of samples spiked after heat treatment with samples spiked before heat treatment. For heat treatment samples were incubated at 90°C and 1,400 rpm for 10 min using Thermomixer comfort (Eppendorf Corporation, Hamburg, Germany).

3. Results and Discussion

3.1. Separation. Best separation conditions of DKPs, their corresponding linear dipeptides, and the amino acids L-Phe and L-Tyr are summarized in Table 1. These two aromatic

Separation of Cyclic Dipeptides (Diketopiperazines) from Their Corresponding Linear Dipeptides by RP-HPLC and Method Validation

45

(a)

(b)

(c)

(d)

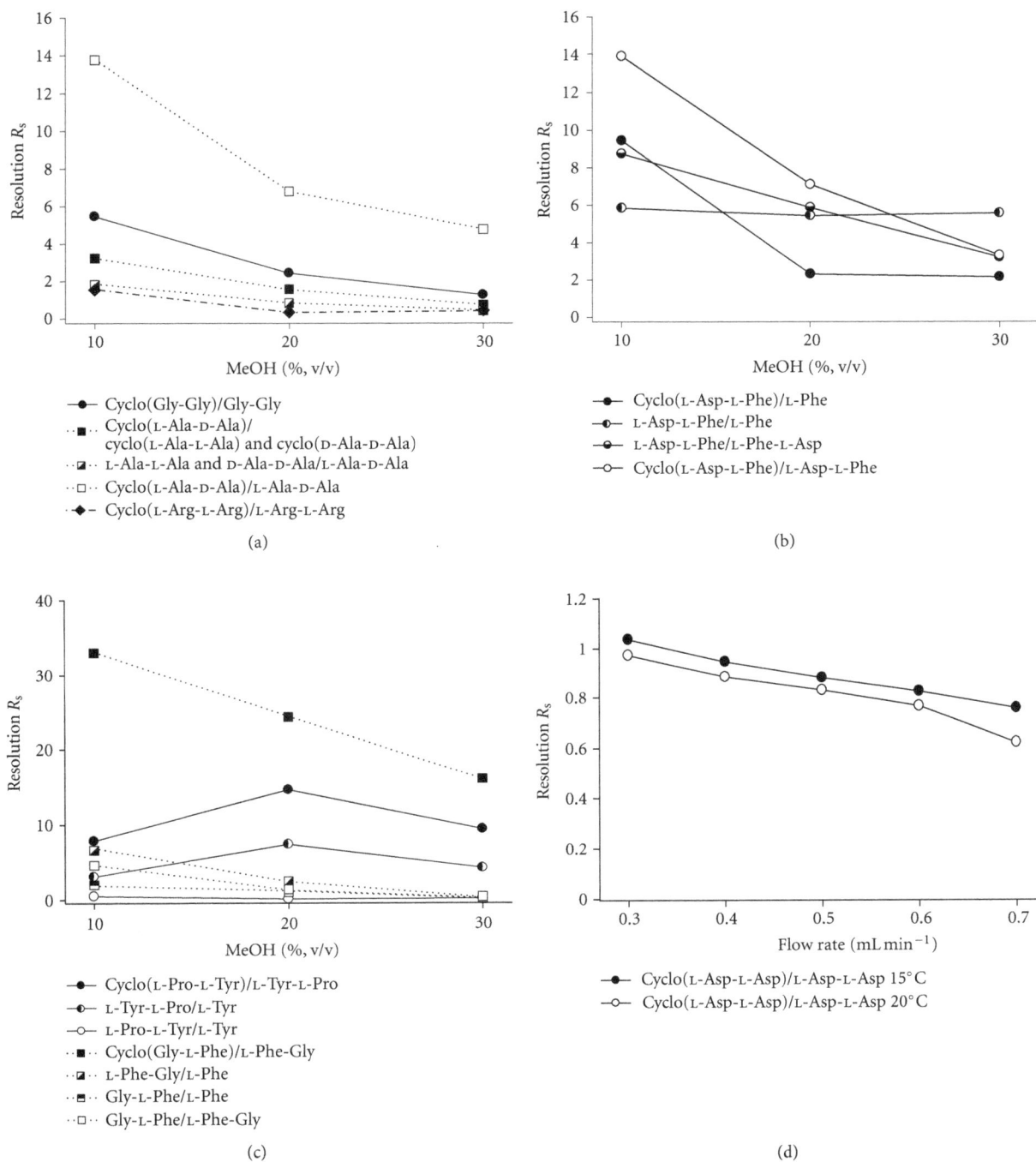

FIGURE 1: Resolution R_s of two adjacent peaks of DKPs, dipeptides, and amino acids depending on MeOH concentration (a–c) or flow rate (d); substances with (a) 10% MeOH, (b) 20% MeOH, and (c) 30% MeOH chosen as best separation conditions; (d) cyclo(L-Asp-L-Asp) and L-Asp-L-Asp at flow rate from 0.3 to 0.7 mL min^{-1} at 15 and 20°C column temperature and 10% MeOH.

amino acids were analyzed due to their detectability at 210 nm.

Separation factors α and peak resolutions R_s were calculated for two adjacent peaks at different MeOH concentrations (Figures 1(a)–1(c)). For some separations the order of peaks changed by altering MeOH concentration, resulting in variation of peaks which are adjacent.

The aim was to identify a method for each DKP, dipeptide and the two amino acids with resolution greater than 1.5 and longest retention time of maximum 15 min.

For separation of cyclo(Gly-Gly) from Gly-Gly and cyclo(L-Arg-L-Arg) from L-Arg-L-Arg best separation conditions were achieved with 10% MeOH mixed with 90% 20 mM sodium phosphate buffer (pH 5.5), 20°C

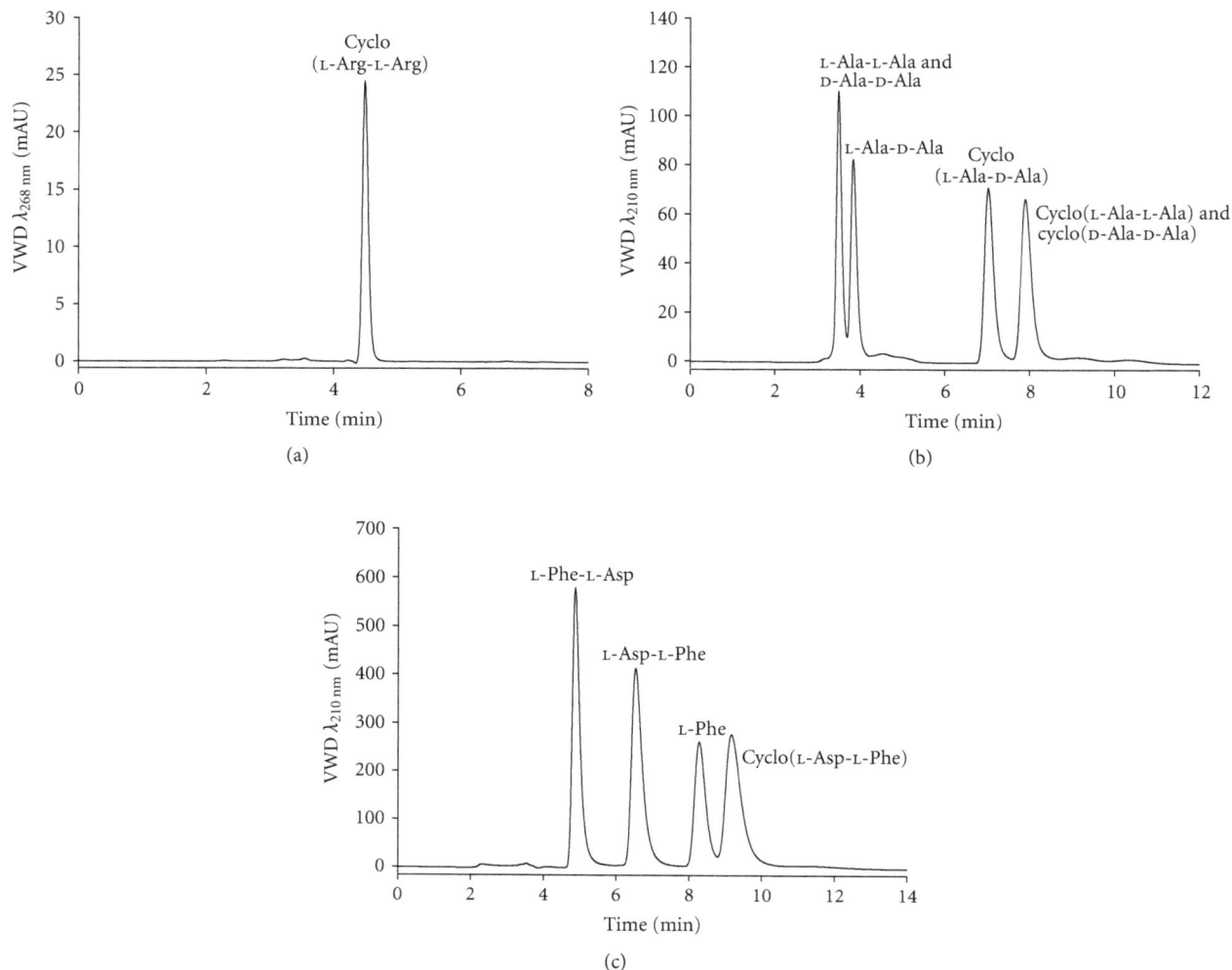

FIGURE 2: HPLC chromatograms of 1 mM standards of (a) cyclo(L-Arg-L-Arg) at 10% MeOH, 268 nm; (b) L-Ala-L-Ala and D-Ala-D-Ala, L-Ala-D-Ala, cyclo(L-Ala-D-Ala) and cyclo(L-Ala-L-Ala) and cyclo(D-Ala-D-Ala) at 10% MeOH, 210 nm; (c) L-Phe-L-Asp, L-Asp-L-Phe, L-Phe and cyclo(L-Asp-L-Phe) at 20% MeOH, 210 nm; all separations were performed at: 20°C column temperature and 0.7 mL min^{-1} flow rate.

column temperature, and a flow rate of 0.7 mL min^{-1} (Table 1, Figure 1(a)). In contrast to all used substances which were measured at a wavelength of 210 nm, cyclo(L-Arg-L-Arg) was detected by absorption of the guanidino groups (λ = 268 nm). Detection of cyclo(L-Arg-L-Arg) at 210 nm resulted in four peaks and it remained unclear which peak belonged to cyclo(L-Arg-L-Arg). On the contrary, detection of cyclo(L-Arg-L-Arg) at 268 nm showed one symmetric peak (Figure 2(a)).

Best separation of cyclo(DL-Ala-DL-Ala) from corresponding dipeptide DL-Ala-DL-Ala was determined with 10% MeOH as well (Table 1, Figure 1(a)). Moreover, for these racemic substances separation of the diastereomers could be performed. Cyclo(L-Ala-D-Ala) was separated from the enantiomeric pair cyclo(L-Ala-L-Ala) and cyclo(D-Ala-D-Ala), and L-Ala-D-Ala was separated from the enantiomeric pair L-Ala-L-Ala and D-Ala-D-Ala (Figure 2(b)). By measuring the enantiomers cyclo(L-Ala-L-Ala) and L-Ala-L-Ala attribution of the peaks was achieved.

It was not possible to separate cyclo(L-Asp-L-Asp) and L-Asp-L-Asp at 10% MeOH, 20°C column temperature, and a flow rate of 0.7 mL min^{-1}. Under these conditions only an R_s value of 0.62 was determined (Figure 1(d)). 10% is the lowest MeOH concentration which can be used for this column, but by decreasing column temperature to 15°C and flow rate to 0.3 mL min^{-1} the resolution was improved to 1.04 (Table 1, Figure 1(d)).

Good separation of cyclo(L-Asp-L-Phe), L-Asp-L-Phe, L-Phe-L-Asp, and L-Phe and retention times lower than 15 min were achieved with 20% MeOH, 20°C column temperature, and a flow rate of 0.7 mL min^{-1} (Table 1, Figure 2(c)). Even better resolution was obtained with 10% MeOH (Figure 1(b)), but under these conditions retention times were longer than 15 min.

For cyclo(L-Pro-L-Tyr), L-Pro-L-Tyr, L-Tyr-L-Pro, and L-Tyr best separation and short retention times were observed with 30% MeOH, 30°C column temperature, and a flow

Separation of Cyclic Dipeptides (Diketopiperazines) from Their Corresponding Linear Dipeptides by RP-HPLC and Method Validation

47

TABLE 2: Linear range, linear equation, correlation coefficient R^2, and limit of quantification (LOQ) for DKPs, dipeptides, and amino acids.

Substance	Linear range (mM)	Linear equation	R^2	LOQ (nmol per 10 μL injection)
Cyclo(Gly-Gly)	0.010–5.00	$y = 1732x + 107$	0.9980	5.00
Gly-Gly	0.010–5.00	$y = 1355x + 124$	0.9989	10.00
Cyclo(L-Ala-D-Ala)	0.025–5.00	$y = 2188x + 93$	0.9991	5.00
Cyclo(L-Ala-L-Ala) and cyclo(D-Ala-D-Ala)	0.025–5.00	$y = 2271x + 96$	0.9991	0.25
L-Ala-L-Ala and D-Ala-D-Ala	0.025–5.00	$y = 1821x + 111$	0.9996	5.00
L-Ala-D-Ala	0.025–5.00	$y = 1642x + 60$	0.9996	5.00
Cyclo(L-Arg-L-Arg)	0.100–50.00	$y = 162x + 5$	1.0000	1.00
L-Arg-L-Arg	0.010–10.00	$y = 1611x + 86$	0.9996	5.00
Cyclo(L-Asp-L-Asp)	0.005–2.50	$y = 6855x + 307$	0.9992	5.00
L-Asp-L-Asp	0.010–5.00	$y = 5192x + 225$	0.9990	5.00
cyclo(L-Asp-L-Phe)	0.005–5.00	$y = 8227x + 162$	0.9998	0.05
L-Asp-L-Phe	0.005–5.00	$y = 8258x + 132$	0.9999	0.05
L-Phe-L-Asp	0.005–2.50	$y = 8389x + 122$	0.9994	0.50
L-Phe (20% MeOH)	0.005–5.00	$y = 5845x + 131$	0.9997	0.10
Cyclo(L-Pro-L-Tyr)	0.005–5.00	$y = 8256x + 116$	0.9999	0.05
L-Pro-L-Tyr	0.005–2.50	$y = 7164x + 87$	0.9997	1.00
L-Tyr-L-Pro	0.010–10.00	$y = 6127x + 172$	0.9997	5.00
L-Tyr	0.010–2.50	$y = 4929x + 43$	1.0000	10.00
Cyclo(Gly-L-Phe)	0.005–5.00	$y = 7662x + 165$	0.9997	0.05
Gly-L-Phe	0.005–2.50	$y = 8360x + 106$	0.9995	0.10
L-Phe-Gly	0.005–2.50	$y = 7392x + 109$	0.9993	0.10
L-Phe (30% MeOH)	0.005–2.50	$y = 6121x + 60$	0.9997	0.10

rate of 0.7 mL min^{-1} (Table 1). Under these conditions one drawback was that L-Pro-L-Tyr was not separated from L-Tyr ($R_s = 0.22$) (Figure 1(c)); nevertheless, the main aim to separate DKP from dipeptides and amino acid was achieved. The same conditions were chosen for separation of cyclo(Gly-L-Phe) from Gly-L-Phe, L-Phe-Gly, and L-Phe (Table 1). Under these conditions separation of cyclo(Gly-L-Phe) from the corresponding dipeptides and L-Phe was possible, but not the separation of the two dipeptides and the amino acid from each other. For separation of all four substances best conditions were 10% MeOH, 20°C column temperature and a flow rate of 0.7 mL min^{-1} with $R_s = 2.01$ for Gly-L-Phe and L-Phe, and $R_s = 4.66$ for Gly-L-Phe and L-Phe-Gly (Figure 1(c)). This resulted in a method time of about 30 min.

All DKPs could be separated from their corresponding dipeptides with a peak resolution R_s higher than 1.5, with the exception of cyclo(L-Asp-L-Asp) and L-Asp-L-Asp with R_s below 1.5.

Method validation was done under best separation conditions (see Table 1). The developed methods were validated by using crude extract of *Paenibacillus chibensis* (DSM 329) spiked with the studied substances. This strain was chosen as biological matrix because of its reported ability to hydrolyze the DKP cyclo(L-Asp-L-Phe) [20].

3.2. Linearity and Sensitivity.

The analytical results for linear range, linear equation, correlation coefficient R^2, limit of detection (LOD) shown as lower limit of linear range and limit of quantification (LOQ) obtained from this investigation are summarized in Table 2.

LOD is in accordance with the lower limit of the linear range and is between 0.005 mM (e.g., cyclo(L-Pro-L-Tyr), cyclo(Gly-L-Phe)) and 0.100 mM for cyclo(L-Arg-L-Arg). The maximum of the linear range varied between 2.50 mM and 50.00 mM depending on the substance. For cyclo(Gly-Gly) a LOD of 5 μg mL^{-1} was described [19]; this equates 0.04 mM, which is higher than the LOD of 0.010 mM for cyclo(Gly-Gly) shown in this study. For cyclo(L-Asp-L-Phe) an LOD of 10 ng per 20 μL injection [13] and 2.5 ng per 10 μL injection [14] was described. These values equate 0.002 mM

TABLE 3: Precision, accuracy, stability, and recovery for DKPs, dipeptides, and amino acids in crude extract of DSM 329.

Substance	Intrarun precision (RSD %)	Interrun precision (RSD %)	Accuracy (RE %)	Short-term stability (%)	Freeze-thaw stability (%)	Recovery (%)
Cyclo(Gly-Gly)/Gly-Gly	1.1/0.9	0.7/2.6	4.7/7.7	100/99	97/98	97/99
Cyclo(L-Ala-L-Ala) and cyclo(D-Ala-D-Ala)/cyclo(L-Ala-D-Ala)	1.1/0.7	3.5/0.9	−0.3/8.1	99/101	99/98	99/99
L-Ala-L-Ala and D-Ala-D-Ala/L-Ala-D-Ala	1.3/1.1	3.9/1.7	6.6/2.9	100/97	101/100	100/100
Cyclo(L-Arg-L-Arg)/L-Arg-L-Arg	1.5/0.5	6.4/6.7	3.9/−0.3	102/99	101/99	100/98
Cyclo(L-Asp-L-Asp)/L-Asp-L-Asp	1.0/1.1	1.3/1.9	5.1/2.1	100/101	100/98	101/98
Cyclo(L-Asp-L-Phe)/L-Phe	1.1/1.5	3.0/3.1	2.2/2.3	102/100	98/98	100/98
L-Asp-L-Phe/L-Phe-L-Asp	2.7/2.1	9.9/1.9	2.3/0.3	94/103	99/98	99/101
Cyclo(L-Pro-L-Tyr)/L-Tyr	4.7/1.9	5.4/4.4	0.7/6.4	100/101	93/94	99/98
L-Pro-L-Tyr/L-Tyr-L-Pro	1.4/1.3	4.5/2.0	4.3/−0.1	98/98	99/98	99/98
Cyclo(Gly-L-Phe)/L-Phe	2.9/1.2	1.4/4.0	−4.2/1.8	97/100	100/100	99/100
Gly-L-Phe/L-Phe-Gly	1.3/0.9	3.4/2.7	0.3/−2.0	98/101	100/100	103/98

and 0.001 mM and thus are slightly lower than 0.005 mM detected in this study, but they are in the same range. We determined a broader linear range for cyclo(L-Asp-L-Phe) (0.005 to 5.00 mM) compared to 0.5 to 10.0 μg mL^{-1} [14] (equates 0.002 to 0.038 mM) or 5 to 100 μg mL^{-1} [15] (0.02 to 0.38 mM).

LOQ was defined as the lowest concentration which could still be measured with a maximum of 20% RSD for precision and ± 20% RE for accuracy. The minimal LOQ was identified for cyclo(L-Asp-L-Phe), L-Asp-L-Phe, cyclo(L-Pro-L-Tyr), and cyclo(Gly-L-Phe) with 0.05 nmol per 10 μL injection and ranged to 10.00 nmol per 10 μL injection for Gly-Gly and L-Tyr (Table 2).

The linear equations calculated in this study strongly varied depending on the substance (Table 2). In general, the slopes for all aromatic substances were higher than for most other used substances, due to the higher UV absorption of the aromatic ring. The correlation coefficient R^2 for each substance was greater than or equal to 0.9990 within the defined linear range, except for cyclo(Gly-Gly) and Gly-Gly with 0.9980 and 0.9989, which are still acceptable values (Table 2).

3.3. *Precision and Accuracy*. The results for intrarun and interrun precision and accuracy are summarized in Table 3. The lowest relative standard deviation for intrarun precision was 0.5% for L-Arg-L-Arg. The highest RSD (%) for intrarun precision was measured for cyclo(L-Pro-L-Tyr) with 4.7%. For interrun precision the lowest RSD (%) value was 0.7%, detected for cyclo(Gly-Gly) and the highest RSD (%) value was 9.9 % for L-Asp-L-Phe.

For substances which showed higher measured concentration than true concentration, the highest relative error for accuracy was 8.1%, detected for cyclo(L-Ala-D-Ala). The best accuracy was 0.3% RE, measured for L-Phe-L-Asp

and Gly-L-Phe. For samples which showed lower measured concentration than true concentration, best accuracy was −0.1% RE obtained for L-Tyr-L-Pro and highest RE (%) value was −4.2%, measured for cyclo(Gly-L-Phe). For all other substances accuracy was between −4.2 and 8.1% RE.

3.4. *Stability and Recovery*. Data measured for short-term and freeze-thaw stability and for recovery are shown in Table 3. After short-term storage at 21°C for 24 h at least 97% of each substance were still detectable except for L-Asp-L-Phe, with 94% stability. For freeze-thaw stability also 97% or more were determined for all substances apart from cyclo(L-Pro-L-Tyr) and L-Tyr, with a stability of 93% and 94%.

For all investigated DKPs, dipeptides and amino acids the recovery rate after heat treatment at 90°C for 10 min was at least 97% or greater. Thus, this procedure is efficient for enzyme inactivation in crude extract and all substances can be detected in sufficient amount.

4. Conclusions

In this work RP-HPLC methods were established for separation of seven DKPs from their corresponding linear dipeptides and in some cases the related amino acids within 15 min analysis time. This approach allows a rapid quantitative analysis of these molecules. Baseline separation with a peak resolution greater than 1.5 was achieved for separation of the six following DKPs from their corresponding linear dipeptides: cyclo(Gly-Gly), cyclo(DL-Ala-DL-Ala), cyclo(L-Arg-L-Arg), cyclo(L-Asp-L-Phe), cyclo(L-Pro-L-Tyr), and cyclo(Gly-L-Phe). Furthermore, for the racemic DKP cyclo(DL-Ala-DL-Ala) and dipeptide DL-Ala-DL-Ala a method to separate the diastereomers was established and could be applied for enantioselective analysis of these

Separation of Cyclic Dipeptides (Diketopiperazines) from Their Corresponding Linear Dipeptides by RP-HPLC and
Method Validation

49

molecules. Cyclo(L-Asp-L-Asp) could also be separated from its corresponding dipeptide, but not with baseline separation.

All methods were successfully validated in terms of sensitivity, linearity, intra- and interrun precision, accuracy, stability, and recovery, demonstrating their usefulness as analytical methods for the detection of DKPs, dipeptides, and amino acids in bacterial crude extract. The methodology could be extended to other analytes of interest and find application, for example, for a screening of strains exhibiting the ability to hydrolyze cyclic dipeptides to their corresponding linear dipeptides.

Acknowledgments

The authors want to thank the "Fachagentur Nachwachsende Rohstoffe e.V. (FNR)" for the financial support of this work within the joint project "PolyTe": polymeric surfactants from renewable resources with optimized performance properties (22012708). They acknowledge the support by "Deutsche Forschungsgemeinschaft" and Open Access Publishing Fund of Karlsruhe Institute of Technology. Moreover they thank Taros Chemicals GmbH & Co. KG for providing DKPs.

References

[1] J. L. Johnson, W. G. Jackson, and T. E. Eble, "Isolation of L-leucyl-L-proline anhydride from microbiological fermentations," *Journal of the American Chemical Society*, vol. 73, no. 6, pp. 2947–2948, 1951.

[2] M. T. G. Holden, S. R. Chhabra, R. De Nys et al., "Quorum-sensing cross talk: isolation and chemical characterization of cyclic dipeptides from *Pseudomonas aeruginosa* and other Gram-negative bacteria," *Molecular Microbiology*, vol. 33, no. 6, pp. 1254–1266, 1999.

[3] K. Ström, J. Sjögren, A. Broberg, and J. Schnürer, "*Lactobacillus plantarum* MiLAB 393 produces the antifungal cyclic dipeptides cyclo(L-Phe-L-Pro) and cyclo(L-Phe-trans-4-OH-L-Pro) and 3-phenyllactic acid," *Applied and Environmental Microbiology*, vol. 68, no. 9, pp. 4322–4327, 2002.

[4] A. Lin, Y. Fang, T. Zhu, Q. Gu, and W. Zhu, "A new diketopiperazine alkaloid isolated from an algicolous *Aspergillus flavus* strain," *Pharmazie*, vol. 63, no. 4, pp. 323–325, 2008.

[5] A. C. Stierle, J. H. Cardellina, and G. A. Strobel, "Maculosin, a host-specific phytotoxin for spotted knapweed from *Alternaria alternata*," *Proceedings of the National Academy of Sciences of the United States of America*, vol. 85, no. 21, pp. 8008–8011, 1988.

[6] E. J. Dumdei, J. S. Simpson, M. J. Garson, K. A. Byriel, and C. H. L. Kennard, "New chlorinated metabolites from the tropical marine sponge *Dysidea herbacea*," *Australian Journal of Chemistry*, vol. 50, no. 2, pp. 139–144, 1997.

[7] F. Fdhila, V. Vázquez, J. L. Sánchez, and R. Riguera, "DD-Diketopiperazines: antibiotics active against *Vibrio anguillarum* isolated from marine bacteria associated with cultures of *Pecten maximus*," *Journal of Natural Products*, vol. 66, no. 10, pp. 1299–1301, 2003.

[8] A. P. Einholm, K. E. Pedersen, T. Wind et al., "Biochemical mechanism of action of a diketopiperazine inactivator of plasminogen activator inhibitor-1," *Biochemical Journal*, vol. 373, no. 3, pp. 723–732, 2003.

[9] A. Lamm, I. Gozlan, A. Rotstein, and D. Avisar, "Detection of amoxicillin-diketopiperazine-2', 5' in wastewater samples.," *Journal of Environmental Science and Health A*, vol. 44, no. 14, pp. 1512–1517, 2009.

[10] U. Kertscher, M. Bienert, E. Krause, N. F. Sepetov, and B. Mehlis, "Spontaneous chemical degradation of substance P in the solid phase and in solution," *International Journal of Peptide and Protein Research*, vol. 41, no. 3, pp. 207–211, 1993.

[11] A. Kocijan, R. Grahek, D. Kocjan, and L. Zupančič-Kralj, "Effect of column temperature on the behaviour of some angiotensin converting enzyme inhibitors during high-performance liquid chromatographic analysis," *Journal of Chromatography B*, vol. 755, no. 1-2, pp. 229–235, 2001.

[12] E. Çubuk Demiralay and G. Özkan, "Optimization strategy for isocratic separation of α-aspartame and its breakdown products by reversed phase liquid chromatography," *Chromatographia*, vol. 60, no. 9-10, pp. 579–582, 2004.

[13] V. George, S. Arora, B. K. Wadhwa, and A. K. Singh, "Analysis of multiple sweeteners and their degradation products in lassi by HPLC and HPTLC plates," *Journal of Food Science and Technology*, vol. 47, no. 4, pp. 408–413, 2010.

[14] K. Saito, M. Horie, Y. Hoshino, N. Nose, H. Nakazawa, and M. Fujita, "Determination of diketopiperazine in soft drinks by high performance liquid chromatography," *Journal of Liquid Chromatography*, vol. 12, no. 4, pp. 571–582, 1989.

[15] H. Y. Aboul-Enein and S. A. Bakr, "Comparative study of the separation and determination of aspartame and its decomposition products in bulk material and diet soft drinks by HPLC and CE," *Journal of Liquid Chromatography and Related Technologies*, vol. 20, no. 9, pp. 1437–1444, 1997.

[16] J.-D. Berset and N. Ochsenbein, "Stability considerations of aspartame in the direct analysis of artificial sweeteners in water samples using high-performance liquid chromatography-tandem mass spectrometry (HPLC-MS/MS)," *Chemosphere*, vol. 88, no. 5, pp. 563–569, 2012.

[17] I. Furda, P. D. Malizia, M. G. Kolor, and P. J. Vernieri, "Decomposition products of L-aspartyl-L-phenylalanine methyl ester and their identification by gas-liquid chromatography," *Journal of Agricultural and Food Chemistry*, vol. 23, no. 2, pp. 340–343, 1975.

[18] C. Liu, H. Wang, Y. Jiang, and Z. Du, "Rapid and simultaneous determination of amoxicillin, penicillin G, and their major metabolites in bovine milk by ultra-high-performance liquid chromatography-tandem mass spectrometry," *Journal of Chromatography B*, vol. 879, no. 7-8, pp. 533–540, 2011.

[19] B. J. Compton, W. C. Purdy, and D. J. Phelps, "A high-performance liquid chromatographic technique for the determination of 2,5-piperazinedione in complex reaction mixtures," *Analytica Chimica Acta*, vol. 105, no. 1, pp. 409–412, 1979.

[20] K. Yokozeki, N. Usui, T. Yukawa, Y. Hirose, and K. Kubota, "Process for producing L-aspartyl-L-phenylalanine and its diketopiperazine," 0220028 B1, 1990.

[21] FDA, http://www.fda.gov/downloads/Drugs/GuidanceComplianceRegulatoryInformation/Guidances/UCM070107.pdf, 2001.

8

Isolation of Low Abundance Proteins and Cells Using Buoyant Glass Microbubble Chromatography

Steingrimur Stefansson,[1] **Daniel L. Adams,**[2] **and Cha-Mei Tang**[2]

[1] *Fuzbien Technology Institute, 9700 Great Seneca Hwy, Suite 302, Rockville, MD 20850, USA*
[2] *Creatv Microtech Inc., 11609 Lake Potomac Drive, Potomac, MD 20854, USA*

Correspondence should be addressed to Steingrimur Stefansson; stennistef@fuzbien.com

Academic Editor: Joselito P. Quirino

Conventional protein affinity chromatography relies on highly porous resins that have large surface areas. These properties are ideal for fast flow separation of proteins from biological samples with maximum yields, but these properties can also lead to increased nonspecific protein binding. In certain applications where the purity of an isolated protein is more important than the yield, using a glass solid phase could be advantageous as glass is nonporous and hydrophilic and has a low surface area and low nonspecific protein binding. As a proof of principle, we used protein A-conjugated hollow glass microbubbles to isolate fluorescently labeled neurofilament heavy chain spiked into serum and compared them to protein A Sepharose and protein A magnetic beads (Dynabeads) using an anti-neurofilament protein antibody. As expected, a greater volume of glass bubbles was required to match the binding capacity of the magnetic beads and Sepharose resins. On the other hand, nonspecific protein binding to glass bubbles was greatly reduced compared to the other resins. Additionally, since the glass bubbles are buoyant and transparent, they are well suited for isolating cells from biological samples and staining them *in situ*.

1. Introduction

Glass is essentially an amorphous 3-dimensional mesh of silica oxides terminating at the surface as silicon hydroxides. Although the surface exposed silica hydroxides readily coordinate with divalent cations such as Ca^{2+} and can promote surface activated plasma coagulation [1], glass is surprisingly resistant to protein adsorption. For example, glass bead chromatography has been used to specifically purify vitronectin, a cell adhesion protein, from serum with a high degree of purity [2]. Vitronectin has an arginine-glycine-aspartic acid (RGD) integrin recognition motif and was initially identified as "serum spreading factor" due to its ability to bind tissue culture plates and mediate cell adhesion and spreading [3]. Additionally, glass beads can be functionalized with a variety of silanes containing amine and thiol functional groups that can be used for biomolecular conjugation [4, 5]. However, development of porous polymer resins with superior flow characteristics and greater surface area have since replaced glass beads as the solid support of choice for most protein purification.

In this study, we revisit glass chromatography by examining the ability of hollow glass microbubbles to isolate proteins, or cells, from serum and blood. Glass bubbles are inexpensive, lightweight, and strong hollow silica spheres that are commonly used in a variety of industrial applications ranging from paints and sealants to adhesives. Glass bubbles are buoyant and float to the surface of liquids, thus separating readily from the bulk solution. Paradoxically, the worst properties of using glass for protein isolation could be its greatest asset, namely, the very low surface area and lack of porosity.

Serum proteomics is currently striving to characterize protein constituents that could enable the discovery of disease biomarkers. However, this is complicated by the fact that there are numerous plasma proteins, with concentrations spanning at least 9 orders of magnitude, while many proteins of diagnostic value present at low concentrations. To analyze valuable serum markers, one must account for the approximately 10 major proteins, including serum albumin and immunoglobulins, which account for more than 90% of the total protein content of serum [6, 7]. Strategies to reduce the interference from these abundant proteins include depleting them from the serum sample or enriching the low abundance proteins of interest. Isolation of proteins from this complex mixture is often performed by immunoprecipitation, but conventional resins (i.e., agarose, sepharose, magnetic beads, etc.) often bind proteins nonspecifically and thus require additional purification steps for samples [8].

As a proof of principle for utilizing glass as a solid support for purifying low abundance proteins, we performed immunoaffinity isolation of purified neurofilament heavy chain spiked into low-IgG fetal bovine serum. Neurofilament proteins belong to the intermediate filament family and are major constituents of the neuronal cytoskeleton. They are mostly expressed by large neurons and myelinated axons and play an important role in neuronal structure and intracellular trafficking [9]. The neurofilament scaffolding is mostly composed of a mixture of three neurofilament proteins known as light, medium, and heavychains that may also include smaller amounts of ancillary proteins (i.e., peripherin, nestin, and vimentin) [10]. Because of their specific localization in the central nervous system (CNS), their presence outside of this compartment may indicate damage and/or disease of the CNS [11]. The presence of neurofilament heavy chain in the cerebrospinal fluid (CSF) is indicative of axonal damage in disorders such as Parkinson's disease, Alzheimer's disease, and HIV dementia [12–15].

Neurofilament heavy chain was fluorescently labeled to follow the purification process. Immunoprecipitation was performed using neurofilament heavy chain antibodies immobilized on protein A-functionalized glass bubbles, protein A-conjugated Sepharose, and protein A-conjugated Dynabeads. Our results show that though glass bubbles have much lower binding capacity than Sepharose and Dynabeads, the glass bubbles had less nonspecific protein binding, as seen by the amount of total protein eluted and by SDS-PAGE protein profile.

In addition to protein purification, glass bubble chromatography was used for cell isolation. Isolation of prostate cancer cells (PC-3) by glass bubbles was shown to be highly specific and allowed for clear fluorescent staining and imaging of cells as the glass bubbles are transparent.

2. Materials and Methods

2.1. Materials. Protein A Sepharose, fluoresceinbiotin, and (3-glycidyloxypropyl)-trimethoxysilane were purchased from Sigma Chemical Company. NeutrAvidin was purchased from Pierce. Polydisperse hollow silica microspheres (5–40 μm diameter) were purchased from 3M. Dynabeads protein A and Ultra Low IgG fetal bovine serum (low IgG-FBS) were purchased from Invitrogen. FluoroLink-Ab Cy5 labelling kit was purchased from GE Healthcare. Silver Stain Kit was purchased from Pierce. Monoclonal antibody against the 200 kD neurofilament heavy chain (no. ab7795)

was purchased from Abcam. Porcine 200 kD neurofilament heavy chain was purchased from Millipore.

2.2. Conjugation of Glass Microbubbles. The glass bubbles were first washed in concentrated sulphuric acid followed by washing with diH$_2$O. These steps clean the surface of the bubbles and remove both broken and smaller, less buoyant glass bubbles. After drying the glass bubbles, silane and either protein A or Neutravidin were reacted together in methanol for 2-3 hrs at RT. This was followed by washing away unbound reactants and then incubating with a blocking agent.

2.3. Protein Labeling and Immunoprecipitation. Porcine neurofilament heavy chain was labelled with Cy5, and the labeled neurofilament heavy chain was serially diluted into 1 mL of low IgG-FBS starting at 1.24 ng/mL. All protein A resins and glass bubble slurries were incubated with 1 μg/mL anti-neurofilament antibody in PBS for 1 hr at RT. After the incubation, excess antibody was washed away with BS, and the slurries were added to the serum containing the Cy5-labeled neurofilament protein. The amount of slurry used in all experiments was normalized to their neurofilament heavy chain binding capacity. Typically, 5–10-fold greater volume of protein A glass bubble slurry was required over that of protein A Sepharose and Dynabeads protein A to obtain equal binding capacity (Figure 2). All serum samples were incubated for 1 hr at RT followed by three washing steps of 1 mL PBS containing 0.1% Tween-20 for 30 min with end-over-end mixing. Proteins were then eluted with 250 μL of 75% ethanol with end-over-end mixing for 30 min at RT. Alternatively, the glass bubbles were destroyed with three 1 sec ultrasonication pulses in 250 μL of 75% ethanol. Total protein in eluted fractions from resins and glass bubbles was determined by bicinchoninic acid (BCA) (Pierce). Empty polypropylene tubes containing only serum were treated in the same way as the slurries, and the protein concentration eluted from the empty tubes was subtracted from protein eluted from the resins and glass bubbles (Figure 3). Samples of the concentrated proteins were also analyzed by SDS-PAGE and stained using a Silver Stain Kit (Pierce).

2.4. Cell Capture Using Glass Microbubbles. The prostate cancer cell line PC-3 (ATCC) was grown to confluency, and 100 μL of cell supernatant was used at first passage. An r-phycoerythrin- (PE-) labeled antibody against epithelial cell adhesion molecule (EpCAM) (eBioscience) was spiked into the cell suspension at 1.25 μg/mL and incubated for 1 hr. Cells were then suspended in 1 mL of low IgG-FBS and protein A-conjugated glass bubble slurry and an FITC-labeled pan-cytokeratin antibody (Miltenyi Biotech) at 5 μg/mL was added. Slurry was incubated for 1 hr at RT with end-over-end mixing followed by three washing steps of 1 mL PBS containing 0.1% Tween-20. The bubbles were then mounted overnight on a microscope slide with a fluoromount/4′,6-diamidino-2-phenylindole (DAPI) solution (Southern Biotech), which stains DNA, and imaged on a 40X magnification on a fluorescent microscope (Olympus).

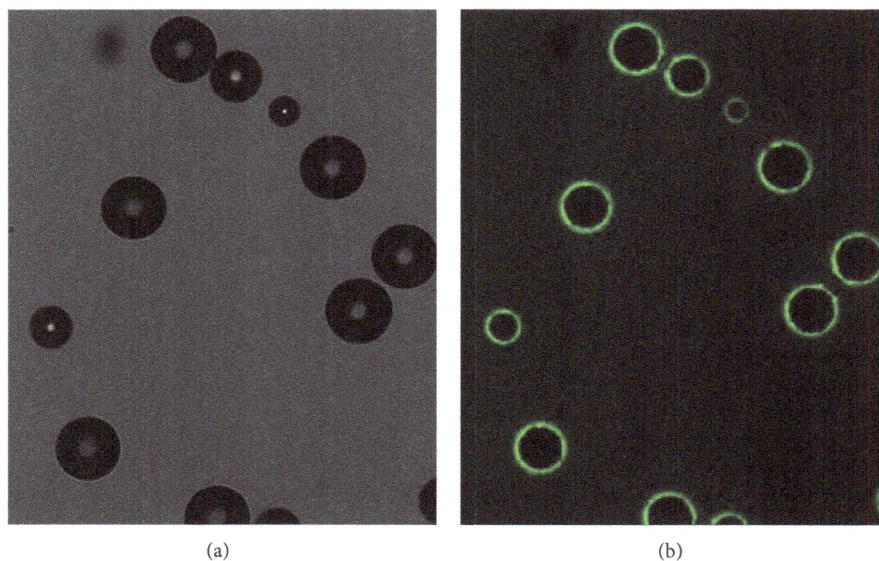

(a)

(b)

FIGURE 1: Neutravidin conjugated glass microbubbles with bound biotin-fluorescein. Panel A shows a 40X light field of the glass bubbles and panel B shows the surface bound biotin-fluorescein.

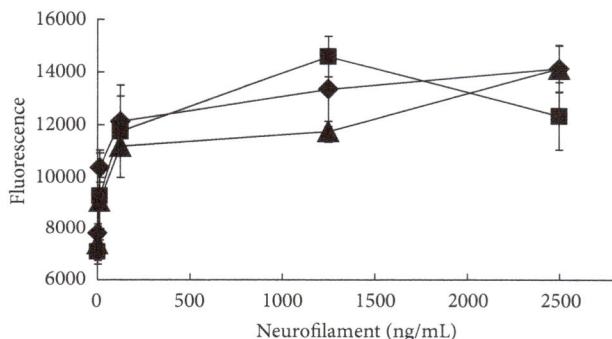

FIGURE 2: Normalization of Protein A Resin and Glass Bubble Binding Capacities. Increasing concentrations of Cy5 labeled neurofilament heavy chain was added to 1 ml aliquots of low-IgG FBS as described in Materials and Methods. Slurries of protein A Sepharose (20 μL), Dynabeads protein A (10 μL) or glass bubbles conjugated with protein A (100 μL) and saturated with anti neurofilament antibodies were added to the tubes and processed as described in Materials and Methods. Closed squares (■) show the protein A glass bubbles, closed diamonds (◆) show the protein A Sepharose and closed triangles (▲) show the protein A Dynabeads.

2.5. Fluorescent Assays. Fluorescent assays were performed using the ultrasensitive Signalyte-II fluorometer (Creatv Microtech, MD) described in more detail in [16–18]. The fluorometer used for these studies was equipped with 4 light emitting diodes (LED) with excitation maxima of 470, 530, 590, and 635 nm and long pass emission filters with cut-on wavelengths of 515, 570, 630, and 665 nm, respectively. The fluorometer cuvettes are glass capillary PCR tubes that can hold a sample size of 20–50 μL. All fluorescent binding assays were performed in PBS unless otherwise stated. Fluorescence measurements of Cy5-labeled neurofilament heavy chain

were performed with an excitation wavelength of 635 nm, and samples were measured from 100 msec. to 3 sec. per tube.

3. Results and Discussion

Figure 1 shows glass bubbles conjugated with Neutravidin and loaded with fluorescein-labeled biotin. The transparency of the glass makes it possible to observe the uniform fluorescence over the entire surface. To examine the potential of using glass bubbles to isolate low abundance proteins, we used protein A-conjugated glass bubbles to isolate fluorescent neurofilament heavy chain spiked into serum.

The glass microbubbles are very robust and remain intact in batch chromatography with end-over-end mixing, which allows for optimal contact of the bubbles with the analyte. Although glass surfaces can be abrasive, the rate of protein leaching off the glass bubbles with end-over-end mixing was similar to that of protein A Sepharose (data not shown). At the end of the purification procedure, the glass bubbles can be easily destroyed with ultrasonication with the glass shards sinking to the bottom.

Figure 2 shows the binding capacity of the protein A resins and glass bubbles for Cy5-labeled neurofilament heavy chain spiked into serum at various concentrations. Immuno-isolation of the labeled protein was performed using glass bubbles, Dynabeads, and Sepharose protein A resins. Removing the serum and PBS washes from the tubes containing glass bubbles was in many respects easier than the resins since no centrifugation was required. Additionally, when a pipette tip was used to aspirate liquid from the bottom of the tube, the glass bubbles clung to the side of the tubes. This probably led to more efficient washing steps as residual liquid associated with the glass bubbles drained to the bottom of the tube.

Cy5-labeled neurofilament heavy chain was eluted from the glass bubbles and protein A resins and fluorescence

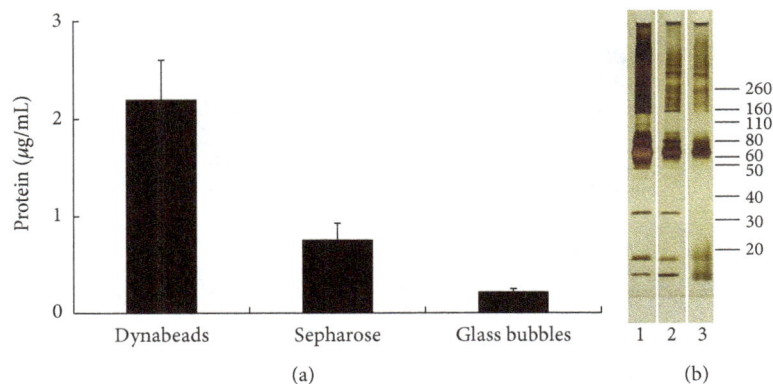

FIGURE 3: Nonspecific serum protein binding to Dynabeads, Sepharose and glass bubbles. Slurries with equivalent neurofilament heavy chain binding capacities were added to serum and processed as described in the Materials and Methods. Panel A shows the total protein eluted from the resins and glass bubbles and panel B shows the protein profile of eluted proteins by SDS-PAGE. Lanes 1, 2 and 3 are proteins eluted from protein A Dynabeads, Sepharose and glass bubbles, respectively. Each lane was loaded with 100 ng of eluted protein.

determined as described in Section 2. Figure 2 demonstrates that to achieve equal binding capacity, approximately 5- to 10-fold greater volume of glass bubble slurry had to be used to equal the binding capacity of the resins, which is consistent with the glass bubbles having a much lower surface area available for immobilizing protein A.

The glass bubbles and resins were washed 3 times using PBS-0.1% Tween-20, which is a rather low-stringency washing condition. Figure 3(a) shows the total protein eluted with 75% ethanol and demonstrates that despite the increased volume of glass bubbles used, the amount of total protein eluted from the glass bubbles was lower compared to the resins. The profiles of the eluted proteins were visualized by SDS-PAGE followed by silver staining. Since most SDS-PAGE protein silver staining is more qualitative than quantitative [19], Figure 3(b) demonstrates that glass bubbles have fewer contaminant species than the Sepharose or Dynabeads resins. However, many protein bands appear to be common for all three. Other researchers have also made similar observations, and Trinkle-Mulcahy et al. [20] developed a methodology to exclude proteins that copurify with the most commonly used resins (Sepharose, agarose, and magnetic beads). Their studies identified over 200 proteins, collectively termed "bead proteome", as they are specifically associated with the resins. Their results also showed that none of the three resins was superior in terms of reducing background binding. For cytoplasmic extracts, the highest background levels were obtained with magnetic beads [21], and our results show that magnetic beads also had the highest background levels for serum (Figure 3).

The major protein contaminant for the glass bubbles and resins in Figure 3(b) is likely serum albumin, based on the molecular weight. Interestingly, some of these nonspecific proteins in the gel were observed to come from the polypropylene tubes used in the batch absorption process (data not shown).

In addition to isolating proteins from serum, these glass bubbles can also be used to isolate cells from biological samples. The advantage of using the glass bubbles for this task is that they remain buoyant even when they are centrifuged. Cells that are not bound to the bubbles are pelleted and thus easily separated. Furthermore, cells need not be separated from the bubbles for imaging as they are transparent. Figure 4 shows a representative sample of PC-3 prostate carcinoma cells captured by protein A-conjugated glass bubbles. Whole cultured cells were first incubated with PE-labeled EpCAM antibody, which also served as the capture antibody for the protein A microbubbles. Only cells with EpCAM surface antigen were captured while the remaining non-EpCAM cells were removed from the population. Cells stained with EpCAM (in red) show a typical cell surface staining pattern. The potential benefits of using transparent glass bubbles for capturing cells can be seen in panels 1(a)–1(e) where a cell that is not directly in the field of vision can be observed through the glass.

The PC-3 cells captured on glass bubbles, visible light (row (a)), were also stained with FITC-labeled cytokeratin antibody (row (b)). The cytokeratin antibody (in green) shows a filamented cytoplasmic staining, EpCAM antibody labeled red (row (c)). Cells were then mounted with a DAPI mounting solution, which stains the nucleus (row (d)). Finally, row (e) is the merged images of the cells. An additional advantage of using the glass bubbles was the low fluorescent background, which could be due to their low nonspecific protein binding. This allowed for long exposures and high quality images of the staining patterns of cells isolated using glass bubbles.

4. Conclusions

Previously, our group and others have examined the utility of using buoyant glass microbubbles to capture cells [21, 22]. The advantage of using glass bubbles to capture cells is that they are easily separated from liquid and their transparency is a benefit if captured cells are stained with fluorescent markers. Circulating tumor cells (CTCs) are an example where high purity cell isolation from a complex mixture may benefit from glass bubble chromatography. In this study, we expand

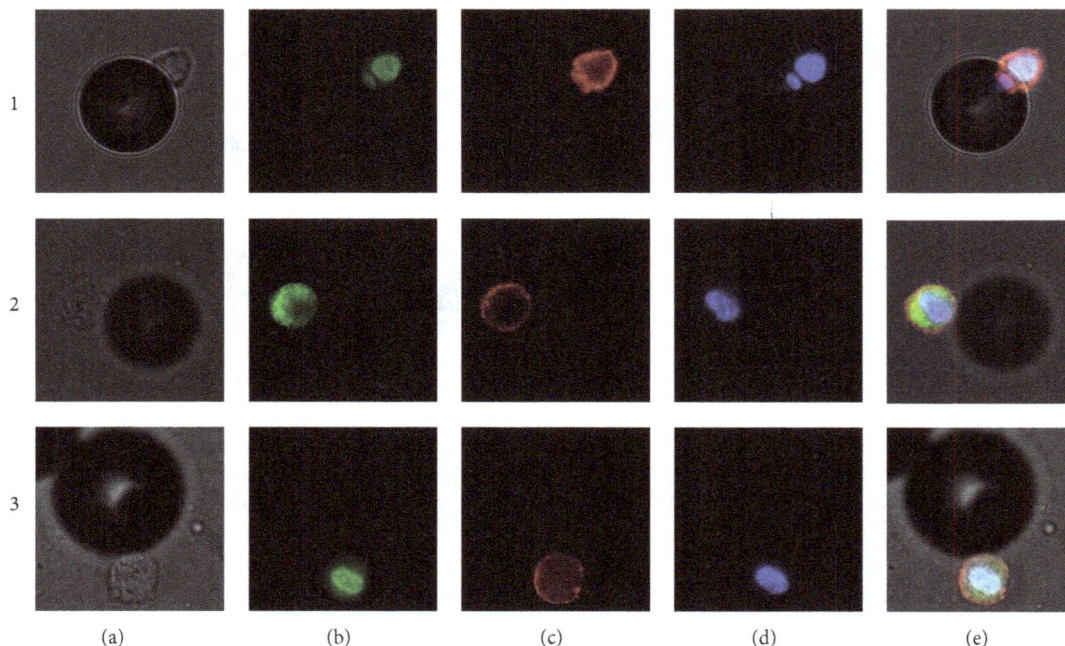

FIGURE 4: PC-3 cells binding to protein A conjugated glass microbubbles. Prostate cancer cells (PC-3), with PE labeled EpCAM, bound to protein A glass microbubbles (row A in visible light) were stained with FITC labeled anti cytokeratin (row B), PE labeled anti EpCAM (row C) and DAPI (row D). Row E shows merged images of the stained cells (40X magnification).

the scope of glass bubble chromatography to include low abundance proteins. By their very nature, glass bubbles are inferior to conventional chromatography resins with respect to binding capacity and flow characteristics. However, in certain applications where the purity of an isolated protein is more important than the yield, glass microbubbles could be a useful alternative to traditional resins.

Conflict of Interests

The authors declare no conflict of interests.

Acknowledgment

The research was supported in part by an NIH Phase I small business Grant 1R43MH082619-01A1 from the National Institute of Mental Health.

References

[1] J. Margolis, "Activation of plasma by contact with glass: evidence for a common reaction which releases plasma kinin and initiates coagulation," *The Journal of Physiology*, vol. 144, no. 1, pp. 1–22, 1958.

[2] D. W. Barnes and J. Silnutzer, "Isolation of human serum spreading factor," *Journal of Biological Chemistry*, vol. 258, no. 20, pp. 12548–12552, 1983.

[3] E. G. Hayman, M. D. Pierschbacher, Y. Ohgren, and E. Ruoslahti, "Serum spreading factor (vitronectin) is present at the cell surface and in tissues," *Proceedings of the National Academy of Sciences of the United States of America*, vol. 80, no. 13, pp. 4003–4007, 1983.

[4] P. Cuatrecasas, "Affinity chromatography," *Annual Review of Biochemistry*, vol. 40, pp. 259–278, 1971.

[5] S. K. Bhatia, L. C. Shriver-Lake, K. J. Prior et al., "Use of thiol-terminal silanes and heterobifunctional crosslinkers for immobilization of antibodies on silica surfaces," *Analytical Biochemistry*, vol. 178, no. 2, pp. 408–413, 1989.

[6] J. N. Adkins, S. M. Varnum, K. J. Auberry et al., "Toward a human blood serum proteome: analysis by multidimensional separation coupled with mass spectrometry," *Molecular & Cellular Proteomics*, vol. 1, no. 12, pp. 947–955, 2002.

[7] N. L. Anderson and N. G. Anderson, "The human plasma proteome: history, character, and diagnostic prospects," *Molecular & Cellular Proteomics*, vol. 1, no. 11, pp. 845–867, 2002.

[8] N. Tang, P. Tornatore, and S. R. Weinberger, "Current developments in SELDI affinity technology," *Mass Spectrometry Reviews*, vol. 23, no. 1, pp. 34–44, 2004.

[9] Q. Liu, F. Xie, A. Alvarado-Diaz et al., "Neurofilamentopathy in neurodegenerative diseases," *Open Neurology Journal*, vol. 5, no. 1, pp. 58–62, 2011.

[10] D. M. Toivola, P. Strnad, A. Habtezion, and M. B. Omary, "Intermediate filaments take the heat as stress proteins," *Trends in Cell Biology*, vol. 20, no. 2, pp. 79–91, 2010.

[11] N. Norgren, L. Rosengren, and T. Stigbrand, "Elevated neurofilament levels in neurological diseases," *Brain Research*, vol. 987, no. 1, pp. 25–31, 2003.

[12] J. Brettschneider, A. Petzold, S. D. Süßmuth et al., "Neurofilament heavy-chain NfHSMI35 in cerebrospinal fluid supports the differential diagnosis of parkinsonian syndromes," *Movement Disorders*, vol. 21, no. 12, pp. 2224–2227, 2006.

[13] J. Brettschneider, A. Petzold, D. Schöttle, A. Claus, M. Riepe, and H. Tumani, "The neurofilament heavy chain (NfH^{SM135}) in the cerebrospinal fluid diagnosis of Alzheimer's disease," *Dementia and Geriatric Cognitive Disorders*, vol. 21, no. 5-6, pp. 291–295, 2006.

[14] A. Antinori, G. Arendt, J. T. Becker, B. J. Brew, D. A. Byrd, and D. B. Clifford, "Biomarkers of HIV-associated neurocognitive disorders," in *Proceedings of the HIV Infection and the Central Nervous System: Developed and Resource Limited Settings*, Rome, Italy, June 2005.

[15] M. Gisslén, L. Rosengren, L. Hagberg, S. G. Deeks, and R. W. Price, "Cerebrospinal fluid signs of neuronal damage after antiretroviral treatment interruption in HIV-1 infection," *AIDS Research and Therapy*, vol. 2, no. 1, article 6, 2005.

[16] P. Zhu, D. R. Shelton, S. Li et al., "Detection of *E. coli* O157:H7 by immunomagnetic separation coupled with fluorescence immunoassay," *Biosensors and Bioelectronics*, vol. 30, no. 1, pp. 337–341, 2011.

[17] S. Li, Y. Zhang, P. Amstutz III, and C.-M. Tang, "Multiplex integrating waveguide sensor: signalyte-II," *Methods in Molecular Biology*, vol. 503, pp. 423–434, 2009.

[18] S. Li, L. Kiefer, P. Zhu et al., "Ultra-sensitive detection using integrated waveguide technologies," in *Biosensors For Health, Environment and Biosecurity/Book 3*, Intech, Rijeka, Croatia, 2011.

[19] H. M. Poehling and V. Neuhoff, "Visualization of proteins with a silver "stain": a critical analysis," *Electrophoresis*, vol. 2, pp. 141–147, 1981.

[20] L. Trinkle-Mulcahy, S. Boulon, Y. W. Lam et al., "Identifying specific protein interaction partners using quantitative mass spectrometry and bead proteomes," *Journal of Cell Biology*, vol. 183, no. 2, pp. 223–239, 2008.

[21] S. Stefansson, D. Adams, and C. M. Tang, "Sample preparation using buoyant silica microspheres," in *Proceedings of the 6th Early Detection Research Network Workshop (EDRN '09)*, Bethesda, Md, USA, August-September 2009.

[22] C. H. Hsu, C. C. Chen, D. Irimia, and M. Toner, "Isolating cells from blood using Buoyancy Activated Cell Sorting (BACS) with glass microbubbles," in *Proceedings of the 14th International Conference on Miniaturized Systems for Chemistry and Life Sciences*, Groningen, The Netherlands, October 2010.

9

Stability-Indicating RP-TLC/Densitometry Determination of Raloxifene Hydrochloride in Bulk Material and in Tablets

A. A. Shirkhedkar, J. K. Rajput, D. K. Rajput, and S. J. Surana

Department of Pharmaceutical Chemistry, R.C. Patel Institute of Pharmaceutical Education and Research, Karwand Naka, Shirpur, Dhule District 425 405, India

Correspondence should be addressed to A. A. Shirkhedkar, atulshirkhedkar@rediffmail.com

Academic Editor: Sibel A. Ozkan

A stability-indicating RP-TLC/Densitometry method for analysis of Raloxifene hydrochloride both in bulk material and in tablets was developed and validated. Densitometric analysis of Raloxifene hydrochloride was carried out at 311 nm on TLC aluminium plates precoated with silica gel 60RP-18 F_{254}S as the stationary phase and methanol : water : ammonia (95 : 05 : 0.1 v/v) as mobile phase. Raloxifene hydrochloride was well resolved at R_f 0.55 ± 0.02. The linear regression analysis data for the calibration plots showed good linear relationship with $r^2 = 0.9969 ± 0.0015$ with respect to peak area in the concentration range 100–600 ng per band. The mean value ± SD of slope and intercept was found to be 15.05 ± 0.44 and 201.9 ± 29.58 with respect to peak area. The limits of detection and quantification were 9.27 ng and 27.10 ng, respectively. Raloxifene hydrochloride was subjected to acid and alkali hydrolysis, oxidation, dry heat, and photodegradation. The drug underwent degradation under basic and oxidation conditions. This indicates that the drug is susceptible to alkali hydrolysis and oxidation. The proposed developed RP-TLC/Densitometry method can be applied for identification and quantitative determination of Raloxifene hydrochloride in bulk material and tablets.

1. Introduction

Raloxifene hydrochloride (RLX), [6-Hydroxy-2-(4-hydroxy-phenyl) benzo [b] thien-3-yl] [4-[2-(1-piperidinyl)-ethoxy] phenyl]-methanone-, hydrochloride (Figure 1) is a selective estrogen receptor modulator (SERM) used in the treatment of osteoporosis in postmenopausal women [1]. Clinically, it is effective in the treatment of breast cancer [2, 3].

Literature survey revealed that RLX was analyzed by HPLC [4–9], Stability-indicating UPLC [10], and several UV-spectrophotometric [11–14] in pharmaceutical formulations. Few methods such as LC-MS-MS [15] and HPLC [16] have been reported for estimation of RLX in biological samples. Although the RP-HPLC and UPLC procedures are accurate and effective means of assaying RLX, they are time and solvent consuming, and therefore, disadvantageous for serial estimation for a large number of samples [17]. However, the prominent application of HPTLC is that many samples can be run simultaneously using a small quantity of mobile phase unlike HPLC, thus reducing the analysis time and cost per analysis. In-reverse phase chromatography, polar mobile phase is used and the stationary phase is nonpolar. It is increasingly being experienced that different components of formulation which could not be resolved using normal-phase TLC could easily be resolved by reverse-phase TLC. Further, in RP-TLC, the impurities either at starting line or near the solvent front can be detected [18].

In view of the above factors, an HPTLC method was well thought-out, to be cheaper, faster, and sometimes more efficient than RP-HPLC and UPLC. From the literature survey, it is revealed that no stability-indicating RP-TLC/Densitometry method has been reported in the literature for analysis of RLX as bulk material or in pharmaceutical formulations.

Hence, the objective of the present investigation was to develop a simple stability-indicating RP-TLC/Densitometry method offering lower analysis time and less cost per analysis for estimation of RLX in bulk material and in tablets and to validate the method according to the ICH guidelines [19, 20].

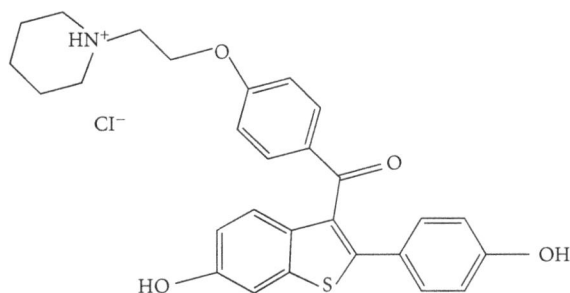

FIGURE 1: Chemical structure of Raloxifene hydrochloride (RLX).

2. Experimental

2.1. Chemicals and Reagents.
Raloxifene hydrochloride (RLX) was procured from Cipla India Ltd, Mumbai, India. All chemicals and reagents used were of Analytical grade and were purchased from Merck Chemicals, India.

2.2. HPTLC Instrumentation.
Chromatography was performed on aluminium plates precoated with Silica gel 60 RP-18 F_{254} S (20 × 10 cm, E. Merck, Germany). The plates were prewashed with methanol and activated at 100°C for 10 min prior to chromatography. The samples were spotted in the form of bands of 6 mm width with a Camag microlitre syringe using a Camag Linomat 5 applicator with a constant rate of application, 150 nL per second. Linear ascending development with methanol : water : ammonia (95 : 05 : 0.1 v/v) as mobile phase was performed in a 20 × 10 cm twin-trough glass chamber (Camag), with tightly fitting lid, previously saturated with mobile phase vapour for 25 min at room temperature (25 ± 2°C). The development distance was 8 cm. After development, the plates were dried in current of air by an air dryer. Densitometric scanning was then performed at 311 nm with a Camag TLC Scanner 3 in absorbance mode operated by winCATS software. The source of radiation was a deuterium lamp. Slit dimensions were 5 mm × 0.45 mm and the scanning speed 20 mm per second.

2.3. Preparation of Stock Standard Solution and Linearity Study.
Stock standard solution was prepared by dissolving 10 mg of RLX in 10 mL methanol. From it, appropriate volume 0.2–1.2 mL was transferred into six separate 10 mL volumetric flask and volume was made up to the mark with methanol. With the help of linomat 5 applicator, 5 μL of each solution was applied on RP-TLC plates to obtain concentration in the range of 100 to 600 ng per band, developed and scanned as described above.

2.4. Preparation of Sample Solution.
Twenty tablets (RALISTA, label claim: 60 mg of RLX per tablet) were weighed and crushed into fine powder. The quantity of powdered drug equivalent to 50 mg of RLX was weighed and transferred in 100 mL volumetric flask containing 50 mL methanol, sonicated for 10 min, volume was adjusted to mark and filtered using 0.45 μm filter (Millifilter, Milford, MA). From the filtrate 1.0 mL was further diluted to 10 mL with methanol. Appropriate volume 8 μL was spotted for assay of RLX. The RP-TLC plates were developed and scanned as described above.

2.5. Method Validation

2.5.1. Precision.
Repeatability of sample application and measurement of peak area were performed using six replicates of the same spot (400 ng per band of RLX). The intra, and interday variation for the estimation of RLX was carried out at three different concentration levels of 200, 300, and 500 ng per band.

2.5.2. Limit of Detection (LOD) and Limit of Quantification (LOQ).
In order to determine detection and quantification limit, RLX concentrations in the lower part of the linear range of the calibration curve were used. From the stock standard solution RLX 100, 120, 140, 160, 180, and 200 ng per band was applied in triplicate on RP-TLC plate. The LOD and LOQ were calculated using equation LOD = 3.3 × N/B and LOQ = 10 × N/B, where "N" is standard deviation of the peak areas of the drugs ($n = 3$), taken as a measure of noise, and "B" is the slope of the corresponding calibration curve.

2.5.3. Specificity.
The specificity of the method was checked by analyzing drug standard and sample. The band for RLX in sample was confirmed by comparing the R_f values and spectra of the band with that of standard. The peak purity of RLX was assessed by comparing the spectra at three different levels, that is, peak-start (S), peak-apex (M), and peak-end (E) positions of the band.

2.5.4. Ruggedness.
Ruggedness of the method was performed by spotting 400 ng of RLX by two different analysts keeping same experimental and environmental conditions.

2.5.5. Accuracy.
The preanalysed samples (200 ng per band) were spiked with extra 80, 100, and 120% of the standard RLX, and the mixture was then reanalysed by the proposed method. At each level of the amount, three determinations were performed. This was done to check the recovery of the drug at different levels in the formulations.

2.5.6. Robustness.
By introducing small deliberate changes in the mobile-phase composition, the effects on the results were examined. Mobile phases having different composition of methanol : water : ammonia (96 : 4 : 0.1 v/v) and (94 : 6 : 0.1 v/v) were tried and chromatograms were run. The amount of mobile phase was varied in the range of ±2 mL. The plates were prewashed by methanol and activated at 100 ± 5°C for 5 and 15 min prior to chromatography. Time from spotting to chromatography and from chromatography to scanning was varied from 0, 20, 40 min.

2.5.7. Stability of Sample Solution in Methanol and Mobile Phase.
To assess the stability of RLX in methanol and mobile

TABLE 1: Repeatability and Intraday, Interday precision.

Parameters	Concentration ng per band	% Amount found	% RSD
Repeatability ($n = 6$)	400	100.94	1.67
	200	99.98	1.27
Intraday ($n = 3$)	300	101.44	0.94
	400	99.70	0.41
	200	99.50	1.37
Interday ($n = 3$)	300	99.68	1.33
	400	101.97	1.35

n: number of determinations.

TABLE 2: Recovery studies.

Drug	Initial amount (ng per band)	Amount of drug standard added (%)	% Drug recovered	% RSD ($n = 3$)
	200	0	100.20	1.2
RLX	200	80	99.78	1.2
	200	100	99.74	1.4
	200	120	100.57	1.5

n: number of determinations.

phase; the sample solutions were separately prepared in methanol and mobile phase and stored at room temperature for 24 (h). The sample solutions were assayed at an interval of 6 (h) for 24 (h).

2.5.8. Stability of Sample Solution on RP-TLC Plate.
The sample solution was applied on RP-TLC plate, kept for 72 (h), and scanned at an interval of 12 (h) as described above.

2.6. Forced Degradation of RLX

2.6.1. Acid, Base and Oxidation Degradation.
Accurately weighed quantity 10 mg of RLX was separately dissolved in 10 mL methanolic solution of 0.5 M HCl and 0.5 M NaOH and 3% (v/v) hydrogen peroxide, respectively; solutions were kept for period of 12 (h) at room temperature in dark to avoid likely degradative effect of light. An appropriate volume 1.0 mL of above solution was taken, neutralized, and diluted up to 10 mL with methanol. The resultant solution was applied on RP-TLC plates in triplicates (5 μL each, i.e, 500 ng per band). The chromatogram was developed and scanned as described above.

2.6.2. Dry Heat Degradation.
Accurately weighed quantity 10 mg of RLX stored at 80°C for 24 (h) in an oven. It was transferred to 10 mL volumetric flask containing methanol and volume was made up to the mark. The 1.0 mL of above solution was taken and diluted up to 10 mL with methanol. The resultant solution was applied on RP-TLC plate in triplicate (5 μL each, i.e, 500 ng per band). The chromatogram was developed and scanned as described above.

2.6.3. Photodegradation.
Accurately weighed quantity 10 mg of RLX was dissolved in 10 mL methanol and solutions was kept for period of 24 (h) in light. An appropriate volume 1.0 mL of above solution was taken and diluted up to 10 mL

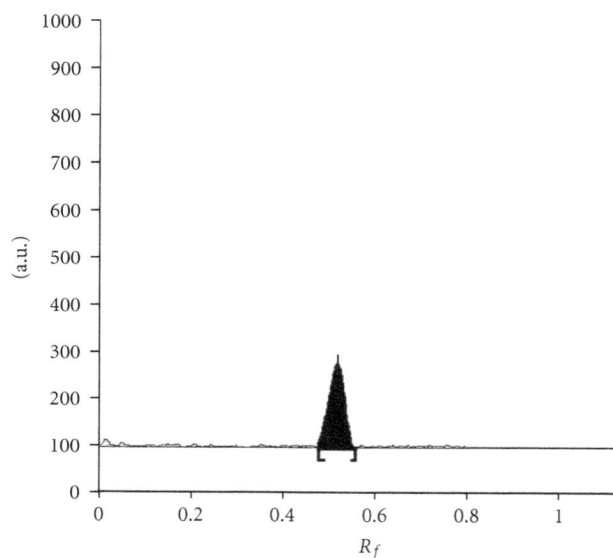

FIGURE 2: Chromatogram of RLX standard (R_f: 0.55 ± 0.02) at 311 nm, in mobile phase methanol : water : ammonia (95 : 05 : 0.1 v/v).

with methanol. The resultant solution was applied on RP-TLC plate in triplicate (5 μL each, i.e, 500 ng per band). The chromatogram was developed and scanned as described above.

3. Results and Discussion

3.1. Development of Optimum Mobile Phase.
For the selection of appropriate mobile phase for RLX, several runs were exercised using mobile phases containing solvents of varying polarity, at different concentration levels. Among the different mobile-phase combinations employed, the mobile phase

TABLE 3: Robustness of the method.

Parameter	±SD of peak area	% RSD ($n = 6$)
Mobile-phase composition: methanol : water : ammonia (96 : 04 : 0.1 v/v)	50.89	1.5
Mobile-phase composition: methanol : water : ammonia (94 : 06 : 0.1 v/v)	52.74	1.6
Mobile-phase volume (±2 mL)	47.91	1.4
Development distance (±0.5 cm)	44.03	1.3
Activation of TLC plate (±5 min)	48.84	0.9
Duration of saturation (±5 min)	44.97	1.4
Time from spotting to chromatography (±10 min)	40.16	0.9
Time from chromatography to scanning (±10 min)	49.34	1.2

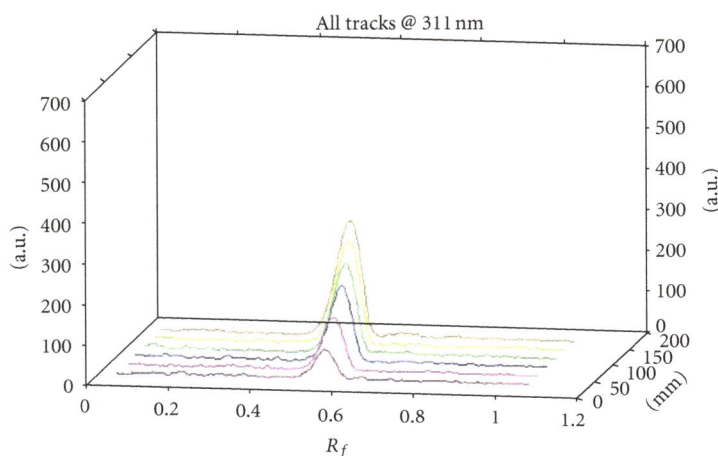

FIGURE 3: Three dimensional chromatograms of RLX sample (100–600 ng per band).

TABLE 4: Summery of validation parameter.

Parameter data	RLX
Linearity range (ng per band)	100–600
Correlation coefficient (r^2)	0.9969
Limit of detection (ng)	9.27
Limit of quantification (ng)	27.10
Recovery ($n = 3$)	99.74–100.57
Ruggedness (% RSD)	
Analyst-I ($n = 6$)	1.53
Analyst-II ($n = 6$)	1.33
Precision (% RSD)	
Repeatability of application ($n = 6$)	0.96–1.49
Interday ($n = 3$)	1.33–1.37
Intraday ($n = 3$)	0.41–1.27
Robustness	Robust
Specificity	Specific

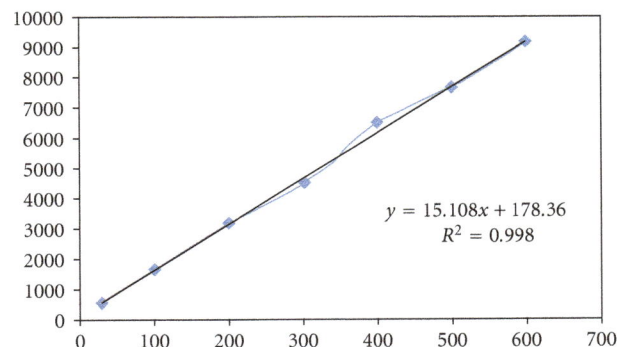

FIGURE 4: Calibration curve containing LOQ value as the lowest point.

3.2. Calibration Curve.
The linear regression data for the calibration curves showed good linear relationship over the concentration range 100–600 ng per band. Linear regression equation was found to be $Y = 15.05X + 201.9$, $r^2 = 0.9969$ (Figure 3).

3.3. Validation of Method

3.3.1. Precision.
The precision of the developed method was represented in terms of % relative standard deviation

consisting of methanol : water : ammonia (95 : 05 : 0.1 v/v) gave a sharp and well-defined peak at R_f value of 0.55 ± 0.02 (Figure 2). Well distinct bands were found when the chamber was saturated with the mobile phase for 25 min at room temperature.

TABLE 5: Summary of forced degradation studies.

Stress conditions	Time (h)	Recovery (%)	R_f of degradants
0.5 M HCl	12	99.5	No degradants formed
0.5 M NaOH	12	80.2	0.32,0.44,0.70
3% (v/v) H_2O_2	12	91.45	0.64
Day light (8 (h)/day)	24	99.9	No degradants formed
Heat (80°C)	24	99.6	No degradants formed

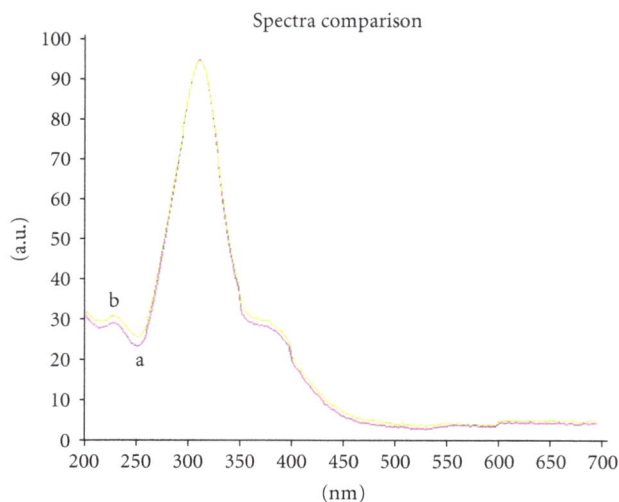

FIGURE 5: Peak purity spectra of RLX standard (a) and RLX extracted from tablets (b) scanned at peak-start, peak-apex and peak-end position.

(% RSD) of the peak area. The results depicted indicated high precision of the method are presented in Table 1.

3.3.2. LOD and LOQ.
The LOD and LOQ were determined from the slope of the lowest part of the calibration plot. The LOD and LOQ were found to be 9.27 ng and 27.10 ng, respectively, which indicates the sensitivity of the method is adequate (Figure 4).

3.3.3. Recovery Studies.
The recovery studies were executed out at 80%, 100%, and 120% of the test concentration as per ICH guidelines. The % recovery of RLX at all the three levels was found to be satisfactory. The amounts of drug added and determined and the % recovery are listed in Table 2.

3.3.4. Specificity.
The peak purity of RLX was assessed by comparing the spectra at peak-start, peak-apex, and peak-end positions of the band, that is, $r^2(S, M) = 0.996$ and $r^2(M, E) = 0.9988$. Good correlation ($r^2 = 0.9989$) was also obtained between drug standard and drug extracted from tablet formulation (Figure 5).

3.3.5. Robustness of the Method.
The standard deviation of peak areas was calculated for each parameter and % R.S.D. was found to be less than 2%. The low values of % RSD

values indicate robustness of the method; results are shown in Table 3.

3.3.6. Solution Stability and Mobile Phase Stability.
The stability study of RLX in methanol demonstrates no significant change in the chromatogram obtained.

Similarly, the stability study of RLX in mobile phase also does not show any noticeable change in the chromatogram.

3.3.7. Stability on Plate.
No major changes were observed in chromagram when plate's were scanned at 0, 12, 24, 36, 48, 72 (h) interval.

The validation of the method is summarized in Table 4.

3.4. Analysis of the Marketed Formulation.
A single spot at R_f 0.55 ± 0.02 was observed in the chromatogram of the drug samples extracted from tablets. There was no interference from the excipients which routinely occur in the tablets. The mean % drug content was found to be 100.28% of the label claim.

3.5. Stability-Indicating Property.
The results of the forced degradation study of RLX are summarized in Table 5. RLX showed degradation in alkali and oxidative conditions. In the base-induced degradation (0.5 M NaOH) study, RLX showed additional peaks at R_f values 0.32, 0.44, 0.70, and in oxidative degradation (3% (v/v) H_2O_2) RLX showed only one additional peak at R_f 0.64 (Figure 6). The spots of the degraded products were well separated from the drug spots. The peak purity spectra of RLX recovered after degradation in 0.5 M NaOH, 3% (v/v) H_2O_2 and RLX standard scanned at peak-start, peak-apex, and peak-end positions of the spot are shown in (Figure 7).

No additional peaks were found in acid, dry heat, and photodegradation. Therefore, RLX is stable in acidic, dry heat, and photoconditions.

4. Conclusion

The developed method was found to be simple, rapid, selective, sensitive, and suitable for determination of Raloxifene hydrochloride in bulk material and pharmaceutical dosage forms without any interference from excipients. As the method is stability-indicating one, it can be used to determine the purity of the drug available from various sources by detecting the related impurities. Furthermore, it can be concluded that the impurities present in the drug

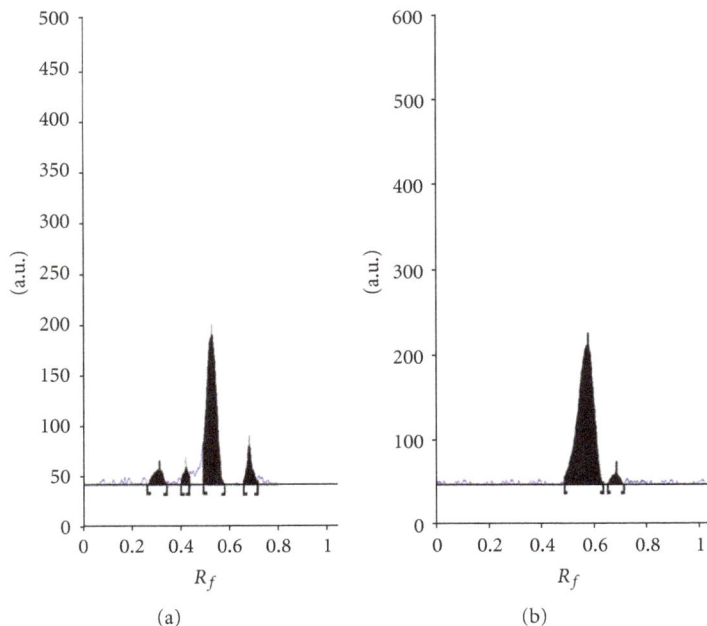

FIGURE 6: RP-TLC chromatogram obtained from forced degradation studies. (a) Base degradation (0.5 M NaOH, 12 (h), RT) showing three degradants at R_f values 0.32, 0.44, and 0.70 for RLX. (b) Oxidative degradation (3% (v/v) H_2O_2, 12 (h), RT) showing one degradant at R_f values 0.64 for RLX.

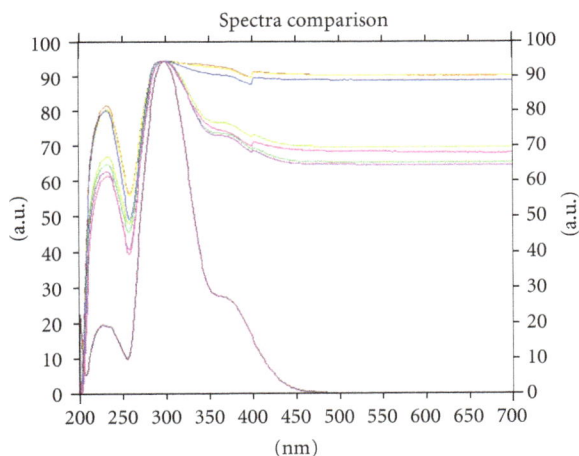

FIGURE 7: Peak purity spectra of RLX recovered after degradation in 0.5 M NaOH, 3% (v/v) H_2O_2, degradants, and RLX standard scanned at peak-start, peak-apex, and peak-end positions.

could be due to hydrolysis or oxidation during processing and storage of the drug.

The proposed procedure fits precision and accuracy usually requested by official methods and can be used as a convenient alternative to HPLC analysis for quantitation of Raloxifene hydrochloride in both bulk and tablet dosage forms. Therefore, the proposed RP-TLC/Densitometry method can be used as an alternative tool in the drug quality control laboratories for quantitative determination of Raloxifene hydrochloride.

Acknowledgment

The authors are thankful to R.C. Patel Institute of Pharmaceutical Education and Research, Shirpur (MS), India, for providing the required facilities to carry out this research work.

References

[1] A. Smith, P. E. Heckelman, J. R. Obenchain, J. A. R. Gallipeau, M. A. D'Arecca, and S. Budavari, *The Merck Index*, Merck Research Laboratories, Whitehouse Station Readington, NJ, USA, 13th edition, 2001.

[2] S. R. Cummings, S. Eckert, K. A. Krueger et al., "The effect of raloxifene on risk of breast cancer in postmenopausal women," *Journal of the American Medical Association*, vol. 281, no. 23, pp. 2189–2197, 1999.

[3] T. Hol, M. B. Cox, H. U. Bryant, and M. W. Draper, "Selective estrogen receptor modulators and postmenopausal women's health," *Journal of Women's Health*, vol. 6, no. 5, pp. 523–531, 1997.

[4] D. Suneetha and A. Lakshmana Rao, "A new validated RP-HPLC method for the estimation of raloxifene in pure and tablet dosage form," *Rasayan Journal of Chemistry*, vol. 3, no. 1, pp. 117–121, 2010.

[5] D. C. Pavithra and L. Sivasubramanian, "RP-HPLC estimation of raloxifene hydrochloride in tablets," *Indian Journal of Pharmaceutical Sciences*, vol. 68, no. 3, pp. 401–402, 2006.

[6] J. Trontelj, T. Vovk, M. Bogataj, and A. Mrhar, "HPLC analysis of raloxifene hydrochloride and its application to drug quality control studies," *Pharmacological Research*, vol. 52, no. 4, pp. 334–339, 2005.

[7] B. Madhu, A. A. Kumara, S. Prashanth et al., "Sensitive and rapid HPLC method for the determination of raloxifene hydrochloride," *Journal of Pharmacy Research*, vol. 4, no. 3, pp. 582–584, 2011.

[8] K. Basavaiah, U. R. A. Kumar, and K. Tharpa, "Gradient HPLC analysis of raloxifene hydrochloride and its application to drug quality control," *Acta Pharmaceutica*, vol. 58, no. 3, pp. 347–356, 2008.

[9] A. Sathyaraj, M. Rao, and V. Satyanarayana, "Gradient RP-HPLC method for the detrmination of purity and assay of raloxifene hydrochloride in bulk drug," *International Journal of Pharmaceutical Chemistry*, vol. 1, no. 3, pp. 372–378, 2011.

[10] G. Srinivas, G. V. Kanumula, P. Madhavan et al., "Development and validation of stability indicating method for the quantitative determination of raloxifene hydrochloride and its related impurities using UPLC," *Journal of Chemical and Pharmaceutical Research*, vol. 3, no. 1, pp. 553–562, 2011.

[11] B. Kalyanaramu and K. Raghubabu, "Development of new analytical method for determination of raloxifene hydrochloride in formulations based on charge-transfer complex formation," *International Journal of Analytical and Bioanalytical Chemistry*, vol. 1, no. 2, pp. 29–33, 2011.

[12] D. C. Pavithra and L. Sivasubramanian, "New spectrophotometric determination of raloxifene hydrochloride in tablets," *Indian Journal of Pharmaceutical Sciences*, vol. 68, no. 3, pp. 375–376, 2006.

[13] K. Basavaiah and U. R. Anilkumar, "New sensitive spectrophotometric methods for the determination of raloxifene hydrochloride in pharmaceuticals using bromate-bromide, methyl orange and indigo carmine," *E-Journal of Chemistry*, vol. 3, no. 13, pp. 242–249, 2006.

[14] M. M. Annapurna, M. E. B. Rao, and B. V. Ravi Kumar, "Spectrophotometric determination of raloxifene hydrochloride in pharmaceutical formulations," *E-Journal of Chemistry*, vol. 4, no. 1, pp. 79–82, 2007.

[15] J. Trontelj, T. Vovic, M. Bogataj, and A. Mrhar, "Development and validation of liquid chromatography-tandem mass spectrometry assay for determination of raloxifene and its metabolites in human plasma," *Journal of Chromatography B*, vol. 855, no. 2, pp. 220–227, 2007.

[16] Z. Y. Yang, Z. F. Zhang, X. B. He, G. Y. Zhao, and Y. Q. Zhang, "Validation of a novel HPLC method for the determination of Raloxifene and its pharmacokinetics in rat plasma," *Chromatographia*, vol. 65, no. 3-4, pp. 197–201, 2007.

[17] S. A. Coran, M. Bambagiotti-Alberti, V. Giannellini, A. Baldi, G. Picchioni, and F. Paoli, "Development of a densitometric method for the determination of cephalexin as an alternative to the standard HPLC procedure," *Journal of Pharmaceutical and Biomedical Analysis*, vol. 18, no. 1-2, pp. 271–274, 1998.

[18] P. D. Sethi, *High Performance Thin Layer Chromatography (Quantitative Analysis of Pharmaceutical Formulations)*, CBS Publishers, New Delhi, India, 1996.

[19] International conference on Harmonization ICH/CPMP guidelines Q2(R1), *Validation of Analytical Procedures: Text and Methodology*, ICH, Geneva, Switzerland, 2005.

[20] International Conference on Harmonization Q1A, *Stability Testing of New Drug Substances and Products*, ICH, Geneva, Switzerland, 1993.

10

Retention Behaviour in Micellar Liquid Chromatography

Maria Rambla-Alegre

Àrea de Química Analítica, QFA, Universitat Jaume I, 12071 Castelló, Spain

Correspondence should be addressed to Maria Rambla-Alegre, mrambla@guest.uji.es

Academic Editor: Samuel Carda-Broch

Retention in micellar liquid chromatography is highly reproducible and can be modelled using empirical or mechanistic models with great accuracy to predict the retention changes when the mobile phase composition varies (surfactant and organic solvent concentrations), thus facilitating the optimisation of separation conditions. In addition, the different equilibria inside the column among the solute, the mobile phase, and the modified stationary phase by monomers of surfactant have been exhaustively studied. In a sequential strategy, the retention of the solutes is not known a priori, and each set of mobile phases is designed by taking into account the retention observed with previous eluents. By contrast, in an interpretative strategy, the experiments are designed before the optimization process and used to fit a model that will allow the prediction of the retention of each solute. This strategy is more efficient and reliable. The sequential strategy will be inadequate when several local and/or secondary maxima exist, as frequently occurs in chromatography, and may not give the best maximum, that is to say, the optimum. More often than not, the complexity of the mixtures of compounds studied and the relevant modification of their chromatographic behaviour when changing the mobile phase composition requires the use of computer-assisted simulations in MLC to follow the modifications in the chromatograms in detail. These simulations can be done with sound reliability thanks to the use of chemometrics tools.

1. Introduction

Most reported procedures for the determination of compounds in MLC make use of micellar mobile phases containing an organic modifier, usually a short-chain alcohol or acetonitrile. These modifiers increase the elution strength and often improve the shape of the chromatographic peaks. The modifiers solvate the bonded stationary phase and reduce the amount of surfactant adsorbed, the effect being larger with increasing concentration and hydrophobicity of the alcohol. Selection of the pH in the mobile phase is also often extremely important for the resolution of complex mixtures, owing to the side acid-base reactions of many solutes. Other variables to be considered are temperature and ionic strength [1].

The chromatographer is concerned with the achievement of the optimum mobile phase that permits the separation of the composition in a mixture, in the minimum time. This task may be really difficult when two or more variables are involved in the optimisation process. The optimisation strategy may be sequential or interpretative. In a sequential strategy, the retention of the solutes is not known *a priori*, and each set of mobile phases is designed by taking into account the retention observed with previous eluents. In contrast, in an interpretative strategy, the experiments are designed before the optimisation process and used to fit a model that will permit the prediction of the retention of each solute. This strategy may be much more efficient and reliable. A sequential strategy will be inadequate when several local (or secondary) maxima exist (as occurs in chromatography) and may not give the best maximum, that is, the optimum.

The necessity for an adequate experimental design becomes especially important when dealing with forms of liquid chromatography suitable for the simultaneous analysis of ionic and nonionic compounds, such as MLC, where several variables should be controlled (e.g., type and concentration of surfactant and organic modifier, pH, temperature, and ionic strength). The method development strategy must provide the chromatographer with an answer to which variables should be used, and how to set up initial experiments to search the appropriate variable space in an effective way.

The separation process in a micellar chromatographic system requires a structured approach in the development of practical applications. Ideally, the resolution of complex mixtures should be made and optimized in a short time, with minimal consumption of reagents.

2. Empirical Models

Retention in a hybrid micellar mobile system can be modelled using a procedure that utilizes the retention data of only five mobile phases [2]: four measurements at the corners of the selected two-dimensional variable space, defined by the concentrations of surfactant and modifier, and the fifth in the centre. The chromatographic data obtained (retention factor, efficiency, and asymmetry factor) are used to fit some equations. The models used for the description of retention behaviour are summarized as in Section 2.1.

2.1. Empirical and Mechanistic Models Used in MLC

Empirical Models. k is the retention factor for a given mobile phase composition; $[M]$ is the concentration of surfactant forming micelles (total concentration of surfactant minus the critical micellar concentration, CMC); φ is the volume fraction of the organic solvent; c_0, c_1, c_2, c_3, and c_{11} are fitting coefficients:

$$\log k = c_0 + c_1[M] + c_2\varphi, \tag{1}$$

$$\frac{1}{k} = c_0 + c_1[M], \tag{2}$$

$$\frac{1}{k} = c_0 + c_1[M] + c_2\varphi, \tag{3}$$

$$\frac{1}{k} = c_0 + c_1[M] + c_2\varphi + c_3[M]\varphi, \tag{4}$$

$$\frac{1}{k} = c_0 + c_1[M] + c_2\varphi + c_3[M]\varphi + c_{11}\varphi^2. \tag{5}$$

Mechanistic Models. K_{AS} is the product of the solute- (A) stationary (S) partition coefficient by the phase ratio, and K_{AM} the solute-micelle (M) association constant; K_{AD} and K_{MD} measure the relative variation produced in the concentration of solute in bulk water and micelle, respectively, in presence of modifier, taking the pure micellar solution as a reference; K_{AD2} corresponds to a quadratic hyperbolic variation in K_{AS} and K_{AM} with φ:

$$k = \frac{K_{AS}}{1 + K_{AM}[M]}, \tag{6}$$

$$k = \frac{K_{AS}(1/(1 + K_{AD}\varphi))}{1 + K_{AM}((1 + K_{MD}\varphi)/(1 + K_{AD}\varphi))[M]}, \tag{7}$$

$$k = \frac{K_{AS}(1/(1 + K_{AD1}\varphi + K_{AD2}\varphi^2))}{1 + K_{AM}((1 + K_{MD}\varphi)/(1 + K_{AD1}\varphi + K_{AD2}\varphi^2))[M]}, \tag{8}$$

$$k = \frac{K_{AS}((1 + K_{SD}\varphi)/(1 + K_{AD}\varphi))}{1 + K_{AM}((1 + K_{MD}\varphi)/(1 + K_{AD}\varphi))[M]}, \tag{9}$$

$$k$$

$$= \frac{K_{AS}(1/(1 + K_{AD1}\varphi + K_{AD2}\varphi^2))}{1 + K_{AM}((1 + K_{MD1}\varphi + K_{MD2}\varphi^2)/(1 + K_{AD1}\varphi + K_{AD2}\varphi^2))[M]}. \tag{10}$$

Equations (1)–(5) correspond to empirical models. In (1), it is assumed that solute retention is linearly related to the mobile phase variables within a selected portion of the space and fitted to a separate logarithmic linear function. However, the predictions obtained with (1) are not accurate enough. Retention in micellar mobile phases with a fixed amount or without organic solvent [3] has been extensively proved to be described by the hyperbolic relationship shown in (2). To model the retention of a solute at varying concentrations of both surfactant and modifier [4], (3) can be used and has yielded acceptable results for small ranges of concentrations of surfactant and modifier, but an interaction between both factors, $[M]$ and φ, exists for larger domains.

The capability of a series of empirical equations to describe the retention behaviour for any surfactant and organic solvent content has been studied [5–7]. Several models were considered where $\log k$ or $1/k$ values were related to the micellar concentration and volume fraction of organic modifier. Logarithmic models usually yielded poorer results. Equation (4) is the simplest equation giving good predictions for both polar and moderately polar compounds, such as amino acids, sulfonamides, β-blockers, and diuretics. Equation (4) yields linear plots of $1/k$ versus φ at fixed concentrations of surfactant. However, for highly hydrophobic compounds, such as steroids or polycyclic aromatic hydrocarbons, the plots are nonlinear. An additional term is required to achieve more accurate descriptions, as shown in (5). The parameters in (4) and (5) should be obtained by fitting the data in experimental designs with at least four and five mobile phases, respectively. However, at least one additional measurement should be taken to check the accuracy of the fittings.

3. Mechanistic Models

The parameters of the empirical models of MLC are related to physicochemical constants that describe the interactions of the solutes with the three environments involved in micellar mobile phases: bulk water, micelles, and the stationary phase. A better understanding of the retention mechanism in micellar systems is provided by these models.

The mechanistic models are based on (2), which is the classical equation proposed for micellar mobile phases at a fixed volume fraction of organic modifier [2]. Equation (2) can be written as (6), which relates the retention of a solute to the concentration of monomers of surfactant in the form of micelles.

For hybrid micellar mobile phases, (6) can be expressed as (7), while (5) gives rise to (8) [8], which may suggest an excessive dependence of the retention on the organic solvent concentration and produce high errors when an extrapolation is made in a region of high concentration of modifier.

As a result, (9) was proposed as an alternative model for highly hydrophobic solutes, and it takes into account the additional change in the concentration of solute associated with the stationary phase produced by the presence of modifier [7]. In (9), the constants K_{MD} and K_{AD} account for the displacement of the water-micelle equilibrium, whereas K_{SD} and K_{AD} describe the modification of the water-stationary phase equilibrium. These changes are caused by the decrease in the polarity of water and the modification of the interactions of solute with micelles and stationary phase, when a modifier is added. Equation (9) provides an accurate description of the retention of solutes of a wide range of polarities, when they are eluted with hybrid mobile phases of SDS and alcohol (propanol, butanol or pentanol) [8]. For acetonitrile and tetrahydrofuran [9, 10], (10) fits better.

4. Peak Shape Modelling

The major drawback of applying MLC to practical separations is still the low chromatographic efficiency, which is caused by resistance to mass transfer in the processes involving micelles and a surfactant-modified stationary phase. This is especially important since the increase in micelle concentration causes a decrease in plate number, resulting in a varying efficiency over the variable space. Thus, it is very important to include the expected peak shape in the expression of the chromatographic quality. The complexity of the chromatographic process does not allow the use of simple equations to describe peak profiles. The best peak-profile predictions are achieved using a Gaussian equation where the standard deviation depends polynomially on the distance to the peak time (polynomial modified Gaussian model) [11]:

$$h(t) = H_0 e^{-0.5((t-t_R)/(s_0+s_1(t-t_R)+s_2(t-t_R)^2+\cdots))^2}, \quad (11)$$

where $h(t)$ is the predicted signal at time t, H_0 the maximal peak height, t_R the retention time, and the coefficients s_i are related to the width and asymmetry of the chromatographic peak. For a given solute and mobile phase, t_R and s_i are ideally invariable, whereas H_0 depends on the concentration. Better descriptions of peak profiles are achieved by increasing the degree of the polynomial. The use of a larger number of coefficients improves the fittings but decreases the practical application of the model. A linear standard deviation in (11) approximates real peak profiles satisfactorily. The linear equation is also useful for simulating chromatograms.

5. Strategies to Measure the Peak Resolution

The simulation of chromatograms requires predictions of peak retention and peak profile that are as accurate as possible. Computer optimization attempts to mimic the methodology followed by an experienced chromatographer so as to reduce the time and the effort required. A software application was developed to assist the chromatographer in the selection of the optimal composition of the mobile phase

in MLC [12]. This program allows chromatograms to be simulated quickly and therefore enables the changes in the chromatograms to be observed as the user simulates variations in the composition of the mobile phase. The resolution surfaces and ease of simulation can be applied in a straightforward manner to select the best composition of the mobile phase.

Different measurements of diverse complexity have been proposed to depict chromatographic performance. Optimization criteria based on calculation of an individual or elementary resolution measurement, r_i, for the least resolved peak or peak pair is a very widely used procedure in chromatographic practice, because of its simplicity:

$$R = \text{MIN}(r_i) \quad 1 \le i \le p, \quad (12)$$

where p is the number of peaks or peak pairs, and R is the global resolution.

This criterion is reasonable, but it considers the resolution of only one peak or peak pair, and is insensitive to the remaining peaks. In many cases, a practically identical resolution of the worst peak can be obtained, while the resolution of the other peaks can be improved. The product of peak resolutions solves this drawback, since it optimizes the resolution of all peaks in the chromatogram.

The normalized-by-mean product is conventionally applied. Section 5.1 shows (13), the global resolution functions, and elementary resolution criteria. This treatment normalizes the resolution approximately, using the mean r_i of all the peaks in the chromatogram instead of the extreme values.

The unnormalized product of (14) seems to be a better alternative, although it can be used only with intrinsically normalized resolution measurements. This product varies from 0 (complete overlapping between at least two peaks) to 1 (full resolution of every peak in the chromatogram).

After the selection of the global resolution function, the appropriate elementary resolution criterion (Section 5.1) should be decided. Some criteria have been based on conventional elementary measurements, such as the modified selectivity in (15), peak-to-valley measurements in (16), and overlapping fraction measurements in (17).

In (16), the valley between two consecutive peaks can be measured at the time giving the largest possible distance, measured orthogonally. If the valley is observed orthogonally, this point is obvious even when there is substantial overlap.

The criterion of overlapping fractions takes into account not only positions but also peak profiles; it isolates the contribution of each component in a mixture, associating a value to each individual peak, which is not affected by the identity of its neighboring peaks, and the intrinsic normalisation facilitates understanding of the information obtained in the optimization process.

5.1. Global Resolution Functions and Elemental Resolution Criteria Used in MLC

Resolution Function. k_i and k_{i+1} $(k_{i+1} > k_i)$ are the retention factors of two neighboring peaks, and $\alpha_{i,i+1}$ is the selectivity; h_1 represents the height of the signal at a specific time depicting the valley location, and h_2 is an interpolated height, measured at that time, from baseline to the line obtained by joining the maximums of the two neighboring peaks; o'_i is the area under a given peak overlapped by the chromatogram yielded by the remaining peaks, and o_i the total area of the peak:

$$R = \frac{\prod_{i=1}^{p} r_i}{\left[\sum_{i=1}^{p} r_i/p\right]^p}, \tag{13}$$

$$R = \prod_{i=1}^{p} r_i, \tag{14}$$

$$r_{i,i+1} = 1 - \frac{k_i}{k_{i+1}}, \tag{15}$$

$$r_{i,i+1} = 1 - \frac{h_1}{h_2}, \tag{16}$$

$$r_i = 1 - \frac{o'_i}{o_i}. \tag{17}$$

6. Conclusions

The development of a micellar analytical procedure for the determination of solutes requires optimization of the column type, the type and concentration of the surfactant and organic modifier, pH, flow rate, and temperature. In addition, the optimization strategy can be sequential or interpretive. In a sequential strategy, a set of mobile phases is designed by taking into account the retention observed with previous eluents (the retention of the solutes is not known a priori). When the experiments are designed before the optimization process and used to fit a model that will allow the retention of each solute to be predicted, the strategy is called interpretative, and, of the two, it is more efficient and reliable. The need for an adequate experimental design becomes especially important when dealing with forms of liquid chromatography suitable for the simultaneous analysis of ionic and nonionic compounds, such as MLC, in which several factors should be controlled. The strategy used to develop the method must provide answers to questions concerning the variables to be used as well as the set-up of the initial experiments that will yield an effective search of the appropriate variable space. Retention in a hybrid micellar mobile system can be modeled using a procedure that utilizes the retention data of only five mobile phases: four measurements at the corners of the selected two-dimensional variable space, defined by the concentrations of surfactant and modifier, and the fifth in the center. The chromatographic data obtained (retention factor, efficiency, and asymmetry factor) are used to fit some equations. Thus, computer optimization permits to mimic the methodology followed by an experienced chromatographer, reducing the time and the effort required.

Acknowledgment

The author wishes to thank the MEC for the FPU grant.

References

[1] M. J. Ruiz-Ángel, S. Carda-Broch, J. R. Torres-Lapasió, and M. C. García-Álvarez-Coque, "Retention mechanisms in micellar liquid chromatography," *Journal of Chromatography A*, vol. 1216, no. 10, pp. 1798–1814, 2009.

[2] E. D. Breyer, J. K. Strasters, A. H. Rodgers, and M. G. Khaledi, "Simultaneous optimization of variables influencing selectivity and elution strength in micellar liquid chromatography. Effect of organic modifier and micelle concentration," *Journal of Chromatography*, vol. 511, pp. 17–33, 1990.

[3] M. Arunyanart and L. J. Cline Love, "Model for micellar effects on liquid chromatography capacity factors and for determination of micelle-solute equilibrium constants," *Analytical Chemistry*, vol. 56, no. 9, pp. 1557–1561, 1984.

[4] J. R. Torres-Lapasió, M. J. Medina-Hernández, R. M. Villanueva-Camanas, and M. C. Garcia-Álvarez-Coque, "Description of the retention behaviour of solutes in micellar liquid chromatography with organic modifiers: comparison of two methods," *Chromatographia*, vol. 40, no. 5-6, pp. 279–286, 1995.

[5] J. R. Torres-Lapasió, R. M. Villanueva-Camañas, J. M. Sanchis-Mallols, M. J. Medina-Hernandez, and M. C. Garcia-Álvarez-Coque, "Modelling of the retention behaviour of solutes in micellar liquid chromatography with organic modifiers," *Journal of Chromatography*, vol. 639, no. 2, pp. 87–96, 1993.

[6] J. R. Torres-Lapasió, R. M. Villanueva-Camanas, J. M. Sanchis-Mallols, M. J. Medina-Hernandez, and M. C. Garcia-Álvarez-Coque, "Interpretive strategy for optimization of surfactant and alcohol concentration in micellar liquid chromatography," *Journal of Chromatography A*, vol. 677, no. 2, pp. 239–253, 1994.

[7] S. Torres-Cartas, R. M. Villanueva-Camañas, and M. C. García-Álvarez-Coque, "Modelling and optimization of the separation of steroids eluted with a micellar mobile phase of sodium dodecyl sulphate containing acetonitrile," *Analytica Chimica Acta*, vol. 333, no. 1-2, pp. 31–40, 1996.

[8] M. C. García-Álvarez-Coque, J. R. Torres-Lapasió, and J. J. Baeza-Baeza, "Description of the partitioning behaviour of solutes and data treatment in micellar liquid chromatography with modifiers," *Analytica Chimica Acta*, vol. 324, no. 2-3, pp. 163–173, 1996.

[9] S. López Grío, J. J. Baeza Baeza, and M. C. García Álvarez-Coque, "Modelling of the elution behaviour in hybrid micellar eluents with different organic modifiers," *Analytica Chimica Acta*, vol. 381, no. 2-3, pp. 275–285, 1999.

[10] S. J. López-Grío, G. Vivó-Truyols, J. R. Torres-Lapasió, and M. C. García-Álvarez-Coque, "Resolution assessment and performance of several organic modifiers in hybrid micellar liquid chromatography," *Analytica Chimica Acta*, vol. 433, no. 2, pp. 187–198, 2001.

[11] J. R. Torres-Lapasió, J. J. Baeza-Baeza, and M. C. García-Álvarez-Coque, "A Model for the Description, Simulation, and

Deconvolution of Skewed Chromatographic Peaks," *Analytical Chemistry*, vol. 69, no. 18, pp. 3822–3831, 1997.

[12] J. R. Torres-Lapasió, *Michrom Software*, Marcel-Dekker, New York, NY, USA, 2000.

11

Validated Stability Indicating RP-HPLC Method for Simultaneous Estimation of Codeine Phosphate and Chlorpheniramine Maleate from Their Combined Liquid Dosage Form

Ramakrishna Kommana and Praveen Basappa

Department of Pharmaceutical Analysis and Quality Assurance, Gokaraju Rangaraju College of Pharmacy, Hyderabad 500 090, India

Correspondence should be addressed to Ramakrishna Kommana; rkkommana@gmail.com

Academic Editor: Irene Panderi

The present paper describes the development of quick stability indicating RP-HPLC method for the simultaneous estimation of codeine phosphate and chlorpheniramine maleate in the presence of its degradation products, generated from forced degradation studies. The developed method separates codeine phosphate and chlorpheniramine maleate in impurities/degradation products. Codeine phosphate and chlorpheniramine maleate and their combination drug product were exposed to acid, base, oxidation, dry heat, and photolytic stress conditions, and the stressed samples were analysed by proposed method. The proposed HPLC method utilizes the Shimadzu HPLC system on a Phenomenex C_{18} column (250 mm × 4.6 mm, 5 μ) using a mixture of 1% o-phosphoric acid in water : acetonitrile : methanol (78 : 10 : 12) mobile phase with pH adjusted to 3.0 in an isocratic elution mode at a flow rate of 1 mL/min, at 23°C with a load of 20 μL. The detection was carried out at 254 nm. The retention time of codeine phosphate and chlorpheniramine maleate was found to be around 3.47 min and 9.45 min, respectively. The method has been validated with respect to linearity, robustness, precision, accuracy, limit of detection (LOD), and limit of quantification (LOQ). The developed validated stability indicating HPLC method was found to be simple, accurate, and reproducible for the determination of instability of these drugs in bulk and commercial products.

1. Introduction

Codeine phosphate (CP) chemically is 7,8-didehydro-4, 5-epoxy-3-methoxy-17-methylmorphinan-6-ol dihydrogen phosphate hemihydrate [1], an alkaloid occurring in *Papaver somniferum* or obtained from morphine by methylation. The structure of CP is shown in Figure 1(a). It acts as a narcotic analgesic.

Chlorpheniramine maleate (CPM) chemically is (*RS*)-3-(4-chlorophenyl)-3-(pyrid-2-yl) propyl-dimethylamine hydrogen maleate [1] as shown in Figure 1(b). Chlorpheniramine maleate is a first generation alkylamine antihistamine used in the prevention of the symptoms of allergic conditions such as rhinitis and urticaria.

The literature survey reveals that many analytical methods are reported for the determination of CP and CPM individually and in combination. The literature survey revealed spectroscopic [2–5] and chromatographic [6–20] methods.

The stability of a drug dosage form refers to the ability of a particular form to maintain its physical, chemical, therapeutic, and toxicological specification presented in the monograph on identity, strength, quality, and purity. The stability of a drug product should ordinarily be demonstrated by its manufacturer by methods appropriate for the purpose. Obviously, a stability testing problem is never simple. Stability testing is an important part of the process of drug product development. The purpose of the stability testing is to provide evidence on how the quality of a drug substance or drug product varies with time under the influence of a variety of environmental factors, such as temperature, humidity, and sun light, and enables recommendation of storage conditions, retest periods, and shelf lives to be established.

Validated Stability Indicating RP-HPLC Method for Simultaneous Estimation of Codeine Phosphate and
Chlorpheniramine Maleate from Their Combined Liquid Dosage Form

69

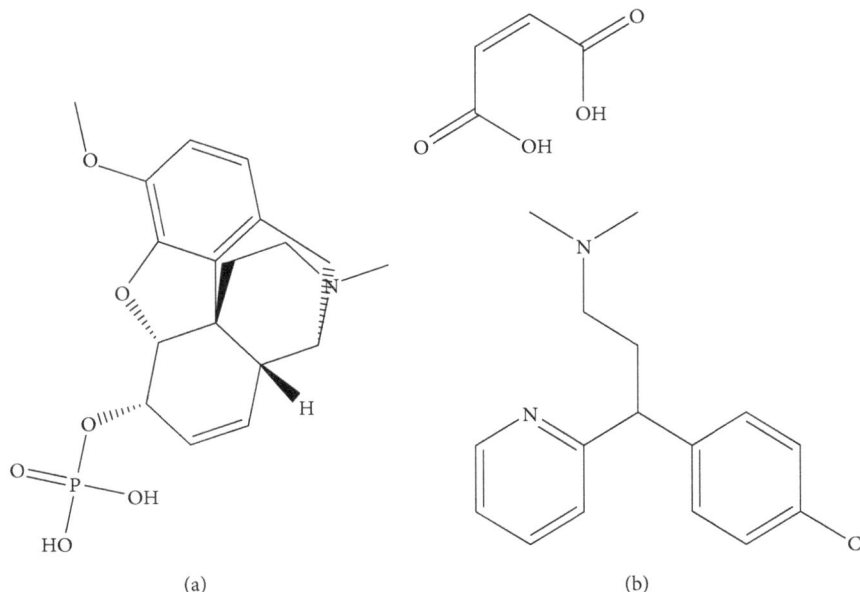

FIGURE 1: Chemical structure of (a) codeine phosphate and (b) chlorpheniramine maleate.

Two main aspects of a drug product that play an important role in shelf life determination are the assay of the active drug and the degradation products generated during the stability study. The drug product in a stability test sample needs to be determined using a stability indicating method, as recommended by the International Conference on Harmonization (ICH) guidelines [21].

The literature survey reveals that many analytical methods are reported for the determination of CP and CPM individually and in combination. The combination of CP and CPM in liquid dosage form has not been reported in any pharmacopoeia. So far, no RP-HPLC stability indicating method has been reported for the simultaneous estimation of CP and CPM in liquid pharmaceutical formulation. Therefore, it is necessary to develop a new rapid and stability indicating method for simultaneous estimation of two compounds (CP and CPM) in liquid pharmaceutical formulation. The proposed method is able to separate CP and CPM with each other and from its impurities/degradation products. Thereafter, this method was validated as per ICH guidelines and successfully applied for separation and quantification of all compounds of interest in the liquid pharmaceutical formulation.

Codeine phosphate impurities (CP-IMPs); A, B, C, and D, are official in British Pharmacopoeia [22]. Chlorpheniramine Maleate impurities (CPM-IMPs); A, B, C, and D, are also official in British Pharmacopoeia [23].

The aim of this work was to develop a quick stability indicating RP-HPLC method for the simultaneous estimation of codeine phosphate and chlorpheniramine maleate in the presence of its degradation products, generated from forced degradation studies. The stability indicating power of the method was established by comparing the chromatograms obtained under optimized conditions before forced degradation with those after forced degradation via acidic, basic, oxidative, thermal, and photolytic stress conditions.

2. Experimental

2.1. Materials and Reagents. Methanol, water, acetonitrile, and *o*-phosphoric acid used were of HPLC (Merck, Mumbai, India). All the glass wares used were of standard quality. Reference samples of CP and CPM were obtained as gift samples from AstraZeneca Pharma India Ltd. Bangalore, India. Liquid dosage form (oral syrup) containing CP 10 mg and CPM 4 mg was used as the sample during the method development process.

Chromatographic Equipments and Conditions. A Shimadzu Prominence binary gradient, a high pressure liquid chromatographic instrument was used for the analysis. The instrument was provided with a Phenomenex C_{18} (250 × 4.6 mm, 5 μ), an LC 20 AD pump and an SPD 20A UV-visible detector were employed. A 20 μL Hamilton syringe was used for sample injection. Data acquisition was done by using LC solutions software.

Mobile Phase Preparation. The mixture of acetonitrile, methanol, and 1% *o*-phosphoric acid in the ratio 78 : 10 : 12 was used as a mobile phase with pH adjusted to 3.0. The mobile phase was filtered through 0.45 μ membrane. The mobile phase was ultrasonicated for 5 min to degas the mixture and then used.

Instrumental Parameters. The separation was done on Phenomenex C_{18} column (250 × 4.6 mm, 5 μ). The flow rate of mobile phase was maintained at 1 mL/min, and the detection was carried out at 254 nm. All determinations were performed at a constant column temperature of 23°C with a load of 20 μL. The retention time of CP and CPM was observed at 3.47 and 9.45 min.

Stress Study. All the stress decomposition studies were performed at a concentration of 50 μg/mL in mobile phase.

Acid hydrolysis was performed in 0.1 M hydrochloric acid. The study in alkaline condition was carried out in 0.1 M sodium hydroxide. Oxidative studies were carried out in 10% hydrogen peroxide. For photolytic degradation studies, pure drug that is a solid state was exposed to UV lamp in UV cabinet at 254 nm. Additionally, the drug powder was exposed to dry heat at 105°C. Samples were withdrawn at appropriate time, cooled and neutralized by adding base or acid, and subjected to RP-HPLC analysis after suitable dilution.

2.2. Preparation of Standard Solutions and Calibration Graphs.
Standard Stock Solutions of CP and CPM were prepared in mobile phase (1 mg/mL).

From CP solution, 10 mL was taken and diluted to 100 mL with mobile phase (100 μg/mL). From CPM solution, 2 mL was taken and diluted to 100 mL with mobile phase (20 μg/mL). Calibration standards at five levels were prepared by appropriately mixing and further diluting stock standard solutions in the concentration range of 50–150 μg/mL for CP and 10–60 μg/mL for CPM. Samples in triplicates were made for each concentration, and peak areas were plotted against the corresponding concentrations to obtain the calibration graph. The regression equation was derived using mean peak area concentration data, and the concentration of the unknown was computed from the regression equation.

2.3. Preparation of Placebo Solution.
An accurately weighed 2 gm of placebo solution was taken into the 100 mL volumetric flask. About 70 mL of diluent was added to this volumetric flask and sonicated in an ultrasonic bath for 5 min. This solution was then diluted up to the mark with diluent and mixed well. It is then filtered through 0.22 μm PVDF syringe filter, and the filtrate was collected after discarding first few milliliters.

2.4. Assay Procedure for Syrup (Combined Formulation of CP and CPM).
Syrup equivalent to 10 mg and 4 mg of CP and CPM was taken in a 100 mL volumetric flask. Volume was made to the mark with mobile phase. The flask was sonicated for 5 min. The above solution was filtered using 0.22 μm PVDF syringe filter, and the filtrate was collected after discarding the first few milliliters. The solution is then injected under above chromatographic conditions, and peak areas were measured. The assay procedure was made triplicate, and weight of sample taken for assay was calculated and summarised in Table 1.

2.5. Method Validation Procedure

2.5.1. Selectivity.
The selectivity of the method was evaluated with regard to interference due to the presence of any other excipients. This shows that drugs were clearly separated from its excipients. Thus, the HPLC method presented in this study was found to be selective, depicted in Figure 2.

2.5.2. Linearity.
The linearity of calibration curves (peak area ratio versus concentration) in pure solution was checked over

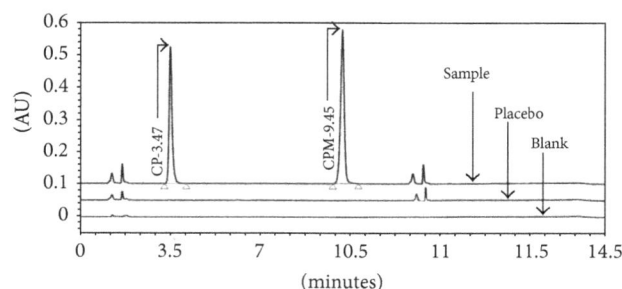

FIGURE 2: Overlay chromatograms of blank, placebo, and sample preparation.

the concentration ranges of about 50–150 μg/mL for CP and 10–60 μg/mL for CPM. Hence, the regression line relating standard concentrations of drug using regression analysis, the calibration curves were linear in the studied range.

2.5.3. Precision.
Three injections of same concentration were given on the same day, and these studies were also repeated on different days to determine interday precision.

2.5.4. Accuracy.
The accuracy of the method was validated by recovery studies and was found to be significant under specification limits.

2.5.5. Limit of Detection (LOD) and Limit of Quantification (LOQ).
Limit of detection was calculated by using the formula:

$$LOD = 3.3SD/S$$

SD = standard deviation of the response

S = slope of calibration curve of the analyte.

Limit of quantification was calculated by using the formula:

$$LOQ = 10SD/S$$

SD = standard deviation of the response

S = slope of calibration curve of the analyte.

2.5.6. Robustness.
To determine robustness of the method the experimental conditions were deliberately changed. The flow rate of the mobile phase, pH of the mobile phase, was varied parameters. The study was performed on same day. The area obtained from each variation was compared with that obtained under optimized conditions.

2.5.7. Filter Compatibility.
Sample compatibility was performed for nylon 0.22 μm syringe filter (Pall life sciences) and PVDF 0.22 μm syringe filter (Millipore). To compare the filter compatibility in proposed analytical method, filtration recovery experiment was carried out by sample filtration technique. Sample was filtered through both syringe filters, and percentage assay was determined and compared against centrifuged sample. Sample solution was not showing any

Validated Stability Indicating RP-HPLC Method for Simultaneous Estimation of Codeine Phosphate and
Chlorpheniramine Maleate from Their Combined Liquid Dosage Form

71

TABLE 1: Assay of formulation.

Formulation	Drug	Label claim	Amount found, (mg) AM ± SD*	% Assay	% RSD
Coscodin	CP	10 mg	10.01 ± 0.05	100.1	0.55
	CPM	4 mg	3.99 ± 0.006	99.75	0.16
Mit's Linctus Codeinae Co.	CP	10 mg	10.05 ± 0.05	100.5	0.53
	CPM	4 mg	4.11 ± 0.006	102.75	0.16

*$n = 6$.

TABLE 2: Filter compatibility results (assay % w/w).

Compound	Centrifuged	PVDF filter 0.22 μm (Millipore)	Nylon filter 0.22 μm (Macherey-Nagel)
CP	99.5	99.3	99.3
CPM	100.3	100.5	100.4

TABLE 3: Optimized chromatographic conditions.

Parameter	Optimized condition
Chromatograph	Shimadzu HPLC with UV detector
Column	Phenomenex C_{18} column (250 × 4.6 mm, 5 μ)
Mobile phase	1% o-phosphoric acid in water, pH adjusted to 3.0 : acetonitrile : methanol (78 : 10 : 12)
Flow rate	1 mL/min
Detection	UV at 254 nm
Injection volume	20 μL
Column temperature	23°C

significant changes in assay percentage with respect to centrifuged sample. Percentage assay results are presented. The result difference in % assay was not observed more than ±0.5, which indicates that both syringe filters are having a good compatibility with sample solution. The filter compatibility results were given in Table 2.

2.5.8. Solution Stability. Stability of sample solution was established by the storage of sample solution at 25°C for 24 hr. Sample solution was reanalyzed after 12 and 24 hr time intervals, and assay was determined for the compounds (CP and CPM) and compared against fresh sample. Sample solution did not show any appreciable change in assay value when stored at ambient temperature up to 24 hr. The results from solution stability experiments confirmed that sample was stable up to 24 hr during assay determination.

2.6. Forced Degradation Studies of API and Formulation. Stress studies were carried by using 50 μg/mL of each solution in different conditions. Acidic and alkaline hydrolysis was carried out in 0.1 M hydrochloric acid (HCl) and 0.1 M sodium hydroxide (NaOH).

The oxidative study was carried out in 10% hydrogen peroxide (H_2O_2) for 24 hr.

Photolytic studies on the drug were carried out by the exposure to UV lamp in UV cabinet at 254 nm.

For thermal stress testing, both the drugs were spread in petridish and placed in the oven at 105°C. Samples were withdrawn periodically and subjected to analysis after suitable dilution.

3. Results and Discussion

The main objective of the RP-HPLC method was to determine a validated stability indicating method for the estimation of CP and CPM simultaneously in liquid pharmaceutical formulation and to obtain well-resolved peaks of CP, CPM, and impurities/degradants. The method should be able to determine degradants/impurities and assay of two compounds in single run and should be accurate, reproducible, robust, stabile indicating, filter compatible, linear, free of interference from blank/placebo/impurities/degradation products, and straightforward enough for routine use in quality control laboratory.

3.1. Method Development and Optimization. Different chromatographic conditions were experimented to achieve better efficiency of the chromatographic system. Parameters such as mobile phase composition, wavelength of detection, column, column temperature and pH of mobile phase were optimized. Several proportions of buffer and solvents were evaluated in order to obtain suitable composition of mobile phase. Different composition of mobile phases containing a mixture (v/v) of 1% o-phosphoric acid in water, acetonitrile, and methanol was tried, but the mixture of 1% o-phosphoric acid in water, pH adjusted to 3.0, acetonitrile, and methanol in the ratio 78 : 10 : 12 was selected as optimal for obtaining well-defined and well-resolved peaks of CP and CPM at a flow rate of 1 mL/min on a Phenomenex C_{18} column. 254 nm was selected as the optimum wavelength for detection and quantitation, at which best detector response for both CP and CPM was obtained. The mean retention time for CP and CPM was found to be 3.47 min and 9.45 min, respectively. The summary of chromatographic condition is given in Table 3.

3.2. Stability Studies. All forced degradation studies were analysed at 50 μg/mL concentration level. The drug stability

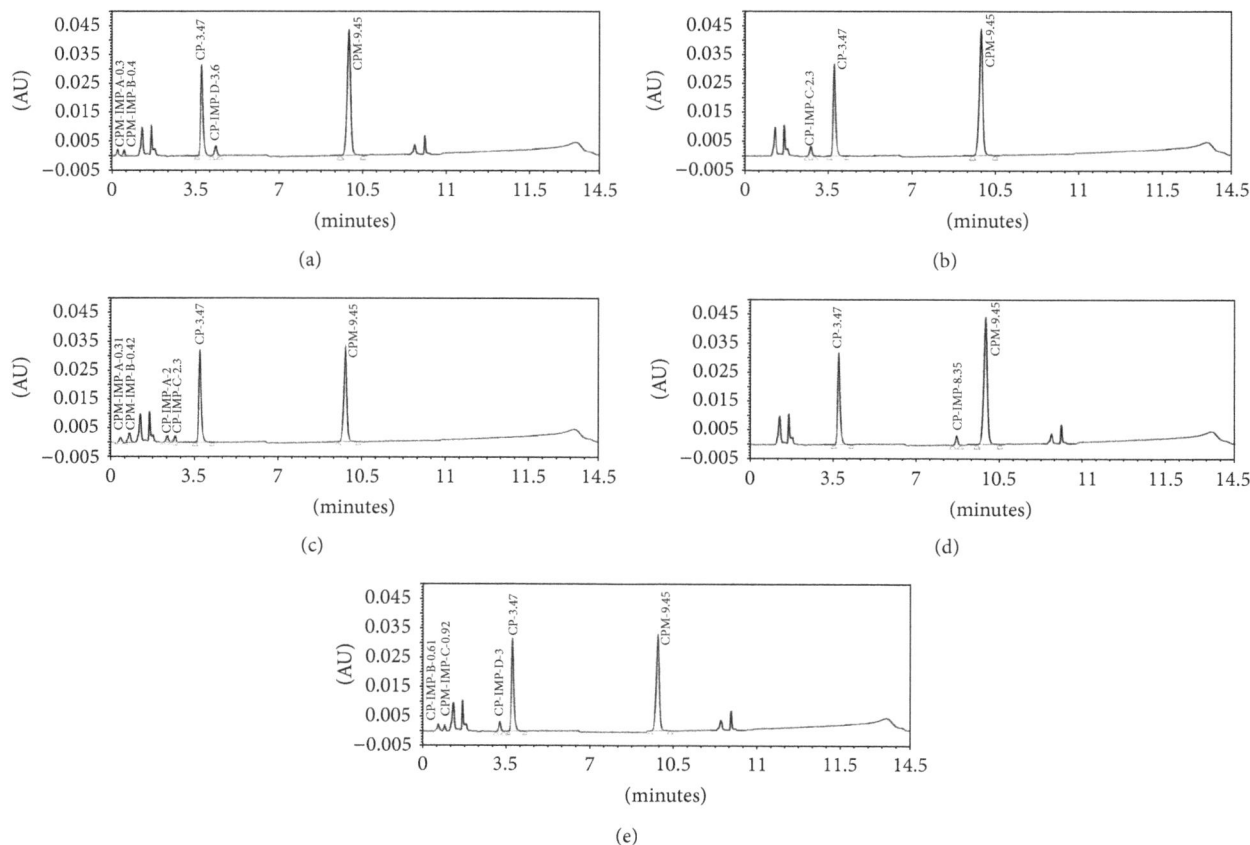

FIGURE 3: (a) Acid degradation, (b) base degradation, (c) oxidative degradation, (d) photodegradation, and (e) thermal degradation.

TABLE 4: Solution stability results.

Time intervals	CP	CPM
% assay initial	100.3	99.5
% assay after 12 hrs	100.7	99.6
% assay after 24 hrs	100.4	99.3

in pharmaceutical formulations is a function of storage conditions and chemical properties of the drug and its impurities. Conditions used in stability experiments should reflect situations likely to be encountered during actual sample handling and analysis. Stability data is required to show that the concentration and purity of analyte in sample at the analysis corresponds to the concentration and purity of the analyte at the time of sampling. CP and CPM were subjected to acid, alkaline, oxidative, photolytic, and dry heat stress conditions. The chromatograms obtained after subjecting to degradation are presented in Figures 3(a), 3(b), 3(c), 3(d), and 3(e).

3.3. Stability of Sample in Diluent.

Drug stability in pharmaceutical formulations is a function of storage conditions and chemical properties of the drug and its impurities. Condition used in stability experiments should reflect situations likely to be encountered during actual sample handling and analysis.

Stability data is required to show that the concentration and purity of analyte in the sample at the analysis corresponds to the concentration and purity of analyte at the time of sampling. Stability of sample solution was established by storage of sample solution at ambient temperature (25°C) for 24 hr. Sample solution was reanalyzed after 12 and 24 hr time intervals, and assay was determined for the compounds (CP and CPM) and compared against fresh sample. Sample solution did not show any appreciable change in assay value when stored at ambient temperature up to 24 hr, which are presented in Table 4. The results from solution stability experiments confirmed that the sample was stable up to 24 hr during assay determination.

Validation of the Method. The described method for the assay of CP and CPM has been validated as per current ICH Q2 (R1).

3.4. Linearity.

Calibration curves were obtained for CP and CPM from which the linear regression equation was computed and found to be $y = 104033x - 777923$, $r^2 = 0.9985$ for CP and $y = 221251x - 956140$, $r^2 = 0.9991$ for CPM.

3.5. Accuracy and Precision.

The percent standard error which is an indicator of accuracy is ≤ 1.0 and is indicative of high accuracy. The calculated percent relative standard

Validated Stability Indicating RP-HPLC Method for Simultaneous Estimation of Codeine Phosphate and
Chlorpheniramine Maleate from Their Combined Liquid Dosage Form

73

TABLE 5: Result of accuracy studies.

Name of drug	Amount in sample (mg)	Amount of drug spiked in mg, (% recovery level)	Amount found, (mg) AM ± SD*	% SE	% RSD
CP	20	16, (80%)	36.37 ± 0.06	0.034	0.18
	20	20, (100%)	40.36 ± 0.10	0.057	0.25
	20	24, (120%)	44.55 ± 0.11	0.063	0.25
CPM	8	6.4, (80%)	14.38 ± 0.01	0.005	0.13
	8	8, (100%)	15.98 ± 0.06	0.034	0.42
	8	9.6, (120%)	17.59 ± 0.06	0.034	0.38

*$n = 3$.

TABLE 6: Results of precision studies.

Drug	Concentration (μg/mL)	Intraday ($n = 3$)		Interday ($n = 3$)	
		Measured concentration (μg/mL) AM ± SD	% CV	Measured concentration (μg/mL) AM ± SD	% CV
CP	100	99.33 ± 0.45	0.45	99 ± 0.45	0.46
CPM	20	19.46 ± 0.12	0.64	19.58 ± 0.1	0.53

TABLE 7: Robustness studies.

Sl. no	Parameter	Modification	Retention time (min)
1	Flow rate	0.8 mL/min	3.2, 9.13
		1.2 mL/min	3.31, 9.32
2	pH	2.8	3.63, 9.7
		3.2	3.3, 8.9

TABLE 8: System suitability parameters.

S. no.	Parameter	CP	CPM
1	Retention time (min)	3.47	9.45
2	Peak asymmetry	1.0	1.0
3	Theoretical plates	11100	17692
4	Resolution	3.05	7.78
5	Limit of detection (LOD), μg/mL	2.263	0.756
6	Limit of quantification (LOQ), μg/mL	6.859	2.293

deviation (%RSD) can be considered to be satisfactory. The percentage RSD values were <1%. The results obtained for the evaluation of accuracy and precision of the method are compiled in Tables 5 and 6.

3.6. Robustness. The robustness of an analytical procedure is a measure of its capacity to remain unaffected by small, but deliberate variations in method parameters, and provides an indication of its reliability during normal usage. The results are summarized in Table 7.

3.7. Solution Stability. Stability of sample solution was established by the storage of sample solution at 25°C for 24 hr. Sample solution was reanalyzed after 12 and 24 hr time intervals, and assay was determined for the compounds (CP and CPM) and compared against fresh sample. Sample solution did not show any appreciable change in assay value

when stored at ambient temperature up to 24 hr. The results from solution stability experiments confirmed that sample was stable up to 24 hr during the assay determination.

3.8. System Suitability. The system suitability was carried out after the method development and validation have been completed. For this, parameters like theoretical plates, resolution, retention time, and peak symmetry of samples were measured and shown in Table 8.

4. Conclusion

The reported RP-HPLC method was proved to be simple, rapid, and reproducible. The validation data indicate good precision, accuracy, and reliability of the method. The developed method offers several advantages in terms of simplicity in mobile phase, isocratic mode of elution, easy sample preparation steps, and comparative short run time which makes the method specific and reliable for its intended use in simultaneous determination of codeine phosphate and chlorpheniramine maleate in liquid oral dosage form. Quick stability indicating RP-HPLC method was developed for the simultaneous estimation of codeine phosphate and chlorpheniramine maleate in the presence of its degradation products, generated from forced degradation studies. The developed method separates codeine phosphate chlorpheniramine maleate in impurities/degradation products. There were no reported stability indicating methods for this combination of drugs in liquid dosage form; hence, this method has an advantage of being unique and novel.

Conflict of Interests

The authors declared that they have no conflict of interests or financial gains in mentioning the companies names or trademarks.

Acknowledgment

The authors acknowledge the management for providing necessary facilities.

References

[1] USP-NF, *The Official Compendia of Standards*, USP Convention, INC., Rock wiley, 2003.

[2] R. Sawant, R. Joshi, P. Lanke, and L. Bhangale, "Simultaneous estimation & validation of paracetamol, phenylephrine hydrochloride and chlorpheniramine maleate in tablets by spectrophotometric method," *Journal of Pharmaceutical Research and Health Care*, vol. 3, no. 2, pp. 23–28, 2011.

[3] D. T. T. An and V. D. Hoang, "Simultaneous determination of paracetamol and codeine phosphate in combined tablets by first-order derivative and ratio spectra first-order derivative UV spectrophotometry," *Asian Journal For Research in Chemistry*, vol. 2, no. 2, pp. 143–147, 2009.

[4] B. Praveen, M. Padmaja, and R. K. Krishna, "Development and validation of UV spectrophotometric method for simultaneous estimation of codeine phosphate and chlorpheniramine maleate in combined liquid dosage form," *International Journal of Pharmacy and Technology*, vol. 3, no. 3, pp. 3390–3400, 2011.

[5] A. Biswas, "Spectrophotometric method for determination of chlorpheniramine maleate in pharmaceutical preparations in the presence of codeine phosphate and ephedrine hydrochloride," *Analyst*, vol. 105, no. 1249, pp. 353–358, 1980.

[6] U. R. Mallu, V. Bobbarala, and S. Penumajji, "Analysis of cough and analgesic range of pharmaceutical active ingredients using RP-HPLC method," *International Journal of Pharma and Bio Sciences*, vol. 2, no. 3, pp. 439–452, 2011.

[7] A. Manassra, M. Khamis, M. el-Dakiky, Z. Abdel-Qader, and F. Al-Rimawi, "Simultaneous HPLC analysis of pseudophedrine hydrochloride, codeine phosphate, and triprolidine hydrochloride in liquid dosage forms," *Journal of Pharmaceutical and Biomedical Analysis*, vol. 51, no. 4, pp. 991–993, 2010.

[8] M. Maithani, R. Raturi, V. Gautam et al., "Development and validation of a RP-HPLC method for the determination of chlorpheniramine maleate and phenylephrine in pharmaceutical dosage form," *International Journal of Comprehensive Pharmacy*, vol. 1, no. 5, pp. 1–4, 2010.

[9] D. J. Hood and H. Y. Cheung, "A chromatographic method for rapid and simultaneous analysis of codeine phosphate, ephedrine HCl and chlorpheniramine maleate in cough-cold syrup formulation," *Journal of Pharmaceutical and Biomedical Analysis*, vol. 30, no. 5, pp. 1595–1601, 2003.

[10] N. Erk and M. Kartal, "Simultaneous high performance liquid chromatographic and derivative ratio spectra spectrophotometry determination of chlorpheniramine maleate and phenylephrine hydrochloride," *Farmaco*, vol. 53, no. 8-9, pp. 617–622, 1998.

[11] G. Santoni, L. Fabbri, P. Gratteri, G. Renzi, and S. Pinzauti, "Simultaneous determination of aspirin, codeine phosphate and propyphenazone in tablets by reversed-phase high-performance liquid chromatography," *International Journal of Pharmaceutics*, vol. 80, no. 2-3, pp. 263–266, 1992.

[12] W. R. Sisco, C. T. Rittenhouse, and W. M. Maggio, "The rapid quantitative analysis of codeine phosphate drug substance by reversed-phase high-performance liquid chromatography," *Chromatographia*, vol. 20, no. 5, pp. 289–292, 1985.

[13] S. Chittrakarn, P. Penjamras, and N. Keawpradub, "Quantitative analysis of mitragynine, codeine, caffeine, chlorpheniramine and phenylephrine in a kratom (*Mitragyna speciosa* Korth.) cocktail using high-performance liquid chromatography," *Forensic Science International*, vol. 217, no. 1–3, pp. 81–86, 2012.

[14] Z. Hu, Q. Zou, J. Tian, L. Sun, and Z. Zhang, "Simultaneous determination of codeine, ephedrine, guaiphenesin and chlorpheniramine in beagle dog plasma using high performance liquid chromatography coupled with tandem mass spectrometric detection: application to a bioequivalence study," *Journal of Chromatography B*, vol. 879, no. 32, pp. 3937–3942, 2011.

[15] D. J. Hood and H. Y. Cheung, "A chromatographic method for rapid and simultaneous analysis of codeine phosphate, ephedrine HCl and chlorpheniramine maleate in cough-cold syrup formulation," *Journal of Pharmaceutical and Biomedical Analysis*, vol. 30, no. 5, pp. 1595–1601, 2003.

[16] R. Ragonese, M. Mulholland, and J. Kalman, "Full and fractionated experimental designs for robustness testing in the high-performance liquid chromatographic analysis of codeine phosphate, pseudoephedrine hydrochloride and chlorpheniramine maleate in a pharmaceutical preparation," *Journal of Chromatography A*, vol. 870, no. 1-2, pp. 45–51, 2000.

[17] K. P. R. Shenoy, K. S. Krishnamurthy, and M. V. Vinod, "Simultaneous determination of codeine phosphate and chlorpheniramine maleate in formulation by reverse phase liquid chromatography," *Indian Drugs*, vol. 36, no. 8, pp. 513–516, 1999.

[18] R. N. Raju, P. Srilakshmi, and M. Muralee Krishna, "Simultaneous determination of chlorpheniramine maleate, ephedrine hydrochloride and codeine phosphate in cough syrups by reverse phase high performance liquid chromatography," *Indian Drugs*, vol. 29, no. 9, pp. 408–411, 1992.

[19] H. Al-Kaysi and T. H. N. M. S. Salem, "Simultaneous quantitative determination of codeine phosphate, chlorpheniramine maleate, phenylephrine hydrochloride, and acetaminophen in pharmaceutical dosage forms using thin layer chromatography densitometry," *Analytical Letters*, vol. 19, no. 7-8, pp. 915–924, 1986.

[20] K. Masumoto, Y. Tashiro, and K. Matsumoto, "Simultaneous determination of codeine and chlorpheniramine in human plasma by capillary column gas chromatography," *Journal of Chromatography*, vol. 381, no. 2, pp. 323–329, 1986.

[21] International Conference on Harmonization, *Validation of Analytical Procedure, Text and Methododlogy Q2 (R1)*, IFMA, Geneva, Switzerland, 2005.

[22] *Codeine Phosphate*, vol. 584 of *(Ph Eur monograph 0074)*, British Pharmacopeia, 2008.

[23] *Chlorpheniramine Maleate*, vol. 498 of *(Ph Eur monograph 0386)*, British Pharmacopeia, 2008.

12

Micellar Liquid Chromatography Determination of Spermine in Fish Sauce after Derivatization with 3,5-Dinitrobenzoyl Chloride

Mei-Liang Chin-Chen,[1] Maria Rambla-Alegre,[2] Samuel Carda-Broch,[1] Josep Esteve-Romero,[1] and Juan Peris-Vicente[1]

[1] QFA, ESTCE, Universitat Jaume I, 12071 Castelló, Spain
[2] Department of Organic Chemistry, Ghent University, 9000 Ghent, Belgium

Correspondence should be addressed to Juan Peris-Vicente, juan.peris@qfa.uji.es

Academic Editor: Maria Jose Ruiz-Angel

A practical liquid chromatographic method has been developed for the selective determination of the levels of spermine in anchovy sauce after derivatization with 3,5 dinitrobenzoyl chloride. The micellar liquid chromatographic separation proposed here uses a C18 column (125×4.6 mm), followed by detection of spermine derivative at 260 nm. Elution of the analyte was performed using a mobile phase of 0.15 M SDS-4% (v/v) 1-pentanol-pH 7 running under isocratic mode at 25°C. Validation parameters were linearity (2–100 μg/mL, $R^2 > 0.999$), detection and quantification limits (0.4 and 1.2 μg/mL, resp.), precision (less than 3.6%), accuracy (93.3–101.1%), and robustness (less than 4.8%). These results are in agreement with the requirements of the FDA guidelines. The proposed method was successfully applied to the monitorization of spermine formation in unsalted and salted fish sauce samples. The suggested methodology was found useful in routine analysis of spermine in fish sauce samples.

1. Introduction

Biogenic amines are biological metabolites present in foods either as natural products or after fermentation, decay microbial contamination, decomposition, or putrefaction processes [1]. They are largely responsible for the foul odour of putrefying flesh, as well as contributing to the odour of such processes as bad breath and bacterial vaginosis. Biogenic amines are also involved in local immune responses, neurotransmission, and chemotaxis of white blood cells. The consumption of an excess of biogenic amines, known as histaminic intoxication, is mainly related to heart, gastrointestinal, and skin diseases, as well as headache [1–3]. Food containing considerable amounts of these amines include alcoholic beverages, beef, chocolate, cheeses, fish, pork, and poultry. Biogenic amines can also be found in semen and some microalgae, together with related molecules like spermine and spermidine. In fact, spermine is formed from spermidine and can be found in a wide variety of organisms and tissues, as it is an essential growth factor in some bacteria. Thus, its detection and quantification is useful to assess the degree of bacterial contamination, mainly caused by incorrect handling or stocking conditions (freezer at $-18°$C), in fish flesh, or derivates as fish sauce. Then the determination of spermine is of the utmost importance to assure that the fish sauce can be eaten without health risk [4, 5].

Several analytical methods have been developed for the determination of spermine. Among them, HPLC with UV-visible absorbance detection using 3,5-dinitrobenzoyl chloride (DNBZ-Cl) as a chromophore has provided excellent results for quantification of biogenic amines in complex food samples. Derivatization reaction is quantitative, quite fast (less than 5 min), and reproducible. The high stability and sensitivity of the obtained derivative makes it an excellent choice for the analysis of spermine [2, 6].

In these methods, the analytes have to undergo a previous extraction step in a suitable organic solvent, evaporation to dryness and redissolution in order to purify and preconcentrate the derivatized amines [6]. However, it introduces the risk of contamination and losing of the sample, improving the possibility of error and even increasing the analysis time [7–9]. Moreover, chromatographic conditions result in either insufficient separation, prolonged analysis times (higher than 1 hour) [6, 10] or need mobile phase running under a gradient program. This requires a stabilization time between two injections, lengthening analysis time and making difficult the analysis of a wide amount of samples [11]. These problems can be avoided by the use of micellar liquid chromatography (MLC), which usually allows direct injection of samples (after filtration), without needing any extraction and purification step [12–16]. Moreover, micellar mobile phases are less toxic, nonflammable, biodegradable, and relatively inexpensive in comparison to aqueous-organic solvents. MLC has proved to be a useful technique in the determination of a wide range of compounds in low time using mobile phases under isocratic conditions, by optimizing separation parameters [12, 13] including food samples [14–16].

The aim of this work was to develop a fast, simple, and selective procedure for the determination of spermine by MLC using a C18 column and UV detection. This analyte was derivatized with DNBZ-Cl to increase sensitivity, and directly injected in the chromatographic system. The suggested methodology was validated in terms of linearity, sensitivity, limits of detection and quantification, accuracy, precision, and recovery, following the FDA guideline [17]. Finally, the method was applied to the study of the anchovy sauce degradation by means of the determination of spermine depending on the storage treatment.

2. Experimental

2.1. Apparatus and Instrumentation. The pH was measured with a Crison GLP 22 (Barcelona, Spain) equipped with a combined Ag/AgCl/glass electrode. The balance used was a Mettler-Toledo AX105 Delta-Range (Greifensee, Switzerland). The vortex shaker and ultrasonification unit were from Selecta (Barcelona). The chromatographic system was an Agilent Technologies Series 1100 (Palo Alto, CA, USA) equipped with a quaternary pump, a thermostated autosampler, and column compartment. The dead time was determined as the mean value of the first significant deviation from the baseline in the chromatograms of the analyte. The signals were acquired by a PC computer connected to the chromatograph through an HP Chemstation.

2.2. Chemicals and Reagents. Spermine and 3,5-dinitrobenzoyl chloride (98% pure) were purchased from Sigma-Aldrich (St. Louis, MO, USA). Sodium dodecyl sulphate (SDS) (99% pure) was from Merck (Darmstadt, Germany); acetonitrile, 1-propanol, 2-propanol, 1-butanol, and 1-pentanol were from Scharlab (Barcelona), sodium dihydrogen phosphate, HCl, and NaOH were from Panreac (Barcelona). All solutions were prepared in ultrapure water (Millipore, S.A.S. Molsheim, France). Mobile phases and samples were filtered through nylon membranes (Millex-HN, Millipore, Bedford, MA, USA). Spermine was dissolved in 0.1 M HCl to provide final concentrations of 100 μg/mL.

2.3. Derivatization of Biogenic Amines with 3,5-Dinitrobenzoyl Chloride. Derivatising reagent 5 mM 3,5-dinitrobenzoyl

chloride (DNBZ-Cl) was dissolved in acetonitrile. Aliquots (400 μL) of spermine (SP) standards, 1 M NaOH (1200 μL), 2-propanol (700 μL) and 3,5-DNBZ-Cl (2100 μL) were mixed in a reaction tube. After 3 min of shaking at 25°C, 2 M HCl (1000 μL) was added to stop the reaction. Finally, after 1 min of shaking, derivatized sample was injected into the chromatographic system. Under these conditions, the formed derivative was (DNBZ)$_4$SP [6]. The fish sauce medium does not affect the derivatization reaction, because the conditions were strongly changed by the addition of organic alcohol and sodium hydroxide. Some matrix compounds are precipitated in 2-propanol/NaOH media, and others are solubilized in the SDS medium [2, 6].

2.4. Chromatographic Conditions. Derivatized spermine elution was performed in a reversed-phase C18 column (125 × 4.6 mm, 5 μm particle size) from Scharlab thermostated at 25°C. Mobile phase was a 0.15 M SDS-4% (v/v) 1-pentanol-NaH$_2$PO$_4$ 0.01 M solution at pH 7, running under isocratic mode. Flowrate, injection volume, and UV wavelength were 1 mL/min, 20 μL and 260 nm, respectively. Samples were thermostated at +15°C in the autosampler module of the HPLC system to avoid decomposition of the spermine derivative [7]. Under these conditions, the retention time was 11.5 min. Chromatographic signals were acquired and processed with an Agilent ChemStation (Rev. B.01.03).

2.5. Sample Preparation. Fish sauce samples of anchovies (*Eugraulidea* spp.) were obtained from a supermarket. A part of the anchovies was mixed with common salt in a relation of 75/25 w/w (a well-known treatment to avoid food spoilage) and another portion was untreated. In both cases, samples were stored at +4°C. For the analyses of fish sauces, 1 g was mixed with 0.5 mL of ethanol that was topped up to 10 mL with 0.1 M SDS. Samples were derivatized as detailed in Section 2.3 and then they were injected directly into the chromatographic system without any other treatment then filtration.

Also, spiked samples were prepared by adding the appropriate volume of spermine standard solution (dissolved in 0.1 M HCl) to 1 g of sample and then the mixture was vigorously shaken to favour homogenization and stored for one day at +4°C to favour the contact between the analyte and the sample, as well as solvent evaporation [18, 19]. Then the spiked sample was mixed with 0.5 mL of ethanol and topped up to 10 mL with 0.1 M SDS solution. An aliquot of the sample (400 μL) was derivatized as explained in Section 2.3, filtered and directly injected into the chromatograph [2, 6].

3. Results and Discussion

3.1. Optimization Strategy and Mobile Phase Selection. Mobile phases were aqueous solutions of SDS buffered at pH 7 with phosphate buffer and the stationary phase was a C18 column [2, 6]. The four pK_a values of spermine lie in the interval 8–11 [20], so its retention behavior will not vary in the whole pH range of the C18 column used (2.5–7.5). In addition, working at pH 7 increases the life of the column.

TABLE 1: Intra- and interday precision (RSD, %) and accuracy (recovery, %) for spermine (*added concentrations expressed in μg/mL).

Intraday precision[a] (RSD, %)			Intraday accuracy[a] (%)		
*2	10	20	2	10	20
3.6	2.5	1.8	93.3	97.4	99.0
Interday precision[b] (RSD, %)			Interday accuracy[b] (%)		
2.5	1.8	0.7	96.2	101.0	99.5

[a] $n = 9$; [b] $n = 5$.

On the other hand, it should be highlighted that spermine is a polar compound, but the derivative (DNBZ)$_4$SP is quite hydrophobic [6]. Initially, the chromatographic parameters for the spermine derivative were obtained using a pure micellar mobile phase of SDS. Several mobile phases containing 0.05, 0.1, and 0.15 M SDS were tested. At these three SDS concentrations, the retention times were found too high. Instead of increasing SDS concentration, the use of an aqueous solution of 0.15 M SDS with a small amount of organic solvent was envisaged to reduce the retention time. Different amounts of 1-propanol and 1-butanol were tested and the retention time was reduced, but it remains very high (more than 60 min), and with an irregular peak shape. Thus, 1-pentanol was tested, and finally a mobile phase of 0.15 M SDS-4% (v/v) 1-pentanol allowed the elution of spermine derivative in an adequate retention time (nearly 11.5 min). The chromatographic parameters for spermine derivative in this mobile phase were k (retention factor) = 9.75; N (efficiency) = 4050 theoretical plates, and B/A (asymmetry factor) = 1.0. Their calculation was performed as in [21].

The possible interference of other compounds in spermine analysis was also studied. A blank of each matrix was performed by analyzing aliquots of unsalted and salted anchovy sauce free of spermine. The front shows a considerable maximal band height and width and no other peak was observed, indicating that all the substances elute at the dead time (Figure 1(a)). In fact, hydrophobic substances present in the matrices are introduced in micelles, thus avoiding their precipitation and allowing their fast elution. Moreover, the addition of 1-pentanol in the mobile phase improves its elution strength, reducing even more the retention time of matrix compounds. Samples of unsalted and salted anchovy sauce were spiked with 10 μg/mL of spermine. As can be seen in Figure 1(b), the chromatographic peak of spermine elutes far from the dead time and without overlapping with any other substances of the unsalted anchovy sauce matrix. The same result was found for the salted anchovy sauce samples. The selectivity of the analytical method has been assessed; it has been optimized by the separation of spermine from other interferences of the matrix. Other biogenic amines do not interfere in the determination of the spermine since they elute earlier [2, 7].

3.2. Method Validation.

The whole validation was performed by spiking samples with spermine, following the Food and Drug Administration (FDA) guideline [17]. Unsalted and salted anchovy sauce samples were studied. Since results were found similar in both cases, only the obtained values for unsalted fish sauce are presented and discussed.

Calibration curves were constructed using the areas of the chromatographic peaks (nine replicates) obtained at seven different concentrations of spermine, in the 2–100 μg/mL range. To study the variability of the calibration parameters, the curves were obtained during 5 days over a period of two months for a different set of standards, and then the average value was considered. Results were similar in the two matrices studied. The slope and intercept were determined by the method of least square linear regression analysis, taking the absorbance in arbitrary units and the concentration in μg/mL. Limits of detection (LOD, *3s criterion*) and quantification (LOQ, *10s criterion*) for derivatized spermine using the proposed method ($n = 10$) were determined. The obtained parameters were slope, 0.15 ± 0.01; intercept, 0.02 ± 0.03; determination coefficient (R^2), 0.999; LOD, 0.4 μg/mL; and LOQ, 1.2 μg/mL.

The intra- and interday precision and accuracy of spermine were determined by analysis of spiked samples at 2, 10, and 20 μg/mL. The intraday values were determined by assaying test solutions nine times on the same day, and interday value was the average of nine measurements of intra-day values taken on 5 days over a 3-month period. The precision was taken as the RSD (%) of the obtained areas, whereas the accuracy was the ratio between the calculated value of the recovered spermine and the spiked one (%). The results obtained were between 0.7–3.6% for precision and 93.3–101% for accuracy (Table 1). Both were in agreement with FDA guideline, which indicates that precision should be less than 15% and accuracy between 80 and 120% [17].

Robustness of the method was examined by replicate injections ($n = 6$) of spermine spiked sample of unsalted anchovy sample at 10 μg/mL under small changes in the chromatographic parameters (SDS concentration, percentage of 1-pentanol, pH and flow rate). Insignificant differences in peak areas (<4.8%) and less variability in retention time (<3.5%) were observed (Table 2). These results indicate that the selected factors remain unaffected by small variations in these parameters.

Stability studies indicated that the degradation of biogenic amines derivatized with DNBZ-Cl took place in 12 h when kept in the fridge and in three hours at room temperature. These results were confirmed by the displacement of the peaks in chromatograms. The biogenic amine samples and the derivatizing reagent DNBZ-Cl were stable for three days in the fridge and four months when kept in a freezer.

3.3. Analysis of Food Samples.

Unsalted and salted anchovy sauces (*Engraulidae* spp.) were analyzed in different days in

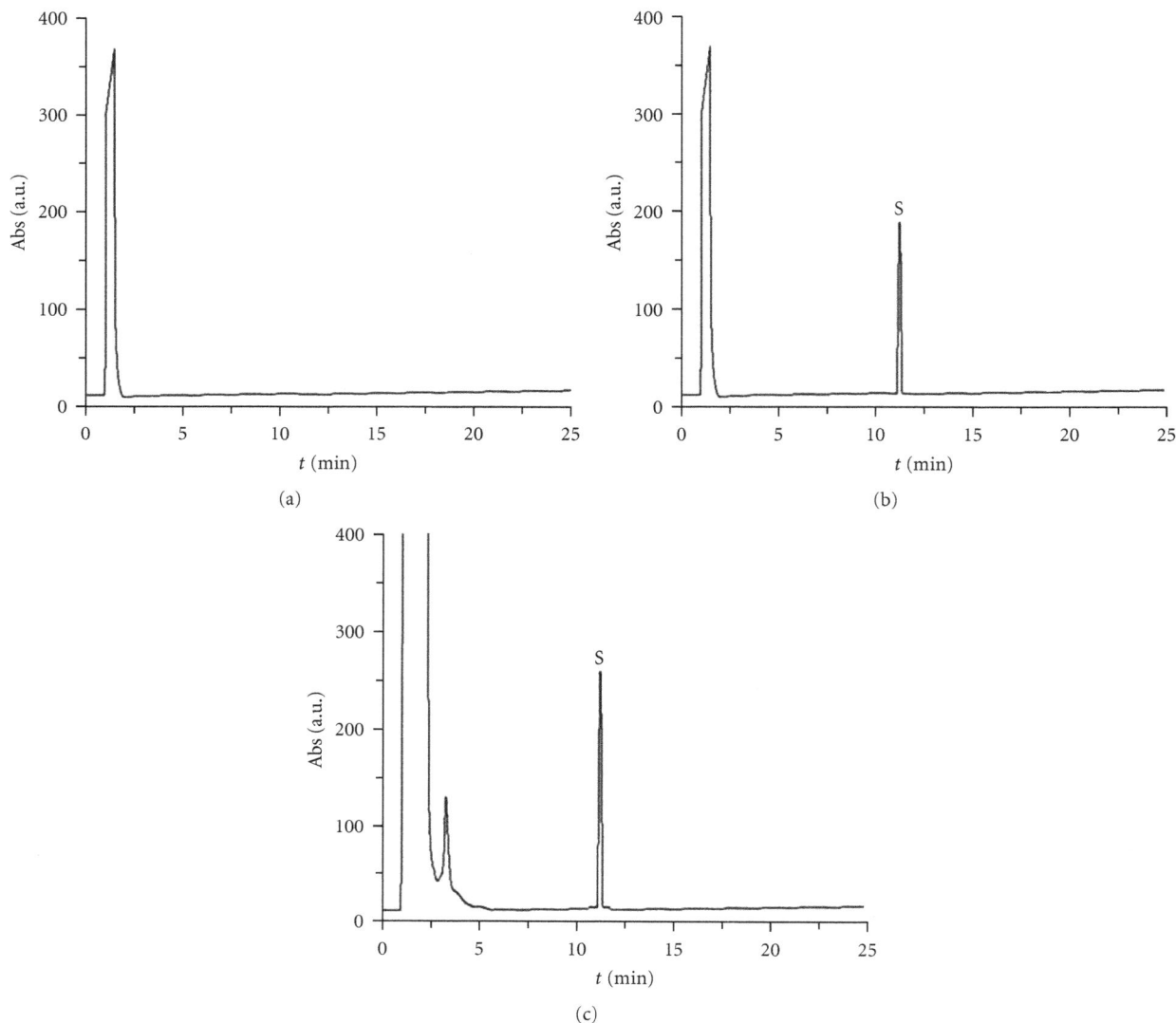

FIGURE 1: Chromatograms obtained by the analysis of unsalted anchovy sauce samples following the proposed methodology: (a) blank, (b) spiked with 10 μg/mL spermine, and (c) a real sample containing 15 μg/mL of spermine after 5-days storage. Chromatographic conditions: 0.15 M SDS-4% (v/v) 1-pentanol-pH 7.

TABLE 2: Evaluation of the MLC method robustness.

Changes in the parameters	Level	Retention time (min) (RSD, %)	Area (RSD, %)
SDS (M)	0.145–0.155	11.6 ± 0.3 (2.6)	1.72 ± 0.05 (2.9)
1-Pentanol (%)	3.9–4.1	11.8 ± 0.4 (3.4)	1.68 ± 0.08 (4.8)
pH	6.9–7.1	11.5 ± 0.2 (1.7)	1.66 ± 0.04 (2.4)
Flow (mL/min)	0.95–1.05	11.3 ± 0.4 (3.5)	1.70 ± 0.05 (2.0)

$n = 6$.

order to evaluate the microbacterial contamination depending on the storage conditions, by means of the amount or spermine. Figure 1(c) shows the chromatogram obtained by analysis of a real sample of unsalted anchovy sauce after 5 days of storage, where 15 μg/mL were detected without interferences. The amount of spermine after 1 month of storage was evaluated. Unsalted sample contained more than 100 μg/mL of spermine, indicating a high microbacterial contamination (Figure 2). The unsalted anchovy sauce is spoilt and is unable to eat. However, the salted sample shows no microbacterial contamination (spermine level under LOD) and can be eaten without risk. These results indicate that salting is an efficient method to prevent microbial contamination during anchovy sauce storage.

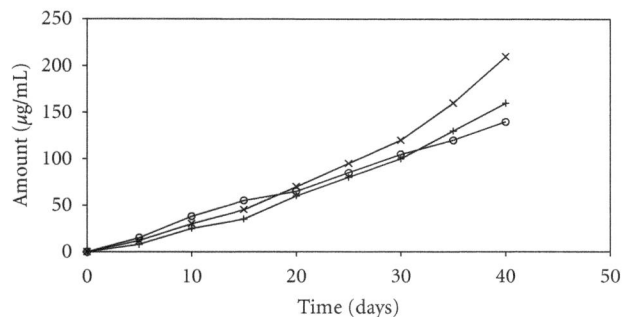

FIGURE 2: Amount found of spermine in three samples of unsalted anchovy sauce stored at +4°C and analyzed by the proposed methodology in different days.

4. Conclusions

In conclusion, the results indicate that the micellar liquid chromatography procedure here developed can be used for the analysis of spermine, with analysis times below 12 min. The analytical method is sensitive enough to be applied for quality control and routine analyses. It is a simple, rapid, effective, and alternative method and does not require any extraction step: the sample is directly injected after the derivatization procedure. The reagent 3,5-dinitrobenzoyl chloride was found to be highly suitable for the analysis of spermine with a very simple method and fast reaction at room temperature and is therefore recommended for use in pollution surveys, to study the degree of microbial contamination, and in the routine practice of food-quality control.

Acknowledgments

This work was supported by Fundació Caixa Castelló-Bancaixa P1-1B2006-12 projects and MEC CTQ 200764473/BQU. M.-L. Chin-Chen also thanks the Foundation for its grant.

References

[1] M. Izquierdo-Pulido, T. Hernández-Jover, A. Mariné-Font, and M. C. Vidal-Carou, "Biogenic amines in European Beers," *Journal of Agricultural and Food Chemistry*, vol. 44, no. 10, pp. 3159–3163, 1996.

[2] M. Chin-Chen, D. Bose, J. Esteve-Romero, J. Peris-Vicente, M. Rambla-Alegre, and S. Carda-Broch, "Determination of putrescine and tyramine in fish by micellar liquid chromatography with UV detection using direct injection," *The Open Analytical Chemistry Journal*, vol. 5, pp. 22–26, 2011.

[3] S. Pons-Sánchez-Cascado, *Study of alternatives to evaluate the freshness and quality of the anchovy*, Doctoral thesis, University of Barcelona, 2004.

[4] S. Rodtong, S. Nawong, and J. Yongsawatdigul, "Histamine accumulation and histamine-forming bacteria in Indian anchovy (Stolephorus indicus)," *Food Microbiology*, vol. 22, no. 5, pp. 475–482, 2005.

[5] J. Yongsawatdigul, Y. J. Choi, and S. Udomporn, "Biogenic amines formation in fish sauce prepared from fresh and temperature-abused Indian anchovy (Stolephorus indicus)," *Journal of Food Science*, vol. 69, no. 4, pp. 312–319, 2004.

[6] J. Kirschbaum, K. Rebscher, and H. Brückner, "Liquid chromatographic determination of biogenic amines in fermented foods after derivatization with 3,5-dinitrobenzoyl chloride," *Journal of Chromatography A*, vol. 881, no. 1-2, pp. 517–530, 2000.

[7] J. Peris Vicente, J. V. Gimeno Adelantado, M. T. Doménech Carbó, R. Mateo Castro, and F. Bosch Reig, "Identification of lipid binders in old oil paintings by separation of 4-bromomethyl-7-methoxycoumarin derivatives of fatty acids by liquid chromatography with fluorescence detection," *Journal of Chromatography A*, vol. 1076, no. 1-2, pp. 44–50, 2005.

[8] J. Peris-Vicente, E. Simó-Alfonso, J. V. Gimeno Adelantado, and M. T. Doménech Carbó, "Direct infusion mass spectrometry as a fingerprint of protein-binding media used in works of art," *Rapid Communications in Mass Spectrometry*, vol. 19, no. 23, pp. 3463–3467, 2005.

[9] J. Peris-Vicente, R. Garrido-Medina, E. Simó-Alfonso, J. V. Gimeno-Adelantado, and M. T. Doménech-Carbó, "Infusion mass spectrometry as a fingerprint to characterize varnishes in oil pictorial artworks," *Rapid Communications in Mass Spectrometry*, vol. 21, no. 6, pp. 851–856, 2007.

[10] E. K. Paleologos, S. D. Chytiri, I. N. Savvaidis, and M. G. Kontominas, "Determination of biogenic amines as their benzoyl derivatives after cloud point extraction with micellar liquid chromatographic separation," *Journal of Chromatography A*, vol. 1010, no. 2, pp. 217–224, 2003.

[11] J. Peris-Vicente, J. V. Gimeno Adelantado, M. T. D. Carbó, R. M. Castro, and F. B. Reig, "Characterization of proteinaceous glues in old paintings by separation of the o-phtalaldehyde derivatives of their amino acids by liquid chromatography with fluorescence detection," *Talanta*, vol. 68, no. 5, pp. 1648–1654, 2006.

[12] J. Esteve-Romero, E. Ochoa-Aranda, D. Bose, M. Rambla-Alegre, J. Peris-Vicente, and A. Martinavarro-Domínguez, "Tamoxifen monitoring studies in breast cancer patients by micellar liquid chromatography," *Analytical and Bioanalytical Chemistry*, vol. 397, no. 4, pp. 1557–1561, 2010.

[13] E. O. Aranda, J. Esteve-Romero, M. Rambla-Alegre, J. Peris-Vicente, and D. Bose, "Development of a methodology to quantify tamoxifen and endoxifen in breast cancer patients by micellar liquid chromatography and validation according to the ICH guidelines," *Talanta*, vol. 84, no. 2, pp. 314–318, 2011.

[14] M. Rambla-Alegre, J. Peris-Vicente, J. Esteve-Romero, and S. Carda-Broch, "Analysis of selected veterinary antibiotics in fish by micellar liquid chromatography with fluorescence detection and validation in accordance with regulation 2002/657/EC," *Food Chemistry*, vol. 123, no. 4, pp. 1294–1302, 2010.

[15] S. Marco-Peiró, B. Beltrán-Martinavarro, M. Rambla-Alegre, J. Peris-Vicente, and J. Esteve-Romero, "Validation of an analytical methodology to quantify melamine in body fluids using micellar liquid chromatography," *Talanta*, vol. 88, pp. 617–622, 2012.

[16] B. Beltrán-Martinavarro, J. Peris-Vicente, S. Marco-Peiró, J. Esteve-Romero, M. Rambla-Alegre, and S. Carda-Broch, "Use of micellar mobile phases for the chromatographic determination of melamine in dietetic supplements," *Analyst*, vol. 137, no. 1, pp. 269–274, 2012.

[17] FDA Guidance for Industry, *Bioanalytical Method Validation*, U.S. Department of Health and Human Services. Food and Drug Administration, Rockville, Md, USA, 2001.

[18] G. Cano-Sancho, S. Marin, A. J. Ramos, J. Peris-Vicente, and V. Sanchis, "Occurrence of aflatoxin M1 and exposure assessment in Catalonia (Spain)," *Revista Iberoamericana de Micologia*, vol. 27, no. 3, pp. 130–135, 2010.

[19] J. Peris Vicente, J. V. Gimeno Adelantado, M. T. Doménech Carbó, R. Mateo Castro, and F. Bosch Reig, "Identification of drying oils used in pictorial works of art by liquid chromatography of the 2-nitrophenylhydrazides derivatives of fatty acids," *Talanta*, vol. 64, no. 2, pp. 326–333, 2004.

[20] C. Frassineti, S. Ghelli, P. Gans, A. Sabatini, M. S. Moruzzi, and A. Vacca, "Nuclear magnetic resonance as a tool for determining protonation constants of natural polyprotic bases in solution," *Analytical Biochemistry*, vol. 231, no. 2, pp. 374–382, 1995.

[21] M. Rambla-Alegre, J. Peris-Vicente, S. Marco-Peiró, B. Beltrán-Martinavarro, and J. Esteve-Romero, "Development of an analytical methodology to quantify melamine in milk using micellar liquid chromatography and validation according to EU Regulation 2002/654/EC," *Talanta*, vol. 81, no. 3, pp. 894–900, 2010.

13

Development and Validation of an RP-HPLC Method for Estimation of Chlorpheniramine Maleate, Ibuprofen, and Phenylephrine Hydrochloride in Combined Pharmaceutical Dosage Form

Pinak M. Sanchaniya, Falgun A. Mehta, and Nirav B. Uchadadiya

Indukaka Ipcowala College of Pharmacy, Beyond GIDC, P.B. No. 53, Vitthal Udyognagar, Gujarat 388121, India

Correspondence should be addressed to Pinak M. Sanchaniya; pinaksanchania@gmail.com

Academic Editor: Susana Casal

The objective of this paper is to develope a simple, precise, accurate, and reproducible reversed phase high performance liquid chromatographic method for the quantitative determination of chlorpheniramine maleate, ibuprofen, and phenylephrine hydrochloride in combined pharmaceutical dosage form. Analysis was carried out using acetonitrile : mathanol : phoshphate buffer (50 : 20 : 30, v/v/v, pH 5.6) mobile phase at 1.0 mL/min flow rate and Sunfire C 18 column (5 μm × 250 mm × 4.6 mm) as stationary phase with detection wavelength of 220 nm. The retention times of chlorpheniramine maleate (CPM), ibuprofen (IBU), and phenylephrine hydrochloride (PHE) were 4.2 min, 13.6 min, and 2.7 min, respectively. The proposed method was validated with respect to linearity, accuracy, precision, specificity, and robustness. The linearity for chlorpheniramine maleate, ibuprofen, and phenylephrine hydrochloride was in the range of 0.5–2.5 μg/mL, 25–125 μg/mL, and 1.25–6.25 μg/mL, respectively. The % recoveries of all the three drugs were found to be 99.44–101.61%, 99.39–101.79%, and 98.66–101.83%. LOD were found to be 32, 120, and 68 ng/mL for CPM, IBU, and PHE, respectively. The method was successfully applied to the estimation of chlorpheniramine maleate, ibuprofen, and phenylephrine hydrochloride in combined pharmaceutical dosage form.

1. Introduction

3-(4-chlorophenyl)-N,N-dimethyl-3-pyridin-2-ylpropan-1-amine is the IUPAC name of the chlorpheniramine maleate (CPM). The empirical formula for CPM is $C_{20}H_{23}ClN_2O_4$ (Figure 1: chemical structure of CPM salt). CPM is a H-1 receptor blocker. CPM is an antihistamine used to relieve symptoms of allergy, hay fever, and the common cold. These symptoms include rash, watery eyes, itchy eyes/nose/throat/skin, cough, runny nose, and sneezing. 2-[4-(2-methyl-propyl)phenyl]propanoic acid is the IUPAC name of the ibuprofen (IBU). The empirical formula for IBU is $C_{13}H_{18}O_2$ (Figure 2: chemical structure of IBU). IBU is a nonselective inhibitor of COX-2, an enzyme involved in prostaglandin synthesis of the arachidonic acid pathway. Its pharmacological effects are believed to be due to inhibition of COX-2 which decrease the synthesis of prostaglandin

involved in mediating inflammation pain, fever, and swelling. (R)-1-(3-hydroxyphenyl)-2-methylamino-ethanol hydrochloride is the IUPAC name of phenylephrine hydrochloride (PHE). The empirical formula for PHE is $C_9H_{13}NO_2HCl$ (Figure 3: chemical structure of PHE). PHE is α-adrenoreceptor agonist, decreases nasal congestion, and increases drainage of sinus cavities. The combination dosage form of CPM, IBU, and PHE is available in the market and it is indicated in the treatment of allergy, congestion relief, and fever reducer [1–5].

These drugs are official in Indian Pharmacopoeia, British Pharmacopoeia, and United states Pharmacopoeia [3–5].

A literature survey regarding quantitative analysis of these drugs revealed that attempts have been made to develop analytical methods for the estimation of chlorpheniramine maleate alone and in combination with other drugs by liquid chromatographic (HPLC) [6–10] and HPTLC [11, 12]. The

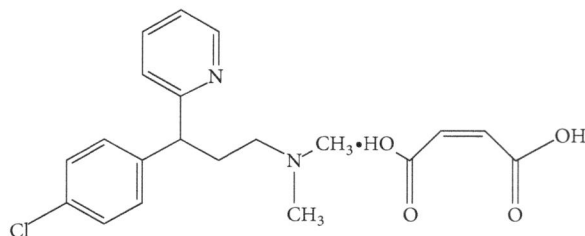

FIGURE 1: Structure of chlorpheniramine maleate.

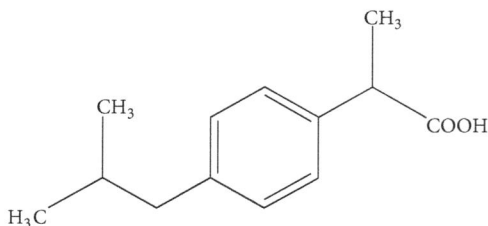

FIGURE 2: Structure of ibuprofen.

liquid chromatographic (HPLC) [13], HPTLC [14, 15], and spectrophotometric methods [16, 17] have been reported for the estimation of ibuprofen alone and in combination with other drugs. The spectrophotometric method [18], liquid chromatography (HPLC) [6, 9], and HPTLC [11, 12, 19, 20] methods have been reported for the estimation of phenylephrine hydrochloride alone and in combination with other drugs.

There is no method reported for the simultaneous estimation of CPM, IBU, and PHE in combined dosage form. The present study involved the development and validation of RP-HPLC method for the estimation of CPM, IBU, and PHE in combined pharmaceutical dosage form.

2. Experimental

2.1. Reagents and Materials. Analytically pure CPM, IBU, and PHE were obtained as gift samples from Elite Pharma Pvt. Ltd., Ahmedabad, India. HPLC grade methanol and acetonitrile were obtained from SRL Ltd., Mumbai, India. The water was distilled and deionised by using Millipore (Vienna, Austria) Milli Q Ultrapure system. Tablet formulation (ADVIL allergy and congestion relief, Pfizer Pharmaceutical, Madison, USA) containing labeled amount of 4 mg of chlorpheniramine maleate, 200 mg of ibuprofen, and 10 mg of phenylephrine hydrochloride was used for the study.

2.2. Apparatus. The liquid chromatographic system consists of Waters series 2998 (Shelton, USA) equipped with a series PDA detector, series 515 quaternary pump, and manual injector rheodyne valve with 20 μL fixed loop. The analytes were monitored at 220 nm. Chromatographic analysis was performed on Sunfire C 18 column (5 μm × 250 mm × 4.6 mm). All the drugs and chemicals were weighed on Shimadzu electronic balance (AX 200, Shimadzu Corp., Japan).

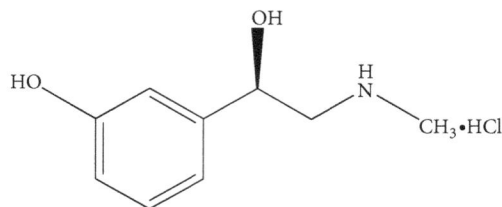

FIGURE 3: Structure of phenylephrine hydrochloride.

2.3. Chromatographic Conditions. The Sunfire C 18 column (5 μm × 250 mm × 4.6 mm) equilibrated with mobile phase acetonitrile : methanol : phosphate buffer (50 : 20 : 30, v/v/v; pH 5.6) and adjusted with 0.01% O-phosphoric acid was used. The flow rate was maintained at 1 mL/min, eluents were monitored with UV detector at 220 nm, and the injection volume was 20 μL. Total run time was kept for 15 min.

2.4. Preparation of Standard Stock Solutions. CPM, IBU, and PHE were weighed (10 mg each) and transferred to three separate 10 mL volumetric flasks and volumes were made up to the mark with mobile phase to yield a solution containing 1000 μg/mL of CPM, IBU, and PHE, respectively. Appropriately diluted with mobile phase to obtain working standard of CPM 100 μg/mL, IBU 1000 μg/mL and PHE 100 μg/mL were used as a working standard.

2.5. Method Validation. The proposed method was subjected to validation for various parameters like linearity and range, precision, accuracy, and robustness in accordance with International Conference on Harmonization Guidelines.

2.5.1. Linearity. Appropriate aliquots of CPM, IBU, and PHE working standard solutions were taken in different 10 mL volumetric flasks and diluted up to the mark with mobile phase to obtain final concentrations of 0.5, 1.0, 1.5, 2.0, and 2.5 μg/mL of CPM, 25, 50, 75, 100, and 125 μg/mL of IBU, and 1.25, 2.50, 3.75, 5.00, and 6.25 μg/mL of PHE, respectively. The solutions were injected using a 20 μL fixed loop system and chromatograms were recorded. Calibration curves were constructed by plotting average peak area versus concentrations and regression equations were computed for all three drugs.

2.5.2. Precision. The repeatability studies were carried out by estimating response of CPM (2 μg/mL), IBU (100 μg/mL), and PHE (5 μg/mL) six times and results were reported in terms of relative standard deviation. The intraday and interday precision studies (intermediate precision) were carried out by estimating the corresponding responses 3 times on the same day and on 3 different days for three different concentrations of CPM (0.5, 1.5, and 2.5 μg/mL), IBU (25, 75, and 125 μg/mL), and PHE (1.25, 3.75, and 6.25 μg/mL), and the results were reported in terms of relative standard deviation.

2.5.3. Accuracy. The accuracy of the method was determined by calculating recoveries of CPM, IBU, and PHE by method

Development and Validation of an RP-HPLC Method for Estimation of Chlorpheniramine Maleate, Ibuprofen, and Phenylephrine Hydrochloride in Combined Pharmaceutical Dosage Form

83

of standard additions. Known amounts of CPM (0, 0.5, 1.0, and 1.5 μg/mL), IBU (0, 25, 50, and 75 μg/mL), and PHE (0, 1.25, 2.50, and 3.75 μg/mL) were added to a prequantified sample solution, and the amounts of CPM, IBU, and PHE were estimated by measuring the peak areas and by fitting these values to the straight-line equation of calibration curve.

2.5.4. Detection Limit and Quantitation Limit. The LOD and LOQ were calculated using the following equation as per ICH guidelines:

$$\text{LOD} = 3.3 \times \frac{\sigma}{S}, \qquad \text{LOQ} = 10 \times \frac{\sigma}{S}, \qquad (1)$$

where σ is the standard deviation of y-intercepts of regression lines and S is the slope of the calibration curve.

2.5.5. Robustness. Robustness of the method was studied by deliberately changing the experimental conditions like flow rate and percentage of mobile phase ratio. The study was carried out by changing 5% of the mobile phase ratio and 0.1 mL/min of flow rate.

2.5.6. Solution Stability. The solutions were prepared and solution stability was checked for 3, 9, 12, and 24 hrs by checking the area over the period of time, using the different analysts and the same instrument.

2.5.7. System Suitability. A system suitability test was an integral part of the method development to verify that the system is adequate for the analysis of CPM, IBU, and PHE to be performed. System suitability test of the chromatography system was performed before each validation run. Five replicate injections of a system suitability standard and one injection of a check standard were made. Area, retention time (RT), tailing factor, asymmetry factor, and theoretical plates for the five suitability injections were determined.

2.6. Analysis of Marketed Formulation. Twenty tablets were accurately weighed and finely powdered. Tablet powder equivalent to 4 mg CPM, 200 mg of IBU, and 10 mg of PHE was taken in 100 mL volumetric flask. Methanol (50 mL) was added to the above flask and the flask was sonicated for 15 minutes. The solution was filtered using Whatman filter paper No. 41 and volume was made up to the mark with the mobile phase.

Appropriate volume of the aliquot was transferred to a 10 mL volumetric flask and the volume was made up to the mark with the mobile phase to obtain a solution containing 1.0 μg/mL of CPM, 50 μg/mL of IBU, and 2.50 μg/mL of PHE. The solution was sonicated for 10 min. It was injected as per the above chromatographic conditions and peak areas were recorded. The quantifications were carried out by keeping these values to the straight line equation of calibration curve.

3. Results and Discussion

3.1. Optimization of Mobile Phase. The optimization of mobile phase was to resolve chromatographic peaks for active drug ingredients with less asymmetric factor.

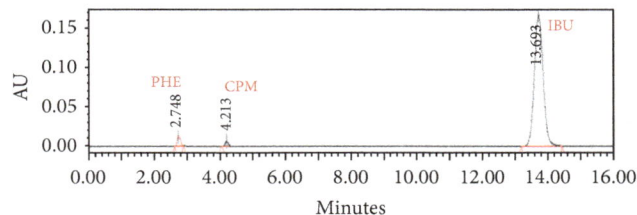

FIGURE 4: Chromatogram of CPM, IBU, and PHE.

There were many mobile phases that tried to resolve all three chromatographic peaks, in that first with simple methanol : water (70 : 30, v/v), methanol : water (50 : 50, v/v), methanol : water (40 : 60, v/v), but we found broadness of peaks and could not get satisfactory results in these chromatograms. To improve these, we worked with phosphate buffer and for sharpness of chromatographic peaks we worked with acetonitrile with slightly acidic condition. So, acetonitrile : phosphate buffer (70 : 30, v/v) (pH: 5.6) was adjusted with 0.01% O-phosphoric acid peaks which were resolved but merged CPM and PHE peak and PHE peak with void volume at 1.9 min. So acetonitrile : phosphate buffer (60 : 40) (pH: 5.6, v/v) was adjusted with 0.01% O-phosphoric acid but PHE peak with void volume. So we added methanol and peak of PHE in Acetonitrile : methanol : phoshphate buffer mobile phase got at above 2.5 min; therefore, acetonitrile : methanol : phosphate buffer (50 : 10 : 40, v/v/v) (pH: 5.6) was adjusted with O-phosphoric acid PHE peak at 2.5 min, CPM peak at 3.8 min, and IBU peak at 24 min. We add more 10% of methanol and, IBU peak at 13.6 min.

Finally, the mobile phase acetonitrile : methanol : phosphate buffer (50 : 20 : 30, v/v/v) (pH: 5.6) adjusted with 0.01% O-phosphoric acid was found to be satisfactory which gave three symmetric and well-resolved peaks for CPM, IBU, and PHE. The retention times for CPM, IBU, and PHE were 4.2 min, 13.6 min, and 2.7 min, respectively (Figure 4). The resolutions between PHE & CPM, PHE & IBU, and CPM & IBU were found to be 4.5, 16.8, and 15.2, respectively, which indicates good separation of all compounds. The mobile phase flow rate was maintained at 1 mL/min. Overlain UV spectra of the drugs showed that CPM, IBU, and PHE absorbed appreciably at 220 nm, so detection was carried out at 220 nm (Figure 5).

3.2. Method Validation. The calibration curve for CPM was found to be linear in the range of 0.5–2.5 μg/mL with a correlation coefficient of 0.998. The calibration curve for IBU was found to be linear in the range of 25–125 μg/mL with a correlation coefficient of 0.996. The calibration curve for PHE was found to be linear in the range of 1.25–6.25 μg/mL with a correlation coefficient of 0.997. Instrument precision was determined by performing injection repeatability test and the RSD values for CPM, IBU, and PHE were found to be 1.38%, 0.57%, and 0.44%, respectively. The intraday and interday precision studies were carried out and the results are reported in Table 1. The low RSD values indicate that the method is precise.

TABLE 1: Validation parameters for CPM, IBU, and PHE.

Parameters	CPM	IBU	PHE
Linearity (range) (μg/mL)	0.5–2.5	25–125	1.25–6.25
Retention time (min)	4.2	13.6	2.7
Detection limit (μg/mL)	0.0321	0.1198	0.0679
Quantitation limit (μg/mL)	0.5	25	1.25
Accuracy (%)	99.44–101.61	99.39–101.79	99.66–101.83
Precision (RSD%)[a]			
Intraday precision ($n = 3$)	0.44–1.28	0.10–0.23	0.38–0.56
Interday precision ($n = 3$)	0.98–1.46	0.66–1.33	0.66–1.53
Instrument precision (RSD%)[a] ($n = 6$)	1.38	0.57	0.12

[a]RSD is relative standard deviation and "n" is number of determinations.

TABLE 2: Robustness study of CPM, IBU, and PHE.

Method parameter/condition	Deliberate changes	% RSD of peak area ($n = 3$)		
		CPM	IBU	PHE
Flow rate	0.9 mL/min	0.89	1.26	1.63
	1.1 mL/min	1.23	1.39	1.73
Mobile phase ratio	45 : 25 : 30	1.24	1.38	1.82
Acetonitrile : methanol : phosphate buffer	55 : 15 : 30	1.28	1.73	1.17

TABLE 3: System suitability parameters.

Parameter	CPM	IBU	PHE
Retention time (min)	4.2	13.6	2.7
Theoretical plates	4155.36	4324.06	3986.31
Tailing factor	0.67	1.03	1.25

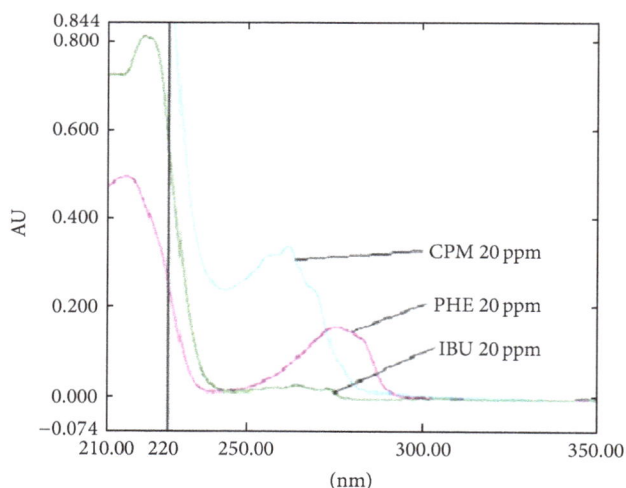

FIGURE 5: Overlain UV spectra of CPM, IBU, and PHE (20 ppm).

The accuracy of the method was determined by calculating recoveries of CPM, IBU, and PHE by method of standard addition. The recoveries were found to be 99.44–101.61%, 99.39–101.79%, and 98.66–101.83% for CPM, IBU, and PHE, respectively. The results are reported in Table 1. The high values indicate that the method is accurate.

The detection limits for CPM, IBU, and PHE were found to be 32 ng/mL, 120 ng/mL, and 68 ng/mL, respectively, while quantitation limits were found to be 0.1 μg/mL, 0.4 μg/mL, and 0.2 μg/mL, respectively. The above data shows that a nanogram quantity of the drugs can be accurately and precisely determined. Robustness study was performed by deliberately changing the experimental conditions like flow rate from 1 mL/min to 0.8 mL/min and 1.2 mL/min. The composition of mobile phase was changed varying the proportion of acetonitrile by 5%. In both conditions the recoveries of both drugs were determined and the RSD was found to be less than 2%. The results are reported in Table 2.

System suitability parameters such as the number of theoretical plates, resolution, and tailing factor were determined. System suitability test was carried out and the results are summarized in Table 3. Asymmetric factors for CPM, IBU, and PHE are 1.0, 1.11, and 0.944, respectively.

Stability of standard and sample solution of CPM, IBU, and PHE were evaluated at room temperature. The solutions of the three drugs were found to be stable for 0, 3, 6, and 24 hrs. The results are reported in Table 4. All three drugs were found to be stable with a recovery of more than 98%.

3.3. Analysis of Marketed Formulations. The proposed method was successfully applied to the determination of CPM, IBU, and PHE in their combined dosage form. The % recovery ± S.D. was found to be 100.37 ± 1.24, 100.24 ± 1.55, and 100.91 ± 1.25, respectively, for CPM, IBU, and PHE (Table 5) which were comparable with the corresponding labeled amounts.

4. Conclusion

The concentration of CPM, IBU, and PHE in pharmaceutical dosage form could be satisfactorily determined using isocratic RP-HPLC system with PDA detector.

This study had shown that PDA detector was sensitive, accurate, and simple method for the determination of the

Development and Validation of an RP-HPLC Method for Estimation of Chlorpheniramine Maleate, Ibuprofen, and
Phenylephrine Hydrochloride in Combined Pharmaceutical Dosage Form

85

TABLE 4: Solvent stability study.

Time (Hrs.)	Area ($n = 3$)			Result %		
	CPM 2 (μg/mL)	IBU 100 (μg/mL)	PHE 5 (μg/mL)	CPM	IBU	PHE
0	61521	4158429	124685	100	100	100
3	61009	4146734	123619	99.16	99.14	99.14
6	60913	4116520	121706	99.01	98.99	97.61
24	60157	4104629	120648	98.71	98.70	96.76

TABLE 5: Analysis of marketed preparation.

Formulation	Labelled amount (mg)			% Recovery[b]		
	CPM	IBU	PHE	CPM	IBU	PHE
ADVIL allergy and congestive relief	4	200	10	100.37 ± 1.24	100.23 ± 1.55	100.91 ± 1.25

[b]Mean value ± standard deviation of three determinations; tablet formulation, ADVIL allergy and congestion relief, Pfizer Pharmaceutical, Madison, USA containing labeled amount of 4 mg of chlorpheniramine maleate, 200 mg of ibuprofen, and 10 mg of phenylephrine hydrochloride.

active ingredients in ADVIL allergy and congestion relief tablet.

This method has been found suitable for the routine analysis of pharmaceutical dosage forms in QC and R & D Laboratories for product of similar type and composition.

Disclosure

The authors have no conflict of interests or no financial gains in mentioning the company names or trademarks. The usage of this trademark symbol or company name is for proving the genuinity of the work and not for any other purpose. As the authors of the paper, they do not have any financial relation with the commercial identity in the paper.

Acknowledgments

The authors are thankful to Elite Pharma Pvt. Ltd., Ahmedabad, India, for providing gift sample of CPM, IBU, and PHE. The authors are very thankful to Principal, Indukaka Ipcowala College of Pharmacy, New Vallabh Vidyanagar for providing necessary facilities to carry out research work.

References

[1] M. d. J. O'niel and P. E. Hackelman, *Merk Index—An Encyclopaedia of Chemicals, Drugs and Biological*, Merk Research laboratory,, 13th edition, 2006.

[2] K. D. Tripathi, *Essentials of Medicinal Pharmacology*, Jaypee Brothers Medical Publisher LTD, New Delhi, India, 5th edition, 2003.

[3] British Pharmacoeia, *The Stationary Office on Behalf of the Medicine and Healthcare Products Regulatory Agency (MHRA)*, vol. 3, London, UK, 2007.

[4] Indian Pharmacopoeia, "Indian pharmacopoeial commission," Ghaziabad, India, 5th edition 2010, vol 2, pp 1071-72, 1479-83, v 3, pp. 1900.

[5] United States Pharmacopoeia and National Formulary, (22th) Asian Edition, The United States Pharmacopoeia Convention Inc., U.S.A., pp. 427-30, 953-55, 1471-73.

[6] M. Maithani, R. Raturi, G. Vertika, and D. Kumar, "Development & validation of RP-HPLC method for the determination of chlorpheniramine maleate and phenylephrine HCl in pharmaceutical dosage form," *International Research Journal of Pharmacy*, vol. 5, pp. 1-4, 2010.

[7] L. E. Geetha and S. M. Rama, "A novel RP-HPLC method for simultaneous estimation of codeine phosphate, chlorpheniramine maleate and its preservative in syrup formulation," *International Journal of Pharmacy and Pharmaceutical Sciences*, vol. 4, supplement 3, pp. 585–590, 2012.

[8] D. B. Wanjari, V. V. Parashar, S. N. Lulay, M. R. Tajne, and N. J. Gaikwad, "Simultaneous HPLC estimation of acetaminophen, chlopheniramine maleate, dextromethorphan hydrobromide and pseudoephedrine hydrochloride in tablets," *Indian Journal of Pharmaceutical Sciences*, vol. 66, no. 3, pp. 345–347, 2004.

[9] H. Senyuva and T. Özden, "Simultaneous high-performance liquid chromatographic determination of paracetamol, phenylephrine HCl, and chlorpheniramine maleate in pharmaceutical dosage forms," *Journal of Chromatographic Science*, vol. 40, no. 2, pp. 97–100, 2002.

[10] S. Rajurkar, "Simultaneous determination of chlorpheniramine maleate, paracetamol, pseudoephedrine HCl in pharmaceutical dosage form by HPLC method," *International Journal of Life Science and Pharma Research*, vol. 1, no. 1, pp. 94–100, 2011.

[11] S. Subramanyan and S. G. Das, "HPTLC method forsimultaneous estimation of chlorpheniramine maleate, phenylephrine in pharmaceutical dosage form," *Journal of AOAC International*. In press.

[12] N. Hunan and S. Multal, "Densitometric analysis of chlorpheniramine maleate, phenylephrine and acetaminophen by HPTLC method," *Analytical LetterIssue*, vol. 19, no. 7-8, 1986.

[13] P. B. Reddy and M. S. Reddy, "RP-HPLC method for simultaneous estimation of paracetamol and ibuprofen in tablets," *Asian Journal of Research in Chemistry*, vol. 2, no. 1, pp. 70–72, 2009.

[14] S. Chitlange, D. Sakarkar, S. Wankhede, and S. Wadodkar, "High performance thin layer chromatographic method for simultaneous estimation of ibuprofen and pseudoephedrine hydrochloride," *Indian Journal of Pharmaceutical Sciences*, vol. 70, no. 3, pp. 398–400, 2008.

[15] R. V. Rele and S. A. Sawant, "Determination of paracetamol and ibuprofen from combined dosage formulation by HPTLC method," *Analytical Chemistry*, vol. 9, no. 1, pp. 302–305, 2010.

[16] R. Gondalia, R. Mashru, and P. Savaliya, "Development and validation of spectrophotometric methods for simultaneous estimation of IBUPROFEN and PARACETAMOL in soft gelatin capsule by simultaneous equation method," *International Journal of ChemTech Research*, vol. 2, no. 4, pp. 1881–1885, 2010.

[17] I. M. Palabiyik, E. Dinç, and F. Onur, "Simultaneous spectrophotometric determination of pseudoephedrine hydrochloride and ibuprofen in a pharmaceutical preparation using ratio spectra derivative spectrophotometry and multivariate calibration techniques," *Journal of Pharmaceutical and Biomedical Analysis*, vol. 34, no. 3, pp. 473–483, 2004.

[18] L. K. Soni, T. Narsinghani, and C. Saxena, "Development and validation of UV-Spectrophotometric assay protocol for simultaneous estimation of ebastine and phenylephrine hydrochloride in tablet dosage form using simultaneous equation method," *International Journal of ChemTech Research*, vol. 3, no. 4, pp. 1918–1925, 2011.

[19] P. V. Devarajan, M. H. Adani, and A. S. Gandhi, "Simultaneous determination of lignocaine hydrochloride and phenylephrine hydrochloride by HPTLC," *Journal of Pharmaceutical and Biomedical Analysis*, vol. 22, no. 4, pp. 685–690, 2000.

[20] I. Wouters, E. Roets, and J. Hoogmartens, "Analysis of tablets containing acetylsalicylic acid and phenylephrine by high-performance liquid chromatography," *Journal of Pharmaceutical and Biomedical Analysis*, vol. 2, no. 3-4, pp. 481–490, 1984.

14

A Stability Indicating UPLC Method for the Determination of Levofloxacin Hemihydrate in Pharmaceutical Dosage Form: Application to Pharmaceutical Analysis

Batuk Dabhi,[1] Bhavesh Parmar,[1] Nitish Patel,[1] Yashwantsinh Jadeja,[1] Madhavi Patel,[1] Hetal Jebaliya,[1] Denish Karia,[2] and A. K. Shah[1]

[1] *Department of Chemistry, Saurashtra University, Rajkot-360 005, Gujarat, India*
[2] *Arts Commerce and Science College, Borsad 388540, Gujarat, India*

Correspondence should be addressed to Batuk Dabhi; batukdabhi@gmail.com

Academic Editor: Antonio Martín-Esteban

A reliable and sensitive isocratic stability indicating RP-UPLC method has been developed and validated for quantitative analysis and content uniformity study of levofloxacin hemihydrate in tablets. An isocratic method for analysis of levofloxacin hemihydrate was archived on ACQUITY UPLC BEH C_{18} (100*2.1) mm particle size 1.7 μ columns within shorter runtime of 4 min with a flow rate of 0.400 mL/min and using a photodiode array detector to monitor the eluate at 294 nm. The mobile phase consisted of acetonitrile-buffer (23 : 77 v/v), (buffer: 20 mM K_2HPO_4 + 1 mL triethylamine in 1 L water, pH = 2.50 by orthophosphoric acid). Response was a liner function of drug concentration in the range of 0.5–80 μg/mL (r^2 = 0.999) with a limit of detection and quantification of 0.1 and 0.5 μg/mL, respectively. Accuracy (recovery) was between 99.77% and 101.55%. The drug was subjected to oxidation, hydrolysis, photolysis, and thermal degradation. Degradation products resulting from the stress studies did not interfere with the detection of levofloxacin hemihydrate, and the assay is stability indicating.

1. Introduction

Levofloxacin hemihydrate (Figure 1) is a synthetic chemotherapeutic antibiotic of the fluoroquinolone drug class and is used to treat severe life-threatening bacterial infection or bacterial infection that have failed to respond to other antibiotic classes.

IUPAC name is (S)-9-fluoro-2,3-dihydro-3-methyl-10-(4-methylpiperazin-1-yl)-7-oxo-7H-pyrido[1,2,3-de]-1,4-benzoxazine-6-carboxylic acid.

Levofloxacin hemihydrate is highly water soluble in nature. It is also soluble in organic solvents, soluble in glacial acetic acid and chloroform, sparingly soluble in methanol, slightly soluble in ethanol, and practically insoluble in ether. Levofloxacin hemihydrate is odourless drug. Methods for quantitative analysis of Levofloxacin by HPLC [1–7], by UV [8–10] spectroscopy in single as well as in combination, are available in the literature. There are no methods available for quantitative analysis by UPLC. The method was developed and validated as per ICH [11–13] and USP [14] guideline.

2. Experimental

2.1. Chemicals and Reagents. Levofloxacin hemihydrates reference standard (claim 99.48%) was provided by Cipla pharmaceuticals. Tablets (500 mg) of Levofloxacin were produced from a pharmacy. HPLC grade methanol and ortho phosphoric acid were purchased from Merck India Limited, Mumbai, India. Analytical grade hydrochloric acid, sodium hydroxide pellets, and hydrogen peroxide solution 30% (v/v) were obtained from Ranbaxy Fine Chemicals, New Delhi, India. High purity water was obtained using Milli-Q (Millipore, Milford, MA, USA) water purification system.

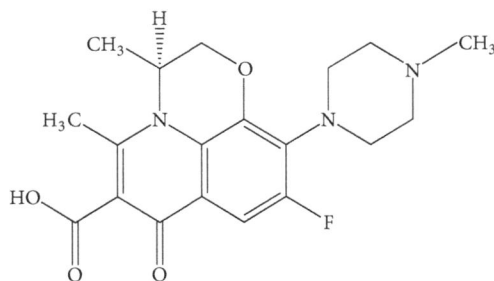

FIGURE 1: Levofloxacin hemihydrate.

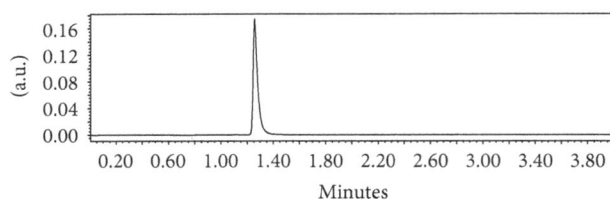

FIGURE 2: Chromatogram of levofloxacin hemihydrates reference standard.

2.2. Instruments and Condition.

The fast liquid chromatography was performed using waters UPLC system with photodiode array detector. Chromatogram and data were recorded by means of Empower software. The chromatographic system was performed using an Acquity BEH C_{18} (100 × 2.1 mm) id., 1.7 μm column. Separation was achieved using a mobile phase consisted of buffer : acetonitrile (77 : 23 v/v) (buffer: 20 mM KH_2PO_4 + 1 mL triethylamine in 1 litre water, pH = 2.5 adjust with orthophosphoric acid) at a flow rate of 0.4 mL/min in only 4 minute runtime. The column temperature was maintained at 30°C, injection volume was 2 μL, and detection wavelength was set at 294 nm for determination of levofloxacin hemihydrate (Figure 2).

2.3. Preparation of Standard Solution.

The stock solution of levofloxacin hemihydrate (500 μg/mL) was prepared by accurately weighing 25 mg of levofloxacin hemihydrate reference standard and transferring to 50 mL standard volumetric flask containing approximately 25 mL of water. The flask was sonicated for 10 minutes to dissolve the solids. Volumes were made up to the mark with water (500 μg/mL stock solution-A). Transferred 5 mL of above stock solution-A to a 50 mL volumetric flask and diluted up to the mark with mobile phase to obtained working standard solution (50 μg/mL) of drug.

2.4. Preparation of Sample Solution.

For analysis of the tablet dosage form, ten tablets were weighed individually, and their average weight was determined. The tablets were crushed to fine homogenous powder, and a quantity equivalent to 25 mg levofloxacin was transferred in a 50 mL volumetric flask. Added about 25 mL of water to the flask, shaken for 10 minutes, and then sonicated for 15 minutes. The solution was allowed to stand at room temperature and filtered through Whatman no. 41 filter paper. The residue was washed with water, and the combined filtrate was made up to the mark with water (stock solution-B 500 μg/mL), transferred quantitatively 5 mL of above stock solution-B in 50 mL volumetric flask and diluted up to the mark with mobile phase (50 μg/mL).

2.5. Specificity/Force Degradation Study.

Force degradation studies were performed to evaluate the stability indicating properties and specificity of the method. All solutions used in stress studies were prepared at an initial concentration of 500 μg/mL of levofloxacin hemihydrate and heat for 1 hr at 80°C. All samples were then diluted in mobile phase to give a final concentration of 50 μg/mL and filtered before injection.

2.5.1. Acid Degradation Study.

Acid decomposition was carried out in 0.1 M HCl at a concentration of 500 μg/mL of levofloxacin hemihydrate and after heat for 1 hr at 80°C. The stressed sample was cooled, neutralized, and diluted with mobile phase.

2.5.2. Base Degradation Study.

Base decomposition was carried out using 0.1 M NaOH at a concentration of 500 μg/mL and heat for 1 hr at 80°C. After cooling, the solution was neutralized and diluted with mobile phase.

2.5.3. Oxidative Degradation.

Oxidation solutions for oxidative stress studies were prepared using 3% H_2O_2 at a concentration of 500 μg/mL of levofloxacin and after heat for 1 hr at 80°C. The sample solution was cooled and diluted accordingly with the mobile phase.

2.5.4. Thermal Degradation.

For thermal stress testing, the drug solution 500 μg/mL was heated for 1 hr at 80°C, cooled, and used for the study.

2.5.5. Photo Degradation.

For Photolytic degradation, the powder drug has been exposed to sunlight for 48 hrs and used for the study.

3. Method Validation

The method was validated as per ICH guideline. The method was validated by performing system suitability, linearity, limit of Detection (LOD), limit of quantification (LOQ), precision, accuracy, selectivity, and robustness.

3.1. Precision.

Method precision of the analytical method was determined by analyzing six sets of sample solution preparation. Assay of all six replicate sample preparations was determined, and mean percentage of assay value, standard deviation and percentage of relative standard deviation for the same were calculated. Intermediate precision of the analytical method was determined by performing method precision on another day by another analyst using different make of raw materials under same experimental condition. Overall assay value of method precision and intermediate precision was compared and percentage of difference and overall percentage of relative standard deviation were calculated.

3.2. LOD and LOQ.
Sensitivity of the method was determined by establishing the LOD and LOQ. The LOD and LOQ were established at signal-to-noise ratio of 3 : 1 and 10 : 1, respectively.

3.3. Linearity.
Linearity test solutions for the assay method were prepared from a stock solution at 7 concentration levels of the assay analyte concentration (20, 30, 40, 50, 60, 70, and 80 μg/mL). The peak area versus concentration data was analyzed with least squares linear regression. The slope and Y-intercept of the calibration curve were reported.

3.4. Accuracy.
Accuracy of the developed method was confirmed by doing recovery study as per ICH guidelines at three different concentration levels 50%, 100%, and 150% by replicate analysis (n = 3). The results indicated that the method is highly accurate for the determination of levofloxacin hemihydrate.

3.5. Robustness.
The robustness of the assay method was established by introducing small changes in the chromatographic condition which included percentage of methanol in mobile phase (58% and 62%), flow rate (0.38 and 0.42 mL/min), and column oven temperature (25°C and 35°C). Robustness of the method was studied using six replicates at a concentration level of 50 μg/mL of levofloxacin hemihydrate.

3.6. Solution Stability.
The solution stability of levofloxacin hemihydrate in the assay method was carried out by leaving both the sample and reference standard solutions in tightly capped volumetric flasks at room temperature for 48 hr. The same sample solution was assayed at 6 hr intervals over the study period. The percentage of RSD of the levofloxacin hemihydrate assay was calculated for solution stability experiments. An additional study was carried out using the stock solution by storing it in a tightly capped volumetric flask at 4°C.

4. Result and Discussion

4.1. Method Validation

4.1.1. Precision.
The precision of the method was determined by repeatability (intraday precision) and intermediate precision (interday precision) of the levofloxacin hemihydrate standard solution. The precision of the assay method was evaluated by carrying out six independent assays. For assay method (n = 6), percentage of RSD for system precision was 0.45% on the same day (intra-day) and 0.41% on different days (inter-day). The mean values of method precision (repeatability) were 101.2% (RSD = 0.26%) for assay on the same day (intra-day) and 101.58% (RSD 0.57%) for assay on different days (inter day). Intermediate precision was

TABLE 1: Result of precision, linearity, LOD, and LOQ study.

Parameter	UPLC
Linearity (μg/mL)	0.5–80
Correlation coefficient (r^2)	0.9994
Slope (m)	46965
Intercept (c)	85486
LOD (μg/mL)	0.1
LOQ (μg/mL)	0.5
Accuracy (% RSD)	0.21–0.76
Interday precision (RSD, n = 6)	0.57
Intraday precision (RSD, n = 6)	0.26

established by determining the overall (inter-day and intraday) method precision for assay. For intermediate precision, overall assay value (n = 12) was 101.24% and RSD = 0.36%. Percentage of RSD value less than 2% indicates that this method is highly precise.

4.1.2. LOD and LOQ.
All the results of LOD and LOQ data were within the acceptance criteria; hence, it can be concluded that the LOD and LOQ of the method were 0.1 μg/mL and 0.5 μg/mL, respectively. The signal-to-noise ratio for the LOD was well within the acceptance criteria which means more than 3.3, and for the LOQ it was more than 10.0. Furthermore, the data of linearity extension chart up to LOQ level also suggest that the levofloxacin can be quantified up to 0.5 μg/mL with well precisely and accurately.

4.1.3. Linearity.
The calibration curve for the levofloxacin hemihydrate was linear over the concentration range of 0.5–80 μg/mL. The data for the peak area versus concentration were treated by linear regression analysis, and the correlation coefficient (r^2) was obtained (0.9994). The regression equation for the calibration curve was found to be $y = 46,965x + 85,486$ (Table 1).

4.1.4. Accuracy.
The percentage recovery of levofloxacin hemihydrate in the drug tablets samples was obtained in a range from 99.97% to 101.55%, respectively (Table 2). Percentage of RSD value of replicated sets was less than 2% which indicates that this method is highly accurate.

4.1.5. Robustness.
The robustness of an analytical procedure refers to its ability to remain unaffected by small and deliberate variations in method parameters and provides an indication of its reliability for routine analysis. The robustness of the method was evaluated by assaying the same sample under different analytical conditions deliberately changing from the original condition. The results obtained from assay of the test solutions were not affected by varying the conditions and were in accordance with the results for original conditions (Table 3). The percentage of RSD value of assay determined for the same sample under original conditions and robustness conditions was less than 2.0%, indicating that the developed method was robust.

TABLE 2: Accuracy study.

Level %	No.	Amount of drug added (μg/mL)	Amount of drug found (μg/mL)	Recovery (%)	Mean recovery (%)	RSD (%)
50	1	25.23	25.15	99.87		
	2	25.29	25.64	101.40	100.63	0.76
	3	25.34	25.63	100.62		
100	1	50.82	50.98	100.32		
	2	50.60	50.40	99.62	99.97	0.35
	3	50.27	50.52	99.97		
150	1	75.21	76.21	101.34		
	2	75.39	76.72	101.77	101.55	0.21
	3	75.06	76.61	101.54		

TABLE 3: Robustness study.

Conditions	Assay (%)	RT (min)	System suitability data	
			Theoretical plates	Asymmetry
Flow mL/min				
0.38	98.89	1.27	5244	1.14
0.40	99.08	1.24	5253	1.11
0.42	99.21	1.25	5276	1.10
Column temperature				
25°C	98.93	1.46	5378	1.12
30°C	99.08	1.24	5253	1.11
35°C	99.19	1.14	5182	1.08
ACN: buffer ratio				
21 : 79 V/V	99.87	1.40	5402	1.10
23 : 77 V/V	99.08	1.24	5253	1.11
25 : 75 V/V	100.27	1.16	5185	1.13

TABLE 4: Specificity Study.

Stress condition	Time	Purity angle	Purity threshold	% Degradation
Acid Hydrolysis	1 hr	0.409	0.874	8.12
Base Hydrolysis	1 hr	1.107	13.28	12.38
Oxidation	1 hr	4.578	12.82	66.44
Thermal	1 hr	0.854	1.057	24.37
Photo	5 days	0.748	1.682	27.82

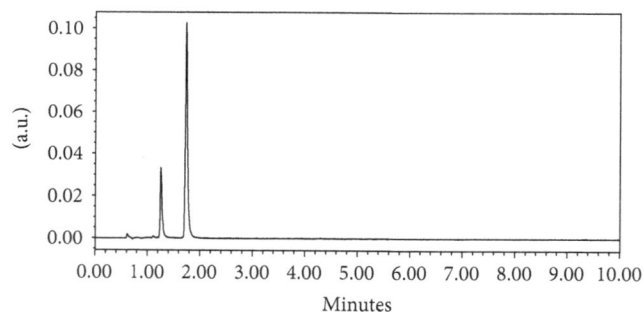

FIGURE 3: H_2O_2 degradation study of levofloxacin.

4.1.6. Solution Stability. *The percentage of RSD of the assay of levofloxacin hemihydrate from the solution stability experiments was within 2%.* The results of solution stability experiments confirm that the sample solutions used during the assay were stable up to 48 hr at room temperature and up to 8 days at 4°C.

4.1.7. Specificity. The specificity of the developed method was determined by injecting sample solutions (50 μg/mL) which were prepared by forcibly degrading under such stress conditions as heat, light, oxidative agent, acid, and base under proposed chromatographic condition. The stability indicating capability of the method was established from the

separation of levofloxacin peak from the degraded samples derived from the empower software. The degradation of levofloxacin hemihydrate was found to be very similar for both the tablets and standard. Typical chromatograms obtained following the assay of stressed samples are shown in Figures 3, 4, and 5.

Levofloxacin hemihydrate standard and tablet powder were found to be quite stable under light and heat conditions. A slight decomposition was seen on exposure of levofloxacin drug solution to acid and heat. On the other hand, the drug

A Stability Indicating UPLC Method for the Determination of Levofloxacin Hemihydrate in Pharmaceutical Dosage Form:
Application to Pharmaceutical Analysis

91

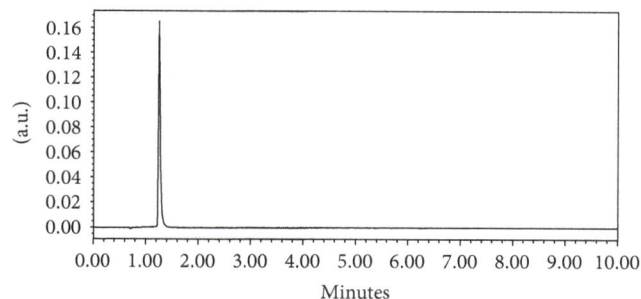

FIGURE 4: Base Degradation Study of Levofloxacin.

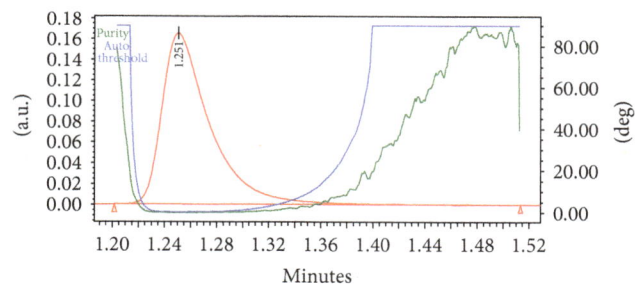

FIGURE 5: Peak purity result of base degradation.

decomposition under oxidative and alkaline degradation was found to be 66.44% and 12.38%, respectively. The results of force degradation study are explained in Table 4.

5. Conclusion

A stability indicating UPLC method was developed, validated, and applied for the determination of levofloxacin hemihydrate in pharmaceutical dosage forms. The developed method was validated as per ICH guidelines and was found to be accurate, precise, robust, and specific. The chromatographic elution step was under taken in a short time (4 min). No interference from any components of pharmaceutical dosage form or degradation products was observed, and the method has been successfully used to perform long term and accelerate stability studies of levofloxacin hemihydrate.

Acknowledgments

The authors are grateful to the Department of Chemistry, Saurashtra University (UGC-SAP Sponsored and DST-FIST Funded) Rajkot, Gujarat, India, for providing the instrumental facilities. Special thanks to the "National Facility for Drug Discovery through New Chemicals Entities (NCE's) Development and Instrumentation Support to Small Manufacturing Pharma Entities" Program under the Drug and Pharma Research Support (DPRS) jointly funded by Department of Science & Technology, New Delhi, Government of Gujarat Industries Commissionerate and Saurashtra University, Rajkot, India.

References

[1] T. Kumar, A. Chitra, V. Amrithraj, and N. Kumar, "New RP-HPLC method development and validation for estimation of levofloxacin in tablet dosage form," *Journal of Global Trends in Pharmaceutical Sciences*, vol. 2, no. 3, pp. 264–276, 2011.

[2] K. Kothekar, J. Balasundaram, A. Khandhar, and R. Mishra, "Quantitative determination of levofloxacin and ambroxol hydrochloride in pharmaceutical dosage form by reversed-phase high performance liquid chromatography," *Eurasian Journal of Analytical Chemistry*, vol. 2, no. 1, 2007.

[3] T. Santhoshi, K. Kumar, V. Rao, and A. Ravipati, "Development and validation of RP-HPLC method for simultaneous estimation of levofloxacin and ornidazole in pharmaceutical dosage form," *Journal of Pharmacy Research*, vol. 4, no. 11, p. 3864, 2011.

[4] N. S. Lakka and N. Goswami, "A novel isocratic RP-HPLC method development and validation for estimation of 5HMF in Levofloxacin Hemihydrate intravenous infusion," *International Journal of Research in Pharmaceutical Sciences*, vol. 2, no. 1, pp. 45–51, 2011.

[5] U. Neckel, C. Joukhadar, M. Frossard, W. Jäger, M. Müller, and B. X. Mayer, "Simultaneous determination of levofloxacin and ciprofloxacin in microdialysates and plasma by high-performance liquid chromatography," *Analytica Chimica Acta*, vol. 463, no. 2, pp. 199–206, 2002.

[6] J. Mehta, Y. Pancholi, V. Patel, N. Kshatri, and N. Vyas, "Development and validation of a sensitive stability indicating method for quantification of levofloxacin related substances and degradation products in pharmaceutical dosage form," *International Journal of PharmTech Research*, vol. 2, no. 3, pp. 1932–1942, 2010.

[7] N. W. Sheikh, A. S. Tripathi, V. Chitra, A. Choudhury, and A. P. Dewani, "Development and validation of RP-HPLC assay for levofloxacin in rat plasma and saliva: application to pharmacokinetic studies," *African Journal of Pharmacy and Pharmacology*, vol. 5, no. 13, pp. 1612–1618, 2011.

[8] S. Rahar, S. Dogra, D. Panchru, P. Singh, and G. Shah, "Development and validation of UV-visible spectroscopic method for the estimation of levofloxacin hemihydrate in bulk and marketed formulation," *International Journal of Institutional Pharmacy and Life Sciences*, vol. 1, no. 2, 2011.

[9] A. A. Shirkhedkar and S. J. Surana, "Quantitative determination of levofloxacin hemihydrate in bulk and tablets by UV-spectrophotometry and first order derivative methods," *Pakistan Journal of Pharmaceutical Sciences*, vol. 22, no. 3, pp. 301–302, 2009.

[10] V. N. Desai, O. E. Afieroho, B. O. Dagunduro, T. J. Okonkwo, and C. C. Ndu, "A simple UV spectrophotometric method for the determination of levofloxacin in dosage formulations," *Tropical Journal of Pharmaceutical Research*, vol. 10, no. 1, pp. 75–79, 2011.

[11] FDA, *ICH-Q1A (R2): Stability Testing of New Drug Substances and Products*, vol. 68, U S Food and Drug Administration, Washington, DC, USA, 2nd edition, 2003.

[12] FDA, *ICH-Q1B: Photo-Stability Testing of New Drug Sub-Stances and Products*, vol. 62, U S Food and Drug Administration, Washington, DC, USA, 1997.

[13] FDA, *ICH-Q2 (R1): Validation of Analytical Procedures: Text and Methodology*, vol. 60, U S Food and Drug Administration, Washington, DC, USA, 1995.

[14] United State Pharmacopoeia, *The U.S. Pharmacopeia Convention*, United State Pharmacopoeia, Rockville, Md, USA, 30th edition, 2007.

15

Pharmacokinetics of Single-Dose and Multi-Dose of Lovastatin/Niacin ER Tablet in Healthy Volunteers

Yan-yan Jia, Song Ying, Chen-tao Lu, Jing Yang, Li-kun Ding, Ai-dong Wen, and Yan-rong Zhu

Department of Pharmacy, Xijing Hospital of the Fourth Military Medical University, Xi'an 710032, China

Correspondence should be addressed to Yan-yan Jia, jiayanyan-2004@hotmail.com and Ai-dong Wen, adwen-2004@hotmail.com

Academic Editor: Meehir Palit

An extended-release (ER) niacin and lovastatin fixed-dose combination has been developed for the treatment of primary hypercholesterolemia and mixed dyslipidemia. The purpose of the present study was to examine the drug interaction between niacin and lovastatin after multi-dose oral administration of lovastatin/niacin ER combination in healthy Chinese volunteers. A single-center, randomized, open-label, 5-period crossover study was conducted in thirty healthy volunteers aged 18 to 45 years with a washout period of 8 days. Subjects were randomized to receive multiple doses of treatment A (1 500 mg niacin ER tablet), B (1 20 mg lovastatin tablet), C (1 20 mg lovastatin and 500 mg niacin-ER tablet), D (2 10 mg lovastatin and 350 mg niacin-ER tablets) or E (2 10 mg lovastatin and 500 mg niacin-ER tablets) in 1 of 5 sequences (ABCDE, BCDEA, CDEAB, DEABC, EABCD) per period. Lovastatin, niacin and its metabolites (nicotinuric acid and nicotinamide) were determined in plasma by LC/MS method. Pharmacokinetic parameters were calculated, and least square mean ratios and 90% confidence intervals for C_{max} and $AUC_{(0-24)}$ were determined for lovastatin/niacin ER versus niacin ER or lovastatin. It revealed that the formulation had no potential drug interaction in healthy Chinese volunteers when the dosage was increased from 500 mg to 1000 mg.

1. Introduction

Niacin (nicotinic acid, 3-pyridine-carboxylic acid, NA), which belongs to the hydrophilic vitamin B complex, is widely used to treat a diverse range of lipid disorders and prevent clinical CVD [1]. It is well known for its effects in reducing total cholesterol, triglycerides (TGs), very low-density lipoprotein (VLDL), low-density lipoprotein (LDL), and lipoprotein (a) (L (pa)), and increasing high-density lipoprotein (HDL) level. NA is metabolized in two pathways: the first is the metabolic route to nicotinuric acid (NUA) through nicotinyl CoA by glycine conjugation, and the second is that to nicotinamide (NAM), which is utilized in NAD synthesis [2]. Lovastatin, as a fungal antibiotic, is a member of the drug class of statins and a specific and nonreversible competitive inhibitor of HMG-CoA reductase, used for lowering cholesterol (hypolipidemic agent) in the patients with hypercholesterolemia and so preventing cardiovascular disease [3]. Many clinical studies have shown that the combination tablet of extended-release (ER) niacin and lovastatin decreases LDL-C and increases HDL-C greater than either treatment alone in patients with dyslipidemia [4–6].

Moreover, lovastatin was primarily metabolized by the cytochrome P450 isoenzyme, especially CYP3A4, with less than 10% being excreted renally [7]. And an in vitro study indicated that NA and NUA inhibited CYP2D6 and NA inhibited CYP3A4, which was responsible for lovastatin metabolism [8]. Genetic variation in those isoenzymes has been surveyed in an ethnically diverse population [9, 10]. Previous work has observed the lack of a pharmacokinetic interaction between niacin and lovastatin after single-dose administration in healthy Hispanic volunteers [11].

The most common adverse events of niacin and lovastatin were flushing, itch of skin, headache, abdominal pain, malaise, dyspepsia, nausea, and hepatic toxicity [3, 4]. And several clinical trials showed that the rates of adverse event with the ER niacin/lovastatin tablet were similar to those with the ER niacin or the lovastatin in Caucasian patients [12–14].

However, it is still unknown whether potential increased risks of adverse events or a pharmacokinetic interaction of lovastatin and niacin exist in Chinese people.

2. Experimental

2.1. Chemicals and Reagents. Lovastatin/niacin ER tablets were supplied by Kangde Pharmaceutical Group Co. Ltd., (Zhejiang, China). Niacin extended-release tablet was purchased from Bejing Second Pharmaceutical Co. Ltd., (China). Lovastatin tablet was obtained from Hisun Pharmaceutical Co. Ltd. (Zhejiang, China). Chemical reference substances of NA, NAM, lovastatin, and simvastatin were obtained from the National Institute for the Control of Pharmaceutical and Biological Products (Beijing, China). NUA was purchased from Sigma (St. Louis, MO), and 6-methyl nicotinic acid as an internal standard was supplied by Aldrich. HPLC grade methanol was obtained from Merck (Darmstadt, Germany), and other chemical reagents were of analytical grade, obtained from Nanjing Chemical Reagent Co., Ltd.(Nanjing, China). Water was deionized and purified by using a Milli-Q system (Millipore, Milford, MA, USA) and was used to prepare all aqueous solutions.

2.2. LC-MS Instrumentation and Analytical Conditions. An agilent (Agilent Technologies, USA) 1100 series LC system equipped with degasser and agilent 1100 MS was used to detect NA and lovastatin during the study. For detecting NA, NAM, and NUA [15], seperation was carried out using a Dikma-C_{18} column (dp 3 μm, 2.1 $\times$ 150 mm ID, Dikma Technologies Inc.), with an isocratic elution system consisting of methanol (containing 0.1% acetic acid) and water (containing 0.3% isopropanol) (2/98, v/v) at a flow rate of 0.2 mL/min. Ion scan mode was with the following settings: the capillary voltage, 100 V; temperature, 350°C; drying gas, 600 L/h; nebulizer pressure, 40 psig. Quantitations of NA, NAM, NUR, and 6-methyl nicotinic acid were achieved by monitoring the ions at $[M + H]^+$, m/z 124.1, 123.1, 181.1, and 138.1, respectively (Figure 1). For detecting lovastatin [16], seperation was carried out using a Lichrospher C_{18} (dp 5 μm), 200 mm $\times$ 4.6 mm ID with a gradient elution system consisting of methanol and water (containing 50 μmol/L sodium acetate) (see Table 1) at a flow rate of 1 mL/min. The column temperature was kept at 25°C. Detection was performed by mass spectrometer (MS) in positive ion mode. Ion scan mode was with the following settings: the capillary voltage, 140 V; temperature, 350°C; drying gas, 600 L/h; nebulizer pressure, 40 psig. Lovastatin and simvastatin, ions at $[M + Na]^+$ m/z 427.2 and 441.3 (Figure 2), were monitored, respectively. All data were collected and analyzed using Agilent Chemstation software.

2.3. Study Design. The pharmacokinetics of Lovastatin/Niacin ER tablet was studied in healthy Chinese subjects in accordance with the Declaration of Helsinki for biomedical research involving human subjects and Good Clinical Practice. The protocol and associated informed consent statements were reviewed and approved by the Committee on Human Rights Related to Human Experimentation, Xijing

TABLE 1: Mobile phrase program for lovastatin.

Time (min)	Water phase (B%)	Flow rate (mL/min)
0	12	1
6.5	12	1
6.6	0	1
10.6	0	1
10.7	12	1
16.0	12	1

Hospital, and the informed consent statements were signed by the volunteers. It was a single-center, randomized, open-label, crossover study with five treatment cycles separated by an eight-day washout cycle. Thirty healthy volunteers who aged from 18 to 45, body mass index (BMI) ranged 19 and 24 Kg/m², were enrolled in this study. All volunteers have passed an obtaining of complete medical history and physical examination before participate in the study. All subjects were fasted for at least 8 hours at last night before our study and were confirmed abstinence from other medications, alcohol, tobacco, and caffeinated products.

The subjects were randomly allocated into five groups (each group have 3 male and 3 female). Each group was randomized to receive multi-dose of treatment A (500 mg niacin ER, one tablet), B (20 mg lovastatin, one tablet), C (one lovastatin/niacin ER tablet (500/20)), D (two lovastatin/niacin ER tablet (350/10)), or E (two lovastatin/niacin ER tablet (500/10)) in 1 of 5 sequences (ABCDE, BCDEA, CDEAB, DEABC, EABCD) per period. Blood samples were collected in heparinized tubes before dosing at days 1, 4, 5, 6, and 7, and on the 1st and 7th day, and blood samples were also collected at 30, 60, and 90 minutes 2, 3, 4, 6, 5, 8, 10, 12, 15, and 24 hours after dosing. All samples were separated immediately by centrifugation at 3500 rpm for 10 min at 4°C and stored at −80°C until analysis.

2.4. Analytical Procedures

2.4.1. Preparation of Stock Solutions and Standard. Stock solutions of NA, NAM, and NUA were prepared by dissolving the drugs in methanol at the concentrations of 0.492, 0.508, 0.0993 mg/mL, respectively. Serial (working) dilutions of NA NAM, and NUA were prepared from the stock solutions by appropriate dilution with methanol at the concentrations of 98.4, 9.84, 0.984, 0.0984 μg/mL for NA; 102, 10.2, 1.02, 0.102 μg/mL for NAM; 9.93, 0.993, 0.0993 μg/mL for NUA, respectively. Stock solution of lovastatin was prepared by dissolving 10.34 mg drug in methanol at a concentration of 1.034 mg/mL. Serial (working) dilutions of lovastatin were prepared with methanol at the concentrations of 103.4, 10.34, 1.034, 0.01034, 0.001034 μg/mL, respectively. Stock solutions of internal standards (IS) were prepared by dissolving the drug in methanol at concentrations of 0.500 mg/mL for 6-methyl nicotinic acid and 1.052 mg/mL for simvastatin. Working solutions of IS were prepared with methanol at concentrations of 2.0 μg/mL for 6-methyl nicotinic acid and 0.2104 μg/mL for simvastatin. All the stock

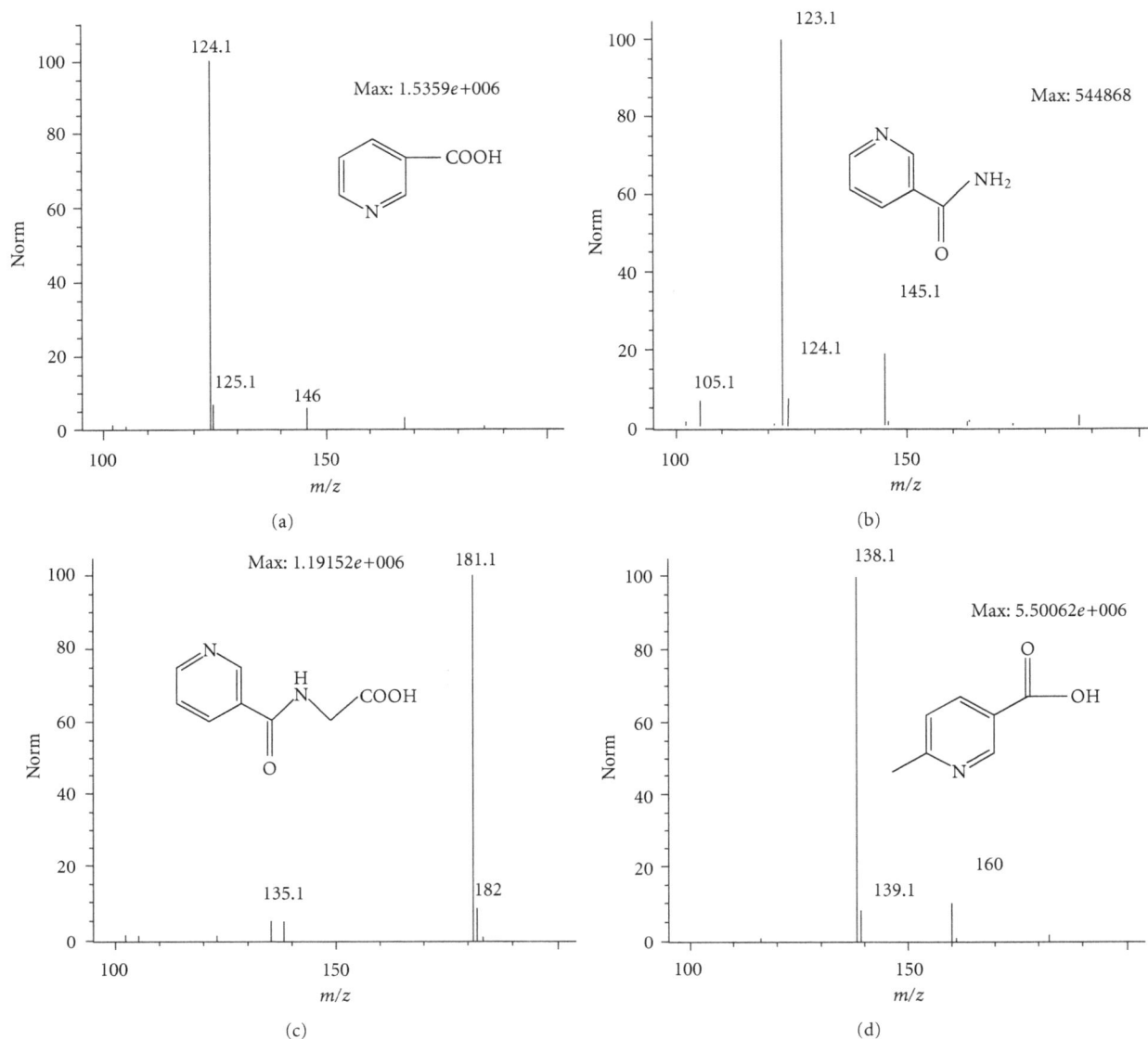

FIGURE 1: Electrospray product ion mass spectrum of the precursor ion of NA (a), NAM (b), NUA (c), and internal standard 6-methyl nicotinic acid (d).

and working solutions were stored at −20°C and prepared for calibration curve and quality controls.

2.4.2. Sample Preparation. To determine the niacin and its metabolites, 50 μL 6-methyl nicotinic acid (2.00 μg/mL, IS) solution was added into 100 μL plasma sample and vortex-mixed for 30 s and then 0.6 mL methanol was added and vortex-mixed for 3 min. After centrifugation (16000 r/min, 6 min), the upper organic layer was separated and evaporated to dryness using a gentle stream of nitrogen. The residuum was reconstituted using the mobile phase and centrifugated 6 min at 16000 r/min. A 5 μL supernatant was autoinjected into the LC/MS system for analysis.

For lovastatin, 1 mL plasmas sample and 75 μL simvastatin solution (0.2104 μg/mL, IS) were accurately added into 10 mL centrifuge tube, and vortex-mixed adequately, then 5 mL redistillate acetidin was added, and, after centrifugation (4000 r/min, 10 min), the upper organic layer was evaporated to dryness using nitrogen in a water bath at 30°C. The residuum was dissolved with 200 μL mobile phase solution, and, after centrifugation (16000 r/min, 6 min), a 20 μL supernatant was transferred into the LC/MS system for analysis.

2.4.3. Calibration Curve. Calibration curves were prepared at the concentration levels of 0.00492, 0.0148, 0.0295, 0.0984, 0.295, 0.984, 1.97, 4.92, and 9.84 μg/mL for NA; 0.00508, 0.0152, 0.0305, 0.102, 0.305, 1.02, 2.03, 5.08, and 10.2 μg/mL for NAM; 0.00497, 0.0149, 0.0298, 0.0993, 0.298, 0.993, 1.99, 4.96, and 9.93 μg/mL for NUA; 0.0517, 0.1551, 0.3102, 1.034, 3.102, 10.34, 20.68, and 41.36 ng/mL for lovastatin by spiking an appropriate amount of the standard solutions

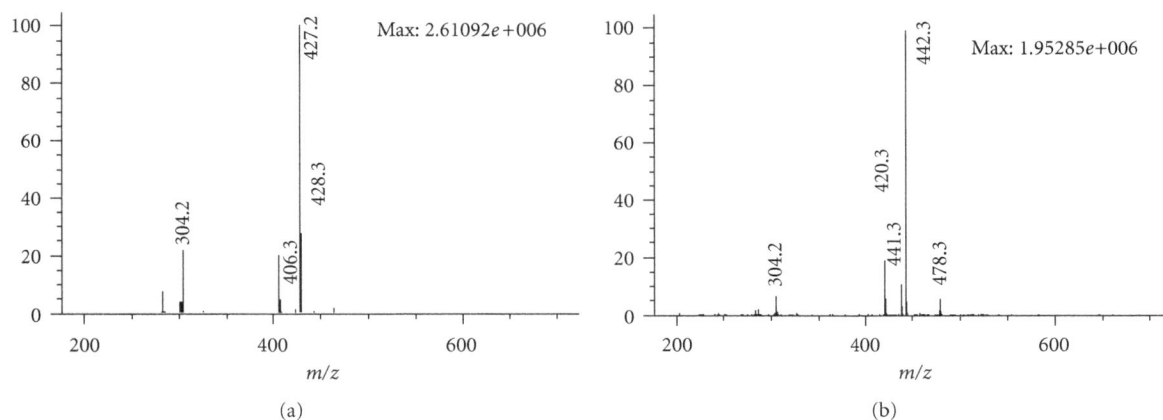

FIGURE 2: Electrospray product ion mass spectrum of the precursor ion of Lovastatin (a), and internal standard simvastatin (b).

in 1 mL blank plasma. The calibration curve was prepared and assayed along with quality control (QC) samples. QC samples were prepared in 1 mL blank plasma at three levels of 0.00984, 0.246, and 8.86 μg/mL for NA; 0.0102, 0.254, and 9.14 μg/mL for NAM; 0.00993, 0.248, and 8.94 μg/mL for NUA; 0.1034, 2.585, 36.19 ng/mL for lovastatin, respectively. The plasma samples were stored at −20°C.

2.4.4. Specificity. The specificity of the method was tested by screening six different batches of blank human plasma. Each blank sample was tested for interferences in the MS channels using the proposed extraction procedure and chromatographic/MS conditions, and the results were compared with those obtained for water solution of the analytes at a concentration near to the lower limit of quantification (LLOQ).

2.4.5. Precision and Accuracy. The intrarun precisions and accuracies were estimated by analyzing five replicates containing NA, NAM, NUA, and lovastatin at three different QC levels. The interrun precisions were determined by analyzing QC samples on three different runs. The criteria for acceptability of the data included accuracy within ±15% deviation (DEV) from the nominal values and a precision of within ±15% relative standard deviation.

2.4.6. Extraction Recovery. The recoveries of NA, NAM, NUA, and lovastatin were determined by comparing the peak area obtained for QC samples that were subjected to the extraction procedure with those obtained from blank plasma extracts that were spiked after extraction to the same nominal concentrations (0.00984, 0.246, and 8.86 μg/mL for NA; 0.0102, 0.254, and 9.14 μg/mL for NAM; 0.00993, 0.248, and 8.94 μg/mL for NUA; 0.1034, 2.585, 36.19 ng/mL for lovastatin).

2.4.7. Stability. The stability of NA, NAM, NUA, and lovastatin in plasma under different temperature and timing conditions was evaluated. Plasma samples were subjected to short-term conditions, to long-term storage conditions (−20°C), and to three freeze-thaw stability studies. The autosampler stability was conducted by reanalyzing extracted

samples kept under the autosampler conditions for 0 and 48 h. All the stability studies were conducted at two concentration levels with three determinations for each.

2.5. Pharmacokinetic Analysis. The noncompartmental model analysis was used in the data processing of NA, NUA, NAM, and lovastatin. The maximum and minimum observed serum concentrations at steady state (C_{max-ss}, C_{min-ss}) and time to C_{max-ss} (T_{max-ss}) were taken from raw data. Ke was determined by linear regression of the terminal linear portion of the concentration-time curve, and $T_{1/2}$ was calculated as $\ln(2)/Ke$. AUC_{ss} (steady-state area under the curve during T (dosing interval) and AUC_{inf} (steady-state area under the curve from 0 to infinity) were calculated by the linear trapezoidal rule. C_{av} (mean concentration between 2 administrations) was calculated as AUC_{ss}/T. The degree of fluctuation (DF) value was calculated as $(C_{max} − C_{min})/C_{av} \times 100\%$ and actual accumulation factor $R = AUC_{ss}/AUC_{inf}$. Clearance (CL/F) was calculated as dose/AUC_{inf}.

2.6. Safety Evaluation. Safety assessments included the recording of all adverse events, vital signs (blood pressure and heart rate), 12-lead electrocardiograms (ECG), laboratory investigations (including biochemistry, haematology, coagulation, and urinalysis), and full physical examinations.

3. Results and Discussion

3.1. Method Validation

3.1.1. Specificity and Selectivity. Good selectivity was observed, and there was no significant interference or ion suppression from endogenous substances observed at the retention time of the analytes. The retention time of NA, NAM, NUA, and 6-methyl nicotinic acid were 4.2, 7, 12, and 4.8 min, respectively. The retention time of lovastatin and simvastatin were 5.6 and 6.6 min, respectively.

3.1.2. Calibration Curve. NA and NAM can be both detected in the blank plasma samples therefore, the background level

TABLE 2: Intra- and interrun precision and accuracy of determination of NA, NAM, NUA, and lovastatin in human plasma.

| | Concentration | | RSD (%) | |
	Nominal	Found C	Intrarun ($n = 5$)	Interrun ($n = 15$)
NA (μg/mL)	0.00984	0.0101	6.0	12.7
	0.246	0.237	4.2	8.3
	8.86	8.71	3.7	8.5
NAM (μg/mL)	0.0102	0.0103	5.5	12.0
	0.254	0.271	3.5	9.3
	9.14	9.58	3.2	13.5
NUA (μg/mL)	0.00993	0.0103	4.8	2.0
	0.248	0.268	2.3	11.3
	8.94	9.52	2.3	13.0
Lovastatin (ng/mL)	0.1034	0.1051	8.2	14.9
	2.585	2.688	5.0	8.6
	36.19	37.09	6.6	9.7

of peak area of NA and NAM will be deducted during the analysis process. Calibration curves of NA, NAM, and NUA in plasma were validated over the concentration ranges of 0.00492–9.84 μg/mL, 0.00508–10.2 μg/mL, and 0.00497–9.93 μg/mL, respectively. The r^2 values for the calibration curves were >0.99. Typical equations of calibration curves were as follows: $y = 0.363 \times C$-0.000002 ($r^2 = 0.9995$, $n = 5$) for NA, $y = 1.17 \times C$-0.00107 ($r^2 = 0.9998$, $n = 5$) for NAM, and $y = 1.07 \times C$-0.000235 ($r^2 = 0.9997$, $n = 5$) for NUA, respectively. Calibration curves of lovastatin in plasma were validated over the concentration ranges of 0.0517–41.36 ng/mL. The limit of quantification (LLOQ), defined as the lowest concentration on the standard curve that can be measured with acceptable accuracy and precision (<20%), was established at 0.00492 μg/mL for NA, 0.00508 μg/mL for NAM, 0.00497 μg/mL for NUA, and 0.0517 ng/mL for lovastatin, respectively (single-to-noise, S/N $\geq$ 10).

3.1.3. Accuracy and Precision. The intra- and interrun precision and accuracy of the assay were assessed by running a single batch of samples containing a calibration curve and five replicates at each QC level. The precision was calculated by using one-way ANOVA. The results, which were summarized in Table 2, demonstrated that the precision and accuracy values were within the acceptable range and the method was accurate and precise.

3.1.4. Extraction Recovery and Matrix Effects. The extraction recoveries of the four analytes were NA 95.2 ± 3.2%, 93.8 ± 2.3%, and 96.7 ± 2.9% at the concentrations of 0.00984, 0.246, and 8.86 μg/mL, respectively; NAM 105.6 ± 4.2%, 94.1 ± 4.5%, and 98.5 ± 4.3% at the concentrations of 0.0102, 0.254, and 9.14 μg/mL, respectively; NUA 92.1 ± 7.6%, 95.7 ± 3.3% and 101.7 ± 5.2% at the concentrations of 0.00993, 0.248 and 8.94 μg/mL respectively; NUA 92.1 ± 7.6%, 95.7 ± 3.3%, and 101.7 ± 5.2% at the concentrations of 0.00993, 0.248, and 8.94 μg/mL respectively; lovastatin 95.5 ± 3.31%, 87.8 ± 1.25%, and 82.8 ± 1.81% at the concentrations of 0.1034, 2.585, 36.19 ng/mL, respectively.

The matrix effect was defined as the direct or indirect alteration or interference in respond due to the presence of unintended or other interfering substances in the samples. It was evaluated by comparing the peak area of the analytes (background subtraction for NA and NAM) dissolved in the blank plasma sample's reconstituted solution (the final solution of the blank plasma after extraction and reconstitution) with that dissolved in mobile phase. Three different concentration levels of analytes were evaluated by analyzing five samples at each level, and the blank plasma used in this study was from five different batches of blank plasma. If the peak area ratio is less than 85% or more than 115%, a matrix effect will be implied. In this study, the peak area ratios of the analytes were NA 87.2 ± 1.6%, 107.8 ± 4.6%, and 89.3 ± 4.7%, NAM 99.1 ± 10.8%, 96.8 ± 3.1% and 104.6 ± 2.9%, NUA 104.9 ± 4.0%, 97.1 ± 3.4%, and 98.2 ± 7.7%, respectively, at concentrations of 0.01, 0.25, and 9 μg/mL; 6-methyl nicotinic acid 95.5 ± 10.5% at the concentration of 2.0 μg/mL; lovastatin 94.5 ± 3.21%, 89.8 ± 1.35%, and 92.8 ± 2.87% at the concentrations of 0.10, 2.6, 36.2 ng/mL, respectively; simvastatin 94.5 ± 7.8% at the concentration of 0.2 μg/mL. The results showed that there was no matrix effect of the analytes and IS from the matrix of plasma in this study.

3.1.5. Stability. The stability of NA, NAM, NUA, and lovastatin in plasma was determined by assessing low- and high-QC samples ($n = 3$ for each concentration). The results are summarized in Table 3. All analytes were found to be stable in plasma samples for at least 12 h at room temperature, for 2 months at −20°C freezing condition, and following three freeze-thaw cycles.

3.2. Pharmacokinetics

3.2.1. NA, NAM, and NUA Plasma Analysis. Mean pharmacokinetic parameters of the treatment A and C were provided in Table 4 for NA, NAM, and NUA. Mean plasma concentration versus time profiles for NA, NAM, and NUA were presented in Figure 3.

TABLE 3: Stability of NA, NAM, NUA, and lovastatin under different storage conditions ($n = 3$).

Storage conditions	Drug	Concentration	
		Nominal	Mean found C
Stability at room temperature for 8 h	NA	0.00984	0.00904 ± 0.00059
		8.86	8.51 ± 0.10
	NAM	0.0102	0.00948 ± 0.00007
		9.14	9.11 ± 0.07
	NUA	0.00993	0.0105 ± 0.0007
		8.94	9.81 ± 0.13
	Lovastatin	0.1034	0.09808 ± 0.0117
		36.19	37.15 ± 3.08
Stability at −80°C for 2 months	NA	0.00984	0.00904 ± 0.00031
		8.86	0.00953 ± 0.00052
	NAM	0.0102	0.00925 ± 0.00072
		9.14	8.49 ± 0.18
	NUA	0.00993	8.53 ± 0.17
		8.94	8.79 ± 0.12
	Lovastatin	0.1034	0.1014 ± 0.0009
		36.19	34.10 ± 1.81
Freeze-thaw stability	NA	0.00984	0.00872 ± 0.00036
		8.86	7.98 ± 0.13
	NAM	0.0102	0.0109 ± 0.0002
		9.14	8.76 ± 0.22
	NUA	0.00993	0.0108 ± 0.0003
		8.94	8.46 ± 0.24
	Lovastatin	0.1034	0.09911 ± 0.0051
		36.19	34.71 ± 1.39
Autosampler stability at 4°C for 24 h	NA	0.00984	0.00955 ± 0.00096
		8.86	8.86 ± 0.16
	NAM	0.0102	0.0100 ± 0.0001
		9.14	9.58 ± 0.30
	NUA	0.00993	0.0100 ± 0.0010
		8.94	9.77 ± 0.139
	Lovastatin	0.1034	0.09931 ± 0.00437
		36.19	40.19 ± 0.31

For the NA study, we found there was significant difference between C_{max}, AUC_{0-24}, and T_{max} for NA on the multiple dose of the treatment A or C, compared with single-dose of treatment A (500 mg niacin ER tablet) or the treatment C (one lovastatin/niacin ER tablet (500/20)). And a higher C_{max} and AUC_{0-24} and longer T_{max} of NA were obtained for the multi-dose treatment A or C. And, for the NAM pharmacokinetic study, the mean NAM C_{max} and AUC_{0-24} values were about 3 times higher for multi-dose administration of 500 mg niacin ER tablet (treatment A or C), comparing with single-dose of 500 mg niacin ER tablet (treatment A or C). It was indicated the metabolism of NA and NAM may exit the accumulation phenomenon in human body. However, there was no significant statistical difference ($P > 0.05$) in the main pharmacokinetic parameters of NA, NAM, NUA

(C_{max-SS}, C_{av}, AUC_{0-24}, T_{max}) between the two treatments (Figure 3). It was suggested that niacin had similar drug delayed release behavior in two treatments, and lovastatin had no effect on pharmacokinetic character of NA. The results initially indicated that no drug interaction existed between NA and lovastatin after multiple oral administration of lovastatin/niacin ER tablet in healthy Chinese volunteers.

For the multi-dose NA pharmacokinetic study, the mean pharmacokinetic parameters of NA, NAM, and NUA after multi-dose three different formulations (treatment C, D, E) were present in Table 5. The NA C_{max} and AUC_{0-24} were appropriately 30 times higher, when the dose was changed from 500 mg to 750 mg, but no big difference when the dose was changed from 750 mg to 1000 mg. It was indicated that there was a liver enzyme saturation phenomenon in NA

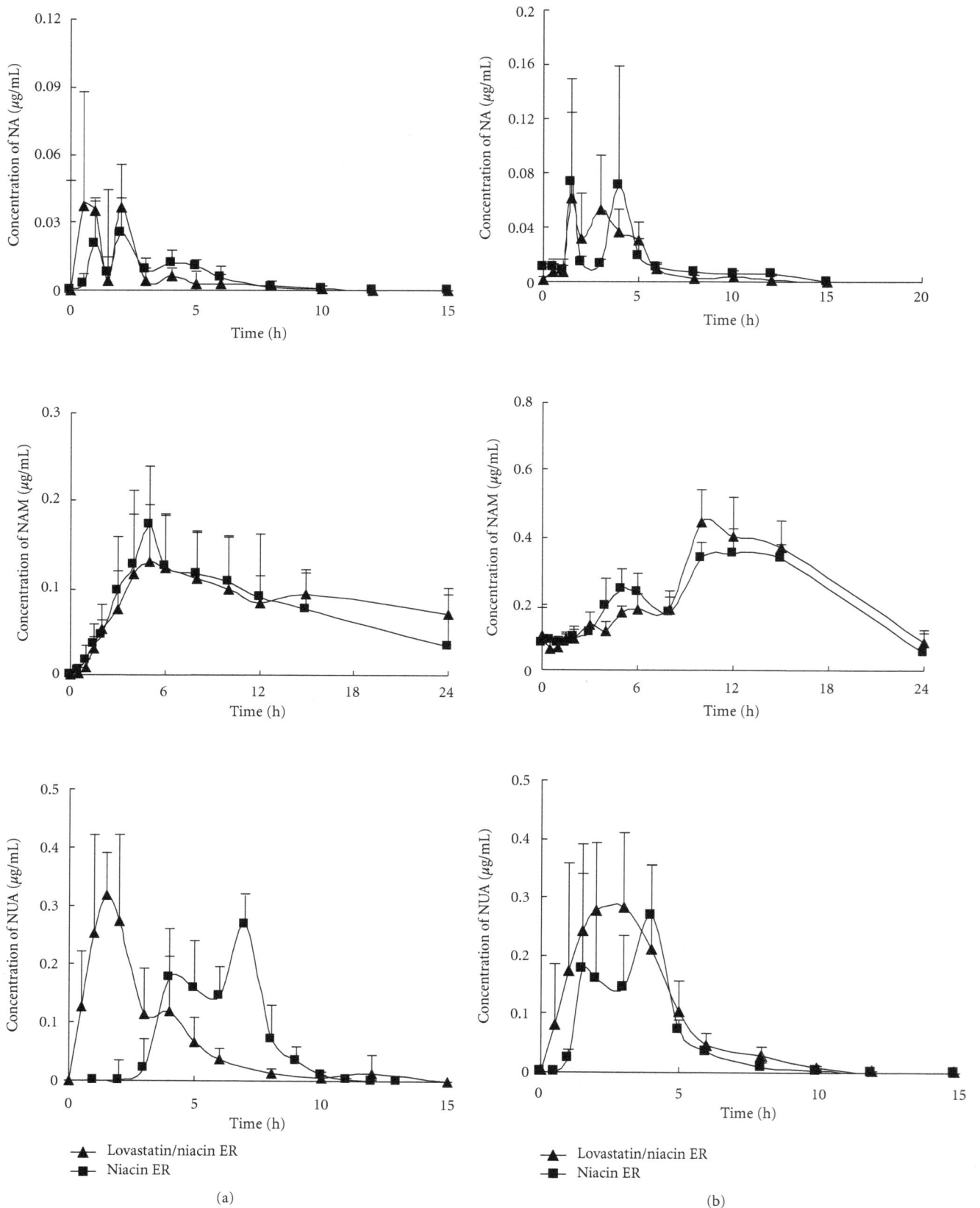

FIGURE 3: Plasma concentration-time curve of NA and NAM and NUA in 10 healthy volunteers after dose administration of lovastatin/niacin ER (20 mg/500 mg) and niacin ER (500 mg) ((a) single-dose; (b) multi-dose).

TABLE 4: The pharmacokinetic variables of NA, NAM, and NUA in volunteers after single- and multi-dose administration of the treatment A (niacin ER tablet, 500 mg) or the treatment C (lovastatin/niacin ER tablet, 20 mg/500 mg) (mean ± SD).

Treatment			NA			NAM			NUA			
			T_{max} (h)	C_{max} (μg/mL)	AUC$_{0-24}$ (μg·h/mL)	T_{max} (h)	C_{max} (μg/mL)	AUC$_{0-24}$ (μg·h/mL)	T_{max} (h)	C_{max} (μg/mL)	AUC$_{0-24}$ (μg·h/mL)	$t_{1/2}$ (h)
C	Single	Mean	1.1	0.079	0.104	7.2	0.148	2.09	1.5	0.404	0.94	1.9
		±s	0.8	0.061	0.067	4.3	0.061	0.79	0.4	0.114	0.32	0.6
	Multi	Mean	2.6	0.097	0.221	10.2	0.448	5.81	2.4	0.375	1.19	1.4
		±s	1.1	0.043	0.055	0.6	0.104	1.29	1.1	0.127	0.38	0.2
A	Single	Mean	2.1	0.0392	0.0761	5.1	0.176	1.82	2.8	0.232	0.743	1.8
		±s	1.6	0.0253	0.0385	0.6	0.068	1.10	1.4	0.062	0.217	0.4
	Multi	Mean	3.3	0.155	0.221	11.6	0.375	5.31	3.15	0.340	0.781	1.3
		±s	1.3	0.121	0.157	3.1	0.050	1.01	1.1	1.18	0.205	0.07
	Single	C/A		1.06	1.14		0.96	1.08		1.12	1.09	
		90%		0.94, 1.21	0.98, 1.29		0.80, 1.25	0.92, 1.24		0.96, 1.30	0.94, 1.24	
	Multi	C/A		1.03	1.06		1.09	1.10		1.11	1.06	
		90%		0.88, 1.22	0.96, 1.16		0.94, 1.26	0.99, 1.20		0.81, 1.21	0.93, 1.19	

TABLE 5: The pharmacokinetic variables of NA, NAM, and NUA in volunteers of group I after multi-dose administration of three different dosage of lovastatin/niacin ER.

Lovastatin/niacin ER	Variable	NA			NAM			NUA		
		AUC$_{0-24}$ (μg·h·L^{-1})	C_{max} (μg/mL)	C_{av} (μg/mL)	AUC$_{0-24}$ (μg·h·L^{-1})	C_{max} (μg/mL)	C_{av} (μg/mL)	AUC$_{0-24}$ (μg·h/L)	C_{max} (μg/mL)	C_{av} (μg/mL)
20 mg/500 mg	Mean	0.221	0.097	0.00919	5.81	0.448	0.242	1.19	0.375	0.0496
	±s	0.055	0.043	0.00227	1.29	0.104	0.054	0.38	0.127	0.0159
20 mg/750 mg	Mean	7.15	3.54	0.298	7.16	0.804	0.298	4.31	1.38	0.179
	±s	3.76	1.90	0.157	1.26	0.168	0.052	0.96	0.35	0.037
20 mg/1000 mg	Mean	5.36	2.73	0.223	7.13	0.730	0.297	4.44	1.13	0.185
	±s	2.96	1.45	0.123	1.73	0.202	0.072	1.21	0.21	0.050

metabolism [11, 12] in Chinese healthy volunteers at the range of 500–1000 mg NA.

3.2.2. Lovastatin Plasma Analysis. For lovastatin study, mean pharmacokinetic parameters of lovastatin (treatment B and C) were provided in Table 6. Mean plasma concentrations versus time profiles for lovastatin were presented in Figure 4. Lovastatin was eliminated with a half-life of approximately 5 h, and peak plasma concentrations of lovastatin were reached within 0.7–1.8 h. These findings are in agreement with the previously reported pharmacokinetic parameters of lovastatin [16–20]. For both treatment B and C comparisons, the ratios ranged from 101% to 124% for C_{max}, 104% to 126% for AUC$_{(0-24)}$. The 90% CI for the ratios of both comparisons was within the 80% to 126% interval. It was indicated that there was no drug interaction between monotherapy and coadministration.

For multi-dose pharmacokinetics study of lovastatin, mean pharmacokinetic parameters of lovastatin (treatment C, D, and E) were provided in Table 7. No significant statistical difference was observed in the pharmacokinetic parameters of lovastatin among the three treatments. The results suggested that different doses of NA have no effect on the pharmacokinetic character of lovastatin and indicated that no drug interaction existed between NA and lovastatin after multiple oral administration of lovastatin/niacin ER tablet in healthy Chinese volunteers.

3.3. Adverse Events. Some subjects in the treatment A (1 subject), C (1 subjects), D (3 subjects), E (4 subjects) reported the adverse events of erubescence, slight fever, pruritus on the skin or mild stomach discomfort. Overall, all adverse events were mild and the volunteers recovered without treatment.

More detail information needs to be collected and analyzed from the phase-two clinical trials of the lovastatin/niacin ER tablet in Chinese patients.

4. Discussion and Conclusion

The study was completed with a sufficient number of subjects to meet the PK objectives. Although there are not observed mean PK differences between monotherapy and coadministration, the overall variability of the study was relatively high,

TABLE 6: The pharmacokinetic variables of lovastatin in volunteers after single-dose administration of lovastatin/niacin ER (20 mg/500 mg) and lovastatin (20 mg) (mean $\pm$ SD).

Variable	Lovastatin/niacin ER (treatment C)		lovastatin (treatment B)	
	Single-dose	Multi-dose	Single-dose	Multi-dose
T_{max} (h)	1.3 ± 0.5	1.3 ± 0.6	1.4 ± 0.2	1.5 ± 0.3
C_{max} (ng/mL)	10.94 ± 3.51	10.25 ± 3.12	9.54 ± 4.47	8.45 ± 2.77
AUC_{0-24} (ng·h/mL)	26.97 ± 6.2	29.85 ± 10.25	27.78 ± 10.33	26.86 ± 8.53
$t_{1/2}$ (h)	3.9 ± 0.8	4.6 ± 1.4	3.8 ± 1.5	5.1 ± 0.5
	Single (C/B, 90%)		Multi (C/B, 90%)	
C_{max}	1.12, (1.00, 1.24)		1.14, (1.04, 1.24)	
AUC_{0-24}	1.13, (1.01, −1.24)		1.15, (1.04, 1.25)	

TABLE 7: The pharmacokinetic variables of lovastatin in volunteers after multi-dose administration of three different formulations of lovastatin/niacin ER (mean $\pm$ SD).

Variable	One Lovastatin/niacin ER tablet (500/20)	Two Lovastatin/niacin ER tablets (375/10)	Two Lovastatin/niacin ER tablets (500/10)
T_{max} (h)	1.3 ± 0.6	0.9 ± 0.3	1.3 ± 0.4
C_{max} (ng/mL)	10.25 ± 3.12	10.45 ± 3.85	12.67 ± 6.09
C_{min} (ng/mL)	0.02719 ± 0.02879	0.03676 ± 0.02553	0.05462 ± 0.02128
C_{av} (ng/mL)	1.244 ± 0.427	1.389 ± 0.411	1.552 ± 0.567
DF	8.42 ± 1.48	7.51 ± 1.37	7.98 ± 1.39
AUC_{0-24} (ng·h/mL)	29.85 ± 10.25	33.33 ± 9.86	37.25 ± 13.61
$t_{1/2}$ (h)	4.6 ± 1.4	4.7 ± 1.1	5.8 ± 1.5
K (h^{-1})	0.167 ± 0.065	0.155 ± 0.043	0.127 ± 0.035
CL/F (L/h)	742.8 ± 247.2	667.3 ± 269.4	616.8 ± 263.6
V_d/F (L)	5001.4 ± 2196.6	4556.9 ± 1911.9	5122.4 ± 2488.6

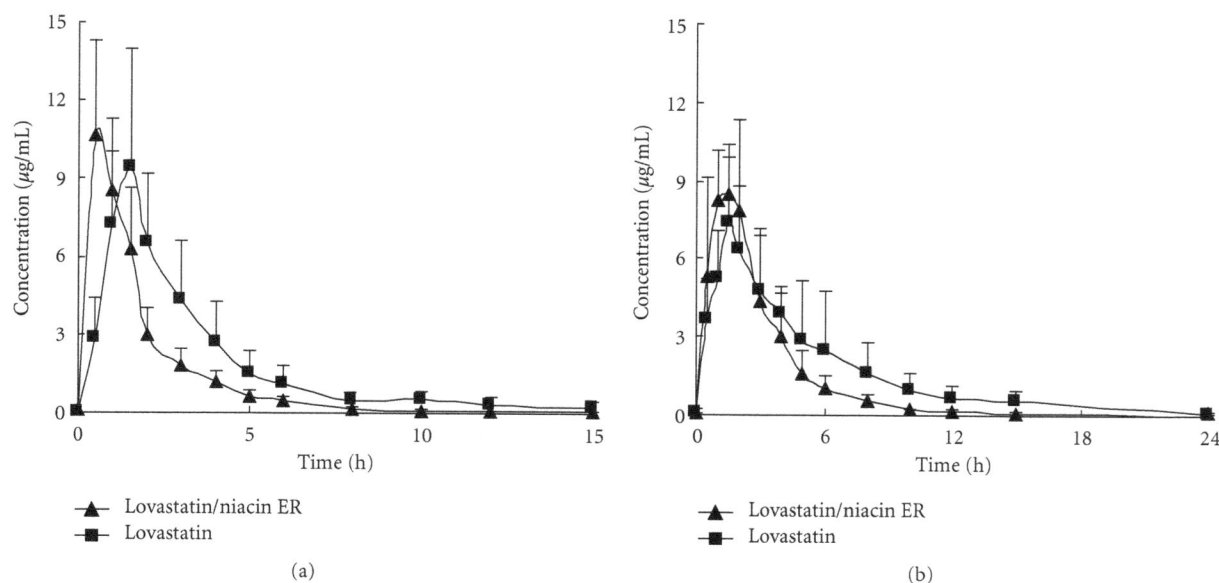

FIGURE 4: Plasma concentration-time curve of lovastatin after single- and multiple-dose administration of lovastatin/niacin ER (500 mg/20 mg) and lovastatin (20 mg) (mean $\pm$ SD) ((a) single-dose; (b) multi-dose).

particularly during ER niacin coadministration. This high variability in conjunction with the small number of subjects and apparently little or no effect on the exposure to the parent drugs makes it difficult to establish cause relationships for the potential interactions.

Lovastatin and ER-niacin in a fixed-dose combination (Advicor) is approved for the treatment of dyslipidemia [12]. In single-dose studies of ADVICOR, rate and extent of niacin and lovastatin absorption were bioequivalent under fed conditions to that from NIASPAN (niacin extended-release tablets) and Mevacor (lovastatin) tablets, respectively. After administration of two ADVICOR 1000 mg/20 mg tablets, peak concentrations averaged about 18 μg/mL and occurred about 5 hours after dosing. And peak lovastatin concentrations averaged about 11 ng/mL and occurred about 2 hours after dosing. It was shown that coadministration of NA and lovastatin did not significantly influence C_{max} and AUC_{0-t} of lovastatin, NA, NUA, and total urinary recovery of niacin and metabolites. Although both drugs are extensively metabolized, genetic variation in those isoenzymes has been surveyed in an ethnically diverse population [9, 10]. Compared with Advicor, the pharmacokinetic profile of NA and lovastatin was similar. But in our study, the mean increase in NA, NAM, and NUA was 30 times, when the dose was changed from 500 mg to 750 mg, but no big difference when the dose was changed from 750 mg to 1000 mg. It was indicated that there was a liver enzyme saturation phenomenon in NA metabolism [11, 12] in Chinese healthy volunteers at the range of 750–1000 mg NA. Meanwhile, the recommended dosage of lovastatin/ER-niacin tablets may be better at the range of 350–500 mg for Chinese volunteers.

In the study of Advicor [12], it was also shown that a 22 to 25% decrease in lovastatin C_{max} was observed, when it was coadministered with NA. Lovastatin appears to be incompletely absorbed after oral administration. Because of extensive hepatic extraction, the amount of lovastatin reaching the systemic circulation as active inhibitors after oral administration is low (<5%) and shows considerable interindividual variation. Peak concentrations of active and total inhibitors occur within 2 to 4 hours after Mevacor administration. But, in our study, there was no significant difference on pharmacokinetic parameters of C_{max}, $AUC_{(0-24)}$ between single use and coadministration and 90% CI of both C_{max} and $AUC_{(0-24)}$ were at the range of 100–125% that is typically established for bioequivalence, considering the small sample size and moderate variability. But it was interesting that the pharmacokinetic parameters for 500 mg, 700 mg, and 1000 mg dose of lovastatin were 10.25 ± 3.12 ng/mL, 10.45 ± 3.85 ng/mL, and 12.67 ± 6.09 ng/mL for C_{max} and 29.85 ± 10.25 ng·h/mL, 33.33 ± 9.86 ng·h/mL, and 37.25 ± 13.61 ng·h/mL for $AUC_{(0-24)}$. It was indicated that lovastatin may have the nonlinear pharmacokinetic profile in Chinese healthy volunteers, when it was coadministrated with NA. It also suggested the metabolism of NA and lovastatin may exit competitive in human body.

On the basis of these results, no dose adjustment for lovastatin should be necessary when lovastatin is administered in combination with sustained-release niacin. The same would be true for the ER niacin. These statements would be true if indeed the patient was equal to a subject displaying near mean plasma levels of drug in this study. Given the wide variability in the results for all the drugs when administered together, caution in the form of close monitoring of the patient by blood tests and safety evaluations would be reasonable.

The treatment emergent adverse events that occurred during coadministration were similar in incidence and severity to those reported during the administration of sustained-release niacin alone or lovastatin alone and to those reported in registration documents, that is, the package insert. The notable changes from baseline in laboratory parameters after coadministration of sustained-release niacin and lovastatin were also expected from previous experience with sustained-release niacin or lovastatin administered alone. The adverse events observed are similar in type and degree to those observed in clinical trials testing the efficacy of combinations of ER niacin and lovastatin. These results, which demonstrate high intra- and intersubject variability due to high-dose ER niacin, have to be tempered by the fact that the individual medications alone and in combination can increase the risk of myopathy and rhabdomyolysis.

In conclusion, the data suggest that there is small PK drug interaction between ER niacin and lovastatin and that, although this is not considered to be clinically significant, the concomitant use of these drugs should be appropriately monitored, especially during the prescribed niacin titration period.

Authors' Contribution

S. Ying and Y. Jia contributed equally to this study.

References

[1] X. P. Song, P. P. Chen, and X. S. Chai, "Effects of puerarin on blood pressure and plasma renin activity in spontaneously hypertensive rats," *Acta Pharmacologica Sinica*, vol. 9, no. 1, pp. 55–58, 1988.

[2] V. S. Kamanna and M. L. Kashyap, "Nicotinic acid (Niacin) receptor agonists: will they be useful therapeutic agents?" *American Journal of Cardiology*, vol. 100, no. 11, pp. S53–S61, 2007.

[3] R. M. Menon, M. A. González, M. H. Adams, D. S. Tolbert, J. H. Leu, and E. A. Cefali, "Effect of the rate of niacin administration on the plasma and urine pharmacokinetics of niacin and its metabolites," *Journal of Clinical Pharmacology*, vol. 47, no. 6, pp. 681–688, 2007.

[4] I. Avisar, J. G. Brook, and E. Wolfovitz, "Atorvastatin monotherapy vs. combination therapy in the management of patients with combined hyperlipidemia," *European Journal of Internal Medicine*, vol. 19, no. 3, pp. 203–208, 2008.

[5] M. L. Kashyap, M. E. McGovern, K. Berra et al., "Long-term safety and efficacy of a once-daily niacin/lovastatin formulation for patients with dyslipidemia," *American Journal of Cardiology*, vol. 89, no. 6, pp. 672–678, 2002.

[6] J. LaFleur, C. J. Thompson, V. N. Joish, S. L. Charland, G. M. Oderda, and D. I. Brixner, "Adherence and persistence with single-dosage form extended-release niacin/lovastatin compared with statins alone or in combination with extended-release niacin," *Annals of Pharmacotherapy*, vol. 40, no. 7-8, pp. 1274–1279, 2006.

[7] H. E. Bays, "Extended-release niacin/lovastatin: the first combination product for dyslipidemia," *Expert Review of Cardiovascular Therapy*, vol. 2, no. 4, pp. 485–501, 2004.

[8] Y. Shitara and Y. Sugiyama, "Pharmacokinetic and pharmacodynamic alterations of 3-hydroxy-3-methylglutaryl coenzyme A (HMG-CoA) reductase inhibitors: drug-drug interactions and interindividual differences in transporter and metabolic enzyme functions," *Pharmacology and Therapeutics*, vol. 112, no. 1, pp. 71–105, 2006.

[9] C. E. Gaudineau and K. Auclair, "Inhibition of human P450 enzymes by nicotinic acid and nicotinamide," *Biochemical and Biophysical Research Communications*, vol. 317, no. 3, pp. 950–956, 2004.

[10] J. F. Solus, B. J. Arietta, J. R. Harris et al., "Genetic variation in eleven phase I drug metabolism genes in an ethnically diverse population," *Pharmacogenomics*, vol. 5, no. 7, pp. 895–931, 2004.

[11] E. Oliveira, S. Marsh, D. J. van Booven, A. Amorim, M. J. Prata, and H. L. McLeod, "Pharmacogenetically relevant polymorphisms in Portugal," *Pharmacogenomics*, vol. 8, no. 7, pp. 703–712, 2007.

[12] R. M. Menon, D. Tolbert, and E. Cefali, "The comparative bioavailability of an extended-release niacin and lovastatin fixed dose combination tablet versus extended-release niacin tablet, lovastatin tablet and a combination of extended-release niacin tablet and lovastatin tablet," *Biopharmaceutics and Drug Disposition*, vol. 28, no. 6, pp. 297–306, 2007.

[13] A. A. Alsheikh-Ali and R. H. Karas, "Safety of lovastatin/extended release niacin compared with lovastatin alone, atorvastatin alone, pravastatin alone, and simvastatin alone (from the United States Food and Drug Administration adverse event reporting system)," *American Journal of Cardiology*, vol. 99, no. 3, pp. 379–381, 2007.

[14] M. Sharma, D. R. Sharma, V. Singh et al., "Evaluation of efficacy and safety of fixed dose lovastatin and niacinER combination in Asian Indian dyslipidemic patients: a multicentric study," *Vascular Health and Risk Management*, vol. 2, no. 1, pp. 87–93, 2006.

[15] W. Insull Jr., M. E. McGovern, H. Schrott et al., "Efficacy of extended-release niacin with lovastatin for hypercholesterolemia: assessing all reasonable doses with innovative surface graph analysis," *Archives of Internal Medicine*, vol. 164, no. 10, pp. 1121–1127, 2004.

[16] P. Pfuhl, U. Kärcher, N. Häring, A. Baumeister, M. A. Tawab, and M. Schubert-Zsilavecz, "Simultaneous determination of niacin, niacinamide and nicotinuric acid in human plasma," *Journal of Pharmaceutical and Biomedical Analysis*, vol. 36, no. 5, pp. 1045–1052, 2005.

[17] J. X. Sun, R. Niecestro, G. Phillips, J. Shen, P. Lukacsko, and L. Friedhoff, "Comparative pharmacokinetics of lovastatin extended-release tablets and lovastatin immediate-release tablets in humans," *Journal of Clinical Pharmacology*, vol. 42, no. 2, pp. 198–204, 2002.

[18] F. Mignini, D. Tomassoni, V. Streccioni, E. Traini, and F. Amenta, "Pharmacokinetics and bioequivalence study of two tablet formulations of lovastatin in healthy volunteers," *Clinical and Experimental Hypertension*, vol. 30, no. 2, pp. 95–108, 2008.

[19] M. Hamidi, N. Zarei, and M. A. Shahbazi, "A simple and sensitive HPLC-UV method for quantitation of lovastatin in human plasma: application to a bioequivalence study," *Biological and Pharmaceutical Bulletin*, vol. 32, no. 9, pp. 1600–1603, 2009.

[20] R. Mullangi and N. R. Srinivas, "Niacin and its metabolites: role of LC-MS/MS bioanalytical methods and update on clinical pharmacology. An overview," *Biomedical Chromatography*, vol. 25, no. 1, pp. 218–237, 2011.

16

Selectivity of Brij-35 in Micellar Liquid Chromatographic Separation of Positional Isomers

Najma Memon, Huma I. Shaikh, and Amber R. Solangi

National Centre of Excellence in Analytical Chemistry, University of Sindh, Jamshoro, Sindh 78060, Pakistan

Correspondence should be addressed to Najma Memon, najmamemon@gmail.com

Academic Editor: Maria Jose Ruiz-Angel

Implementation of Brij-35, a nonionic surfactant, as a mobile phase for separation of positional isomers is investigated. Chromolith C-18 SpeedROD is used as a stationary phase. The effect of surfactant and organic modifier (propanol) concentration on the separation of some selected isomers is studied and evaluated in terms of linear solvation energy relationship (LSER). Shape selectivity is assessed by α value of sorbic and benzoic acid, which is found to be 1.339 by using mobile phase composed of 0.5% aqueous solutions of Brij-35 and propanol in 9 : 1. Isomers of parabens, nitroanilines, nitrophenols, and quinolinols are successfully separated using mobile phases composed of various percentages of surfactant and propanol. System constants for nonionic MLC using LSER analysis show that hydrogen bond basicity and dipolarity may be major contributors to selectivity, while excess molar refraction helps fine-tuning the separation which also imparts unique selectivity to nonionic surfactants as compared to ionic ones.

1. Introduction

Use of surfactants as mobile phases in HPLC above critical micellar concentration (cmc) gave birth to a new branch of chromatography, now known as micellar liquid chromatography (MLC) [1]. Since its first use for separation of PAHs in 1980 [2], a number of publications, books, and monographs appeared in the literature covering the applications and characterization of MLC systems [3–5].

The most commonly used surfactant in liquid chromatography and in overall analytical chemistry is sodium dodecylsulfate (SDS) [6], an anionic surfactant as compared to other cationic and nonionic surfactants [7]. In nonionic surfactants, Brij-35 (polyoxyethylene dodecyl ether) has found applications in liquid chromatography as it is non-UV absorbing, which is a drawback of other nonionic surfactants having aromatic ring into their structure.

Brij-35 is reported as mobile phase for the separation of a number of compounds [7]. In a previous report [8] by our group, we reported that isomers of propyl-parabens could be better separated by Brij-35 as compared to SDS. Takayanagi et al. [9, 10] have also reported the separation

of various positional isomers using Brij-35 as a surfactant in capillary electrophoresis. Vlasenko et al. [11] have studied the dissociation constants of hydroxybenzoic acids and parabens and found that addition of Brij-35 alters the dissociation of these compounds. Inspite of the number of publications on MLC using non-ionic surfactants as mobile phases, none of them is specifically focused on separation of isomers.

Separation of positional isomers is of importance from pharmaceutical perspective, due to differences in their complexation, toxicity, and reactivity in biological systems. Positional isomers cannot normally be separated through reverse phase HPLC with ODS columns commonly available in the laboratories with methanol/water or acetonitrile/water as mobile phases [12]. The objective of current study is to evaluate if Brij-35 could be used as mobile phase for separation of positional isomers and to find out the interaction responsible for selectivity in MLC using Brij-35. Here, we have investigated nonionic surfactants as mobile phase for separation of positional isomers using Chromolith C-18 column. The organic modifier (propanol) and surfactant (Brij-35) concentration on separation of some

TABLE 1: Retention time and resolution data for various positional isomers using hydro-organic and hybrid surfactant mobile phase composition.

Analyte	Solvent composition	$*t_R$ (R)	Solvent composition	$*t_R$ (R)
2-nitroaniline		2.87	88:12	4.57
3-nitroaniline	1:1 MeOH/water	2.28 (1.99)	Brij35 (2% aq. soln.)/propanol	3.7 (2.56)
4-nitroaniline		1.99 (1.14)		3.43 (0.92)
2-quinolinol	1:1MeOH/water	2.2	99:1	2.53
4-quinolinol		1.67 (1.53)	Brij-35 (1.5% aq. soln.)/propanol	1.78 (1.629)
o-cresol		2.03	90:10	1.69
p-cresol	7:3 MeOH/water	1.96	Brij35 (1.75% aq. soln.)/propanol	1.65
m-cresol		1.96		1.65
m-nitrophenol	7:3 MeOH/water	2.27	88:12	1.83
p-nitrophenol		2.07 (0.453)	Brij-35 (2% aq. soln.)/propanol	1.62 (1.35)
Dexamethazone	7:3 MeOH/water	2.49	99:1	4.24
Betamethasone		2.52	Brij-35 (1.5% aq. soln.)/propanol	4.04
Propylparaben	7:3 MeOH/water	2.3	96:4	7.33
iso-propylparaben		2.21 (0.32)	Brij-35 (1.75% aq. soln.)/propanol	6.47 (0.809)
Bezoic acid	4:6 MeOH/water	1.78	90:10	4.92
Sorbic acid		2.24 (0.64)	Brij-35 (1.5% aq. soln.)/propanol	3.88 (1.73)

$*t_R$ is retention time in minutes; in parenthesis "R" is resolution.

positional isomers is studied. The findings are discussed and selectivity is accessed by using benzoic acid/sorbic acid selectivity ratio and reported in terms of LSER parameters, which then are correlated to understand the system selectivity.

2. Experimental

2.1. Instrumentation. A Hitachi 6010 liquid chromatograph fitted with a Hitachi L-4200 variable wavelength UV-Vis detector, a Rheodyne 7125 injector, and a Chromolith performance RP-18 e, 100 mm × 4.6 mm i.d. column (E. Merck, Darmstadt, Germany) was used. CSW32 software (Data Apex) was used for data acquisition and integration. The λ 220 nm (quinolinols), 254 nm (Methazones, parabens, cresols, and nitrophenols), and 320 nm (nitroanilines) were set, and flow rate was 1.0 mL/min for all experiments except nitrophenols and quinolinols, which is 2.0 mL/min.

2.2. Materials and Reagents. Brij-35 was purchased from Sumito Corporation, Tokyo, Japan, prepared in Millipore water and used without degassing and filtration; propanol (Fluka, Sigma-Aldrich, Buchs, St.Gallen, Switzerland) was mixed within system depending on percentage required. Benzoic acid, quinolinols, nitroanilines, parabens (Fluka, Sigma-Aldrich, Buchs, St.Gallen, Switzerland), sorbic acid (Merck, Frankfurt, Germany), beta-and dexamethasone (Glaxo SmithKline, Dungarvan, Ireland) and cresol (Riedel-Hansen, Sigma-Aldrich, Seelze, Lower Saxony, Germany) were used as received their stock solutions were prepared in

propanol, and working solutions were diluted with running mobile phase. HPLC grade methanol was purchased from Fisher Scientific, Loughborough, UK.

3. Results and Discussion

3.1. Separation of Isomers. Table 1 shows the retention time and resolution data for various isomers including benzoic acid and sorbic acid for hydro-organic and hybrid non-ionic micellar mobile phase. It could be clearly observed that same mobile phase is not optimum for all separations and for most solutes studied here; separation is also possible with hydro-organic mobile phases but micellar mobile phase offers better resolution.

These separations were studied with mobile phase composed of aqueous solution of surfactant with concentration (1-2%) and modifier in the ratio 1–20% to the surfactant at pH 3 (adjusted with phosphoric acid). All the solutes were studied for 1.25, 1.5, 1.75, and 2% Brij-35 aqueous solution prepared in millipore water with various percentages of propanol. It was found that the increase in surfactant concentration can reduce the retention time but does not affect separation appreciably. So, for all pair of analytes, surfactant that provides better resolution was selected. Addition of organic modifier (1-propanol) imparts interesting changes in retention times and resolution. Figure 1 shows the effect of propanol concentration on retention times of isomers of nitroanilines, quinolinols, and parabens.

Propanol decreases the retention times in all cases, and more difference in retention time is seen at small

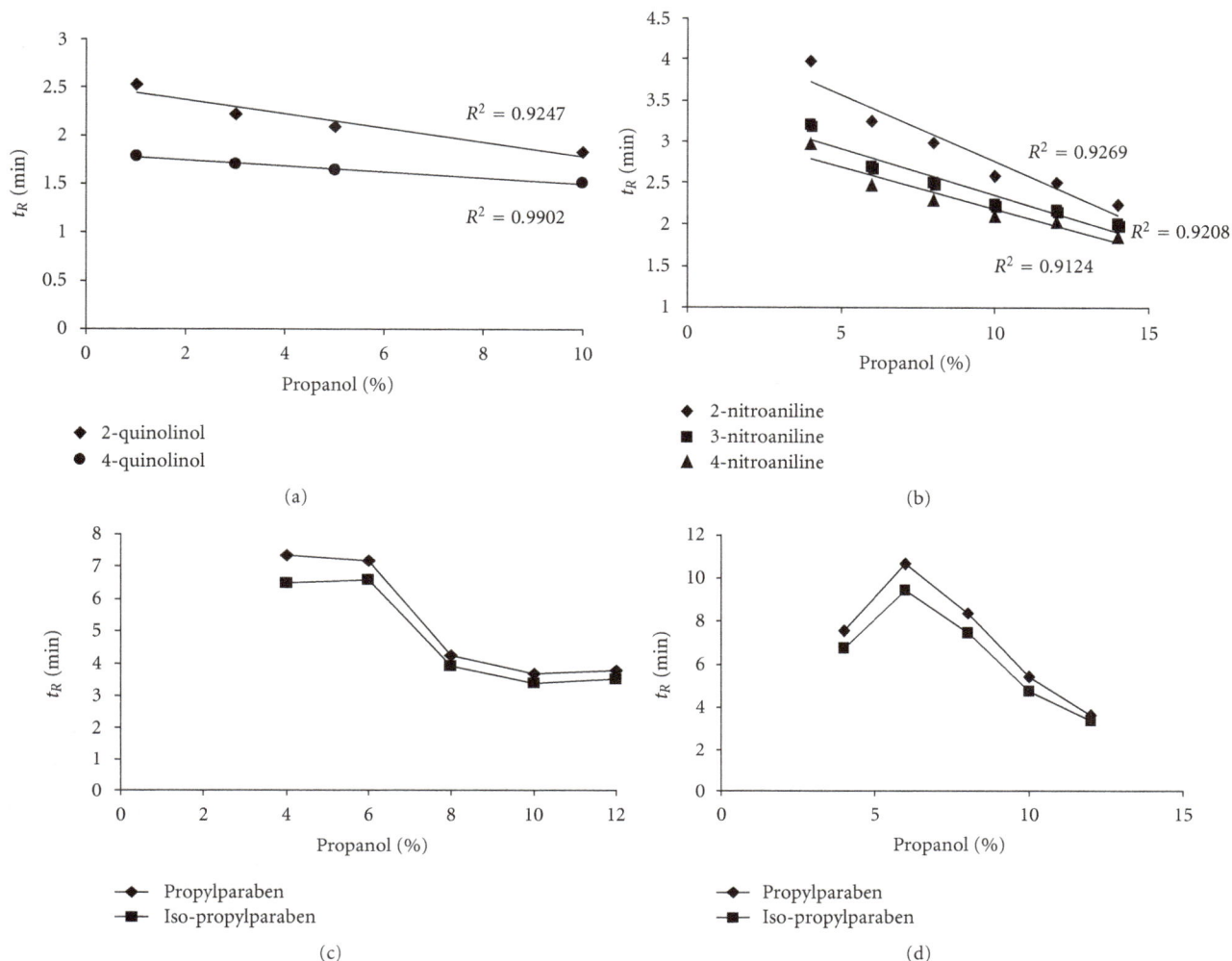

FIGURE 1: Effect of propanol concentration on retention of isomers of (a) quinolinol (Brij-35 1.5% aq. Soln.), (b) nitroanilines (Brij-35 1.75% aq. Soln.), (c) parabenes (Brij-35 1.5% aq. Soln.), and (d) parabens (Brij-35 1.75% aq. Soln.).

organic modifier content but resolution is better at higher concentrations. The decrease in retention is linear for quinolinols (Figure 1(a)) and nitroanilines (Figure 1(b)) though for nitroanilines not as good as for quinolinols. For parabens, a visible shift in retention time trend at 6% propanol (Figures 1(c) and 1(d)) is seen which is also observed with nitroanilines (visible when Figure 1(b) is expanded) and other compounds (data not shown). These findings are different in the previously reported data [7] that up to 20% propanol concentration micelles exist which may still be valid that this trend may be due to stripping of adsorbed surfactant from stationary phase at increasing propanol concentration. This study indicates that retention behaviour in MLC changes at 6% propanol with non-ionic Brij-35 surfactant when hybrid micellar mobile phase in HPLC with C-18 columns is used.

Better separation of isomers is observed at higher propanol concentration in most cases studied here as relatively sharp peaks appeared as compared to those at low propanol content (Figure 2). This may be due to decreased

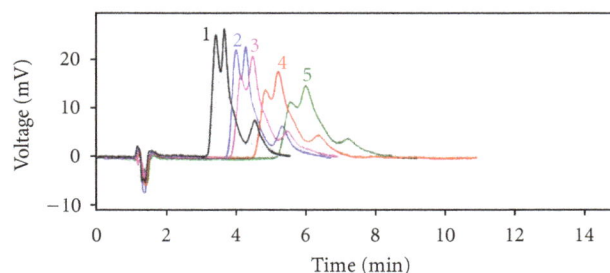

FIGURE 2: Effect of propanol concentration on sharpness of peaks of isomers of nitroanilines. Ratio of Brij-35 (2% aq. Soln.) to propanol (1) 88 : 12, (2) 90 : 10, (3) 92 : 08, (4) 94 : 06, (5) 96 : 04.

amount of adsorbed surfactant on C-18 stationary phase, which helps better mass transfer between stationary and mobile phase hence reduce diffusion of analyte.

It is also interesting to observe that o- and m-nitroanilines are better resolved with methanol/water while m- and p-nitroanilines with hybrid surfactant system. Cresols could not be separated with either system; for rest of the studied isomers, better separations are obvious using hybrid surfactant mobile phases.

Selectivity of system was assessed by calculating α value of benzoic acid/sorbic acid $\alpha_{BA/S}$. Highest $\alpha_{BA/S}$ value of 1.40 was observed with Brij35 (1.5% aqueous solution)/propanol in the ratio of 90 : 10, while with 40 : 60 methanol/water the factor was 0.74. Also the elution order was reversed; with surfactant containing mobile phase sorbic acid elutes first, while with hydro-organic benzoic acid appears first.

3.2. Interactions Responsible for Separation of Isomers. In traditional hydro-organic chromatography, selectivity for separation of isomers is correlated with molecular order within stationary phase. Selectivity is enhanced with increasing hydrocarbon stationary phase, increased chain length, and decreased temperature [17]. Unique selectivity is observed with micellar liquid chromatography as compared to hydro-organic chromatography, that is, retention factor has linear relationship with homologous series in MLC in contrast to LC which is logarithmic. The phenomenon is explained on the basis of absence of free energy change in MLC for the formation of cavity in mobile phase for incorporation of solute with increasing hydrophobicity [18]. The linear increase in "k" is also ascribed for larger number of compounds eluted per unit time [14, 18].

In MLC, not only mobile phase contributes to the separation but surfactant modified stationary phase also plays a role. It is obvious in early reports on MLC that surfactants adsorb on stationary phases consequently modify their properties. Ionic surfactants adsorb on ODS surfaces and reaches plateau when surfactant-modified mobile phases are run. For nonionic surfactants, no plateau is observed and adsorption of surfactant is continuous with micellar mobile even above the cmc [19]. Quiñons-Torrelo et al. [20] have reported that nonionic surfactant impart interesting changes in polarity of stationary phase and termed the type of chromatography as Micellar Biopartitioning Chromatography (MBC). The same author further reported that this cannot afford separation of enantiomers, while in our work we can partially separate enantiomers of steroids; this shows that statement is valid for MBC only.

Linear solvation energy relation (LSER) is used to understand the types and relative strengths of chemical interactions that control retention and selectivity in various separation techniques [21]. LSER equation as proposed by Abraham is $\log k' = c + eE + sS + aA + bB + vV$, where k is retention factor, c is constant, E, S, A, B, and V are solute descriptors independent of mobile and stationary phase used. E is solute excess molar refraction, S is dipolarity/polarizability, A and B are acidity and basicity and V is MsGowan's characteristic molecular volume. The lower case letters c, e, s, a, b, and v are the system constant reflecting the difference in solute interactions between mobile and stationary phase [22].

LSER has also been applied to MLC [13], where additional solute descriptors are added in the equation to explain the behaviour of ionic surfactants. In a recent review by Ruiz-Ángel [14], it is demonstrated that hydrophobicity

coefficient "v" varies in narrow range with change in micellar mobile phase composition hence hydrophobic interactions are scarcely affected. LSER studies of MBC has shown to have pronounce effect by hydrophobicity as given by high coefficients for "v" [20]. This type of effect can also be observed in a report by Mutelet et al. [15], where v values for Brij-35 (0.08 M), SDS (0.1 M), and CTAB (0.01 M), with 15% isopropanol as additive are 1.69 ± 0.13, 0.75 ± 0.08, and 0.56 ± 0.08, respectively.

Nonetheless, characterization of separation of isomers is explained on the basis of hydrophobicity value v since most of the positional isomers have identical v value for example, 2-, 3-, 4-nitroaniline. Berthod et al. [22] have discussed the utility of Abraham's LSER model to characterize chiral recognition behaviour of teicoplanin aglycon chiral phases. Appreciably different patterns were found for chiral phases as compared to others, while it is worth noting in the report that e and s parameters besides b and v also play a role in fine tuning of separation mechanism.

The system constants using LSER approach are system specific, that is, applicable to given stationary and mobile phase. As for non-ionic liquid chromatography, system constants are already reported so to understand the isomer recognition mechanism of MLC Linear Solvation Energy Relationship parameters for various systems reported in the literature gathered (Table 2).

Systems listed in Table 2 are included here on the basis of columns and mobile phases used; ODS column is most commonly used while polar-endcapped and polar embedded columns were included bearing in the mind that non-ionic surfactants that modify the surface of ODS column may have any similarity. TAG columns are chiral selectors with multiple interactions which also could provide clue, and monolith column is the part of current study and is most relevant. The LSER parameter; "e" is maximum for TAG phases with MeOH as mobile phase, 0 with monolithic columns with 10% MeOH and negative for Brij-35, "s" is negative for all systems except TAG phases with MeOH, while "a" and "b" are negative and "v" is positive throughout. Table 2 shows that excess molar refraction "e" is a discriminating factor for non-ionic micellar liquid chromatography, which is negatively related to log k value. Such types of interactions are also reported by Quina et al. [16] while characterizing incorporation of non-ionic solutes in aqueous micelles using LSER analysis [23–25] and Altomare et al. [26] for estimating partitioning parameters of non-ionic surfactants. The "e" value for Brij-35 surfactant is 1.63 as compared to 0.76 for CTAB and 0.32 for SDS. The author has attributed this to polyoxyethylene head groups in Brij-35 that contribute to the polarizability of the micellar solubilization sites. Also the hydrogen bond basicity of micellar solubilization site is enhanced by the presence of ether oxygen of the polyoxyethylene headgroup. Jason et al. [23] have highlighted similar finding for other non-ionic surfactants, where molar excess refraction and basicity are major contributors to partitioning as compared to previous work on other systems. To check if similar factors contribute here, available molecular descriptors for some of the isomers are compiled in Table 3. It could be observed from Table 3 that dipolarizability is a major

TABLE 2: LSER system constants for various stationary and mobile phases reported in the literature.

Column	Mobile phase	System parameters					Ref.
		e	s	a	b	v	
ODS	50% ACN	0.181	−0.527	−0.417	−1.646	1.492	[13]
Polar Endcapped	50% ACN	0.173	−0.444	−0.396	−1.261	1.323	
Polar embedded	50% ACN	0.146	−0.339	−0.079	−1.581	1.320	
TAG	20% ACN	0.345	−0.001	−0.315	−1.202	1.349	[14]
TAG	25% MeOH	0.478	0.267	−0.493	−1.211	1.496	
Monolith C-18	50% ACN	0.07	−0.33	−0.52	−1.55	1.49	[15]
Monolith C-18	50% MeOH	0.30	−0.67	−0.41	−1.77	2.10	
Monolith C-18	10% MeOH	0	−0.51	−0.26	−2.18	3.83	
Monolith C-18	0.04 M Brij-35*	−0.165	−0.087	−0.191	−1.284	1.080	[16]

*Mobile phase is 0.04 M Brij-35 with 0.01 M sodium dihydrogen phosphate at pH 7.4 adjusted with NaOH at 40°C.

TABLE 3: Abraham's LSER molecular descriptors for some of the isomers included in current study.

Analyte	A (acidity)	B (basicity)	S (dipolarity/dipolarizability)	E (excess molar refraction)
4-nitroaniline	0.46	0.35	1.93	1.22
3-nitroaniline	0.4	0.36	1.71	1.2
2-nitroaniline	0.3	0.36	1.37	1.18
o-cresol	0.52	0.30	0.86	0.84
m-cresol	0.57	0.34	0.88	0.82
p-cresol	0.57	0.31	0.87	0.82
p-nitrophenol	0.82	0.26	1.72	1.07
m-nitrophenol	0.79	0.23	1.57	1.05

contributor while molar excess refraction also varies for three isomers. For cresols, no appreciable separation was observed. Only, o-cresol was little separated and this separation may be attributed to the combined effect of "S" and "E". For p- and m-dinitrophenols, "B" and "S" are major contributors.

4. Conclusions

Non-ionic micellar liquid chromatography offers different mode of interaction than hydro-organic or ionic micellar liquid chromatography. Better separation of positional isomers is possible with Brij-35/propanol hybrid mobile phase. Besides basicity, dipolarizability and excess molar refraction are responsible for fine-tuning of separation. This new face of non-ionic MLC opens field for many applications in separation of positional isomers.

Acknowledgment

Corresponding author is thankful to Professor Dr. Willie L. Hinze of Department of Chemistry, Wake Forest University, Winston, Salem, NC, USA, for his suggestions and discussion in using LSER for characterization of molecular interactions.

References

[1] A. Berthod and C. GarcÃa-Álvarez-Coque, *Micellar Liquid Chromatography*, Marcel Dekker, New York, NY, USA, 2000.

[2] D. W. Armstrong and S. J. Henry, "Use of an aqueous micellar mobile phase for separation of phenols and polynuclear aromatic hydrocarbons via HPLC," *Journal of Liquid Chromatography*, vol. 3, no. 5, pp. 657–662, 1980.

[3] A. Detroyer, S. Stokbroekx, H. Bohets et al., "Fast monolithic micellar liquid chromatography: an alternative drug permeability assessing method for high-throughput screening," *Analytical Chemistry*, vol. 76, no. 24, pp. 7304–7309, 2004.

[4] R. Izquierdo-Hornillos, R. Gonzalo-Lumbreras, and A. Santos-Montes, "Method development for cortisol and cortisone by micellar liquid chromatography using sodium dodecyl sulphate: application to urine samples of rugby players," *Journal of Chromatographic Science*, vol. 43, no. 5, pp. 235–240, 2005.

[5] N. Memon and M. I. Bhanger, "Micellar liquid chromatographic determination of aluminum using 8-hydroxyquinoline-5-sulphonic acid," *Acta Chromatographica*, vol. 14, pp. 172–179, 2004.

[6] M. Gil-Agustí, J. Esteve-Romero, and M. H. Abraham, "Solute-solvent interactions in micellar liquid chromatography. Characterization of hybrid micellar systems of sodium dodecyl sulfate-pentanol," *Journal of Chromatography A*, vol. 1117, no. 1, pp. 47–55, 2006.

[7] M. J. Ruiz-Ángel, M. C. Garcia-Álvarez-Coque, and A. Berthod, "New insights and recent developments in micellar liquid chromatography," *Separation and Purification Reviews*, vol. 38, no. 1, pp. 45–96, 2009.

[8] N. Memon, M. I. Bhanger, and M. Y. Khuhawer, "Determination of preservatives in cosmetics and food samples by micellar liquid chromatography," *Journal of Separation Science*, vol. 28, no. 7, pp. 635–638, 2005.

[9] T. Takayanagi and S. Motomizu, "Equilibrium analysis of reactions between aromatic anions and nonionic surfactant micelles by capillary zone electrophoresis," *Journal of Chromatography A*, vol. 853, no. 1-2, pp. 55–61, 1999.

[10] T. Takayanagi, K. Fushimi, and S. Motomizu, "Separation of various positional isomers of aromatic anions by nonionic micellar electrokinetic chromatography coupled with ion association distribution," *Journal of Microcolumn Separations*, vol. 12, no. 2, pp. 107–112, 2000.

[11] A. S. Vlasenko, L. P. Loginova, and E. L. Iwashchenko, "Dissociation constants and micelle-water partition coefficients of hydroxybenzoic acids and parabens in surfactant micellar solutions," *Journal of Molecular Liquids*, vol. 145, no. 3, pp. 182–187, 2009.

[12] T. Chasse, R. Wenslow, and Y. Bereznitski, "Chromatographic selectivity study of 4-fluorophenylacetic acid positional isomers separation," *Journal of Chromatography A*, vol. 1156, no. 1-2, pp. 25–34, 2007.

[13] J. R. Torres-Lapasió, M. J. Ruiz-Ángel, M. C. García-Álvarez-Coque, and M. H. Abraham, "Micellar versus hydro-organic reversed-phase liquid chromatography: a solvation parameter-based perspective," *Journal of Chromatography A*, vol. 1182, no. 2, pp. 176–196, 2008.

[14] M. J. Ruiz-Ángel, S. Carda-Broch, J. R. Torres-Lapasió, and M. C. García-Álvarez-Coque, "Retention mechanisms in micellar liquid chromatography," *Journal of Chromatography A*, vol. 1216, no. 10, pp. 1798–1814, 2009.

[15] F. Mutelet, M. Rogalski, and M. H. Guermouche, "Micellar liquid chromatography of polyaromatic hydrocarbons using anionic, cationic, and nonionic surfactants: armstrong model, LSER interpretation," *Chromatographia*, vol. 57, no. 9-10, pp. 605–610, 2003.

[16] F. H. Quina, E. O. Alonso, and J. P. S. Farah, "Incorporation of nonionic solutes into aqueous micelles: a linear solvation free energy relationship analysis," *Journal of Physical Chemistry*, vol. 99, no. 30, pp. 11708–11714, 1995.

[17] C. A. Rimmer and L. C. Sander, "Shape selectivity in embedded polar group stationary phases for liquid chromatography," *Analytical and Bioanalytical Chemistry*, vol. 394, no. 1, pp. 285–291, 2009.

[18] M. F. Borgerding, F. H. Quina, W. L. Hinze, J. Bowermaster, and H. M. McNair, "Investigation of the retention mechanism in nonionic micellar liquid chromatography using an alkylbenzene homologous series," *Analytical Chemistry*, vol. 60, no. 22, pp. 2520–2527, 1988.

[19] M. F. Borgerding and W. L. Hinze, "Characterization and evaluation of the use of nonionic polyoxyethylene (23) dodecanol micellar mobile phases in reversed-phase high-performance liquid chromatography," *Analytical Chemistry*, vol. 57, no. 12, pp. 2183–2190, 1985.

[20] C. Quiñones-Torrelo, Y. Martin-Biosca, J. J. Martínez-Pla, S. Sagrado, R. M. Villanueva-Camañas, and M. J. Medina-Hernández, "QRAR models for central nervous system drugs using biopartitioning micellar chromatography," *Mini Reviews in Medicinal Chemistry*, vol. 2, no. 2, pp. 145–161, 2002.

[21] M. Vitha and P. W. Carr, "The chemical interpretation and practice of linear solvation energy relationships in chromatography," *Journal of Chromatography A*, vol. 1126, no. 1-2, pp. 143–194, 2006.

[22] A. Berthod, C. R. Mitchell, and D. W. Armstrong, "Could linear solvation energy relationships give insights into chiral recognition mechanisms?. 1. π-π and charge interaction in the reversed versus the normal phase mode," *Journal of Chromatography A*, vol. 1166, no. 1-2, pp. 61–69, 2007.

[23] W. C. Jason, "Comparison of retention on traditional alkyl, Polar endcapped, and Polar embedded group stationary phases," *Journal of Separation Science*, vol. 31, no. 10, pp. 1712–1718, 2008.

[24] Y. Chu and C. F. Poole, "System maps for retention of neutral organic compounds under isocratic conditions on a reversed-phase monolithic column," *Journal of Chromatography A*, vol. 1003, no. 1-2, pp. 113–121, 2003.

[25] R. Lu, J. Sun, Y. Wang et al., "Characterization of biopartitioning micellar chromatography system using monolithic column by linear solvation energy relationship and application to predict blood-brain barrier penetration," *Journal of Chromatography A*, vol. 1216, no. 27, pp. 5190–5198, 2009.

[26] C. Altomare, A. Carotti, G. Trapani, and G. Liso, "Estimation of partitioning parameters of nonionic surfactans using calculated descriptors of molecular size, polarity, and hydrogen bonding," *Journal of Pharmaceutical Sciences*, vol. 86, no. 12, pp. 1417–1425, 1997.

17

HPTLC-Densitometric Analysis of Eperisone Hydrochloride and Paracetamol in Their Combined Tablet Dosage Form

Nirav Uchadadiya, Falgun Mehta, and Pinak Sanchaniya

Department of Pharmaceutical Chemistry and Analysis, Indukaka Ipcowala College of Pharmacy, New Vallabh Vidyanagar, Gujarat 388121, India

Correspondence should be addressed to Nirav Uchadadiya; nikup982@gmail.com

Academic Editor: Irene Panderi

A simple, precise, accurate, and reliable HPTLC method has been developed and validated for the analysis of EPE-Eperisone hydrochloride and PCM-Paracetamol in their combined dosage form. Identification and analysis were performed on 100 mm × 100 mm layer thickness 0.2 mm, precoated silica gel G_{60}-F_{254} aluminum sheet, prewashed with methanol, and dried in an oven at 50°C for 5 min. Toluene : methanol : ethyl acetate : glacial acetic acid (4 : 3.5 : 2.5 : 0.05) (v/v/v/v) was used as mobile phase. Calibration plots were established showing the dependence of response (peak area) on the amount chromatographed. The validated calibration ranges were 200–700 ng/spot and 1300–4550 ng/spot for EPE and PCM with correlation coefficient (R^2) 0.994 and 0.996, respectively. Average % recovery was between 98.61–100.94% and 99.18–100.57% for EPE and PCM, respectively. The spots were scanned at 248 nm in a reflectance mode. The proposed method was validated as per ICH guidelines and successfully applied to the estimation of EPE and PCM in their combined tablet dosage form.

1. Introduction

EPE is chemically (2RS)-1-(4-ethylphenyl)-2-methyl-3-(1-piperidinyl)propan-1-one and hydrochloride (1 : 1) (Figure 1(a)) [1, 2]. Eperisone HCl is a centrally acting muscle relaxant; it acts at the level of spinal cord by blocking sodium channels and calcium channels. Eperisone HCl exerts its spinal reflex inhibitory action predominantly via a presynaptic inhibition of the transmitter release from the primary afferent endings via a combined action on voltage-gated sodium and calcium channels. Eperisone HCl increases the blood supply to skeletal muscles; this action is noteworthy since a muscle contracture may compress the small blood vessels and induce an ischemia leading to release of pain stimulating compounds [3]. EPE is official in Japanese pharmacopoeia [1]. Chemically PCM is N-(4-hydroxyphenyl)acetamide (Figure 1(b)). Paracetamol is a weak inhibitor of PG-Prostaglandin synthesis of COX-1 and COX-2. Cyclooxygenase serves as a pain activator, is responsible for the biosynthesis of prostaglandins, is used for the relief of headaches, fever, pains, and is a major

ingredient in numerous cold and flu remedies [4]. Paracetamol is official in Japanese Pharmacopoeia [1], British Pharmacopoeia [5], United States Pharmacopoeia [6], and Indian Pharmacopoeia [7].

The review of the literature revealed that analytical methods involving spectrophotometry [8], LC-ESI-MS [9], have been reported for EPE in single form and in combination with other drugs. Several analytical methods have been reported for PCM in single form and in combination with other drugs including spectrophotometry [10–14], HPLC [15–17], and HPTLC [18, 19].

To the best of our knowledge, there is no published HPTLC method for this combination. So, the present paper describes a simple, accurate, and precise method for simultaneous estimation of EPE and PCM in combined tablet dosage form by HPTLC method. The developed method was validated in accordance with ICH guidelines [20] and successfully employed for the assay of EPE and PCM in their combined dosage form.

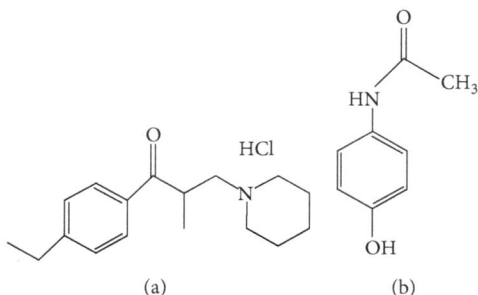

FIGURE 1: Chemical structure of (a) EPE and (b) PCM.

TABLE 1: System suitability test parameters.

System suitability parameter	EPE	PCM
Peak purity	0.994	0.996
R_f value	0.26 ± 0.017224	0.79 ± 0.010328

2. Materials and Methods

2.1. Reagents and Chemicals.

Analytically pure EPE and PCM were kindly provided by Macleods Pharmaceuticals Ltd., Mumbai, Maharashtra, India, and Elysium Pharmaceutical Ltd., Vadodara, Gujarat, India, respectively, as gratis samples. Analytical grade methanol was purchased from SRL limited, Mumbai, India. Tablet of EPE and PCM in combined dosage form, MYOSONE PLUS, with a 50 mg EPE and 325 mg PCM label claim, manufactured by Macleods Pharmaceuticals Ltd., was procured from local market.

2.2. Instrumentation and Conditions.

Chromatography was performed on 100 mm × 100 mm on precoated silica gel G_{60}-F_{254} aluminum sheet (E. Merck, Mumbai, India). Before use the plates were washed with methanol then dried at room temperature. Samples were applied as 6 mm bands by means of a Camag Linomat V (Muttenz, Switzerland) sample applicator equipped with 100 μL syringe; operated with settings of band length, 6 mm; distance between bands, 5 mm; distance from the plate edge, 10 mm; and distance from the bottom of the plate, 10 mm. The constant application rate was 15 s μL^{-1}, and a nitrogen aspirator was used. Ascending development of plate, migration distance 70 mm was performed at ambient temperature with Toluene : methanol : ethyl acetate : glacial acetic acid (4 : 3.5 : 2.5 : 0.05) (v/v/v/v), as mobile phase in a 10 cm × 10 cm Camag twin-trough chamber previously saturated for 15 min. After development the plates were dried with hot-hair dryer and viewed in a CAMAG UV cabinet. Densitometric scanning at 272 nm was then performed with a Camag TLC Scanner 3 equipped with winCATs 3.2.1 software. The scanning rate was 20 mm s^{-1}. The source of radiation used was the deuterium lamp.

2.3. Calibration

2.3.1. Eperisone Hydrochloride (50 μg/mL) and Paracetamol (325 μg/mL) Standard Stock Solution.

Standard EPE 50 mg and PCM 325 mg were weighed and transferred to a 100 mL volumetric flask and dissolved in methanol. The flask was sonicated, and volume was made up to the mark with methanol to give a solution containing 500 μg/mL EPE and 3250 μg/mL PCM. One mL of this aliquot was added to 10 mL volumetric flask, and volume was made up to the mark with methanol to give a solution containing 50 μg/mL EPE and 325 μg/mL PCM.

2.3.2. Calibration Curve for EPE and PCM.

Semiautomatic spotter was used containing a syringe having capacity of 100 μL. Mixed stock solution having concentration of 50 μg/mL of EPE and 325 μg/mL PCM was filled in the syringe, and under nitrogen stream, it was applied in the form of band of desired concentration range for each drug on a single plate having concentration of 200 to 700 ng/spot for EPE and 1300 to 4550 ng/spot for PCM. Plate was developed using the above-mentioned conditions. Plots of peak area versus concentration for both drugs were obtained.

2.4. Analysis of Marketed Formulation.

Twenty tablets were weighed accurately; average weight was found and finely powered. A quantity equivalent to 50 mg EPE and 325 mg PCM was accurately weighed and transferred to volumetric flask of 100 mL capacity. 80 mL of methanol was transferred to this volumetric flask and sonicated for 20 min to dissolve the drug. Resulting solution was filtered through Whatman filter paper (0.45 μ) into a 100 mL volumetric flask. The flask was shaken, and volume was made up to the mark with methanol to give a solution containing 500 μg/mL of EPE and 3250 μg/mL of PCM. One mL of this aliquot was added to 10 mL volumetric flask, and volume was made up to the mark with methanol to give a solution containing 50 μg/mL of EPE and 325 μg/mL of PCM. Now this prepared sample solution was applied on TLC plate, developed, dried in air, and photometrically analyzed as described above. From the peak area obtained in the chromatogram, the amounts of both of the drugs were calculated.

2.5. Validation of the Method

2.5.1. Linearity and Range of the HPTLC Method.

Calibration graphs were constructed by plotting peak areas versus concentrations of EPE and PCM, and the regression equations were calculated. The calibration graphs were plotted over 6 different concentrations in the range of 200–700 ng/spot for EPE and 1300–4550 ng/spot for PCM by applying different volumes stock solution containing EPE and PCM (50 μg/mL of EPE and 325 μg/mL PCM). The calibration graphs were developed by plotting peak area versus concentrations ($n = 6$) with the help of the winCATS software.

2.5.2. Accuracy (Recovery).

Known amounts of standard solution of EPE (300, 400, and 500 ng/spot) and PCM (2050, 2600, and 3250 ng/spot) for the HPTLC method were added

TABLE 2

Concentration (ng/spot)	Area mean ± S.D. ($n = 6$)	CV
Result of calibration readings for EPE by HPTLC method		
200	1194.333 ± 13.80821	1.15
300	1675.967 ± 14.6988	0.87
400	2093.35 ± 17.86894	0.85
500	2504.867 ± 35.20316	1.4
600	3171.433 ± 29.04952	0.91
700	3645.45 ± 23.91458	0.66
Result of calibration readings for PCM by HPTLC method		
1300	9337.983 ± 86.3672	0.92
1950	11511.37 ± 122.5659	1.06
2600	13646.1 ± 86.2213	0.63
3250	16556.83 ± 104.0746	0.62
3900	18071.97 ± 129.3408	0.71
4550	20566.1 ± 110.7175	0.53

TABLE 3: Determination of accuracy.

% Level	Amount added (ng/spot)		Amount recovered (ng/spot) ($n = 3$)		% Recovered ± S.D	
	EPE	PCM	EPE	PCM	EPE	PCM
50	300	2050	302.82	2057.17	100.94 ± 0.89	100.35 ± 0.41
100	400	2600	394.44	2578.68	98.61 ± 0.74	99.18 ± 0.50
150	500	3250	492.95	3268.52	98.59 ± 0.57	100.57 ± 0.93

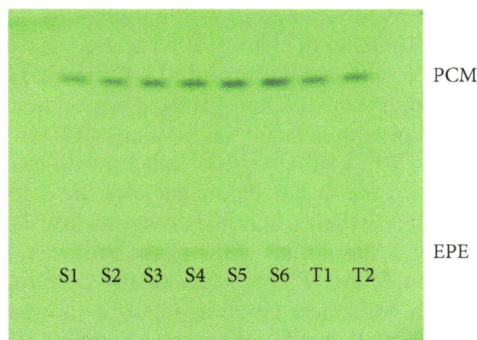

FIGURE 2: Photograph of developed HPTLC plate of EPE and PCM at 248 nm.

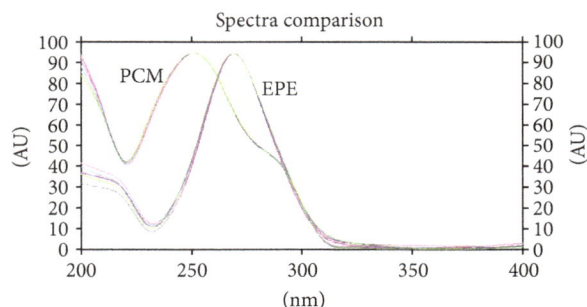

FIGURE 4: Overlain spectrums of EPE and PCM at 248 nm by HPTLC method.

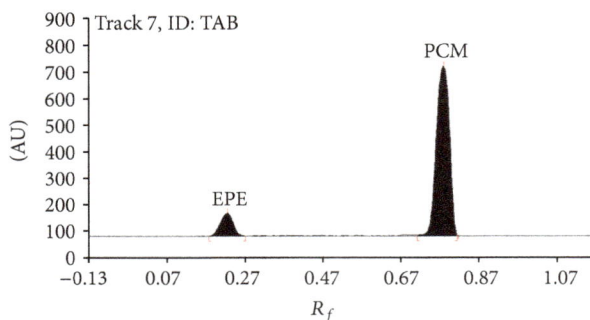

FIGURE 3: Densitogram of market formulation containing EPE and PCM 400 ng/spot and 2600 ng/spot, respectively.

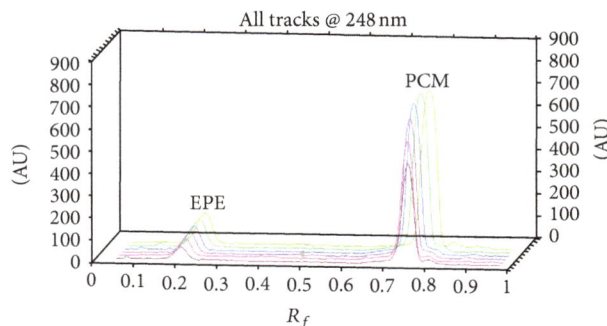

FIGURE 5: 3D overlain spectra of EPE and PCM by HPTLC method.

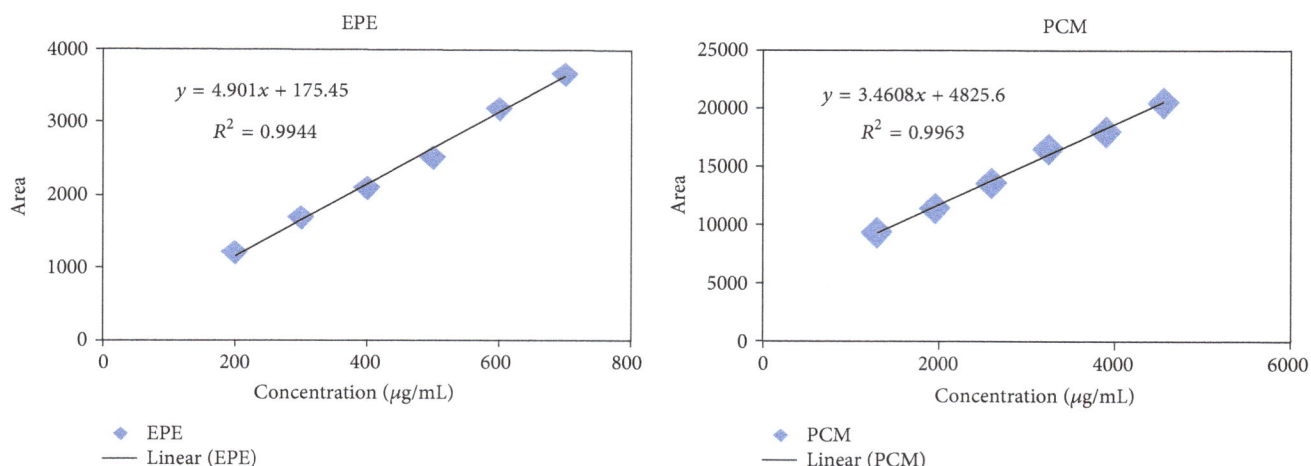

FIGURE 6: Calibration curve of EPE and PCM.

to prequantitated sample solutions of tablet dosage forms. The amounts of EPE and PCM were estimated by applying values of peak area to the regression equations of the calibration graph.

2.5.3. Precision. Precisions of the proposed HPTLC methods were determined by analyzing mixed standard solution of EPE and PCM at 3 different concentrations (300, 400, and 500 ng/spot for EPE and 1950, 2600, and 3250 ng/spot for PCM) 3 times in the same day and in 3 different days. The results are reported in terms of coefficient of variance (CV).

2.5.4. Repeatability. Repeatability of method was assessed by applying the same sample solution 6 times on a plate with the automatic spotter using the same syringe and by taking 6 scans of the sample spot for both EPE and PCM (400 ng/spot of EPE and 2600 ng/spot of PCM) without changing the positions of the plate.

2.5.5. Specificity. The specificity of the method was ascertained by analyzing standard drug and sample. The band of EPE and PCM in sample was confirmed by comparing the R_f and spectra of the band with that of standard. The peak purity of both drugs was assessed by comparing the spectra at 3 different levels, that is, peak start (S), peak apex (M), and peak end (E) position of the band.

2.5.6. Limit of Detection and Limit of Quantification. The limit of detection (LOD) and the limit of quantification (LOQ) of the drug were derived by using the following equations as per International Conference on Harmonization (ICH) guidelines which is based on the calibration curve:

$$LOD = 3.3 \times \frac{\sigma}{S},$$
$$LOQ = 10 \times \frac{\sigma}{S}, \tag{1}$$

where σ is the standard deviation of y-intercepts of regression lines and S is slope of calibration curve.

2.5.7. Robustness. Sample solution was prepared and then analyzed with change in the typical analytical conditions like amount of mobile phase, proportion of mobile phase, saturation time, plate pretreatment, and stability of analytical solution.

3. Results and Discussion

3.1. Method Optimization. Several mobile phases were tried to accomplish good separation of EPE and PCM. Using the mobile phase Toluene : methanol : ethyl acetate : glacial acetic acid (4 : 3.5 : 2.5 : 0.05) (v/v/v/v) and 10 × 10 cm HPTLC silica gel 60 F254 aluminum-backed plates, good separation was attained with retention factor (R_f) values of 0.26 for EPE and 0.79 for PCM. A wavelength of 248 nm was used for the quantification of the drugs. Figure 2 shows the detection of both of the drugs in their combined dosage form at 248 nm by HPTLC method. Resolution of the peaks with clear baseline separation was found. Figure 3 shows the densitogram of mixture which has a clear baseline. Figure 4 showed a good linearity when overlapped and scanned between 200 nm to 400 nm. Figure 5 shows a 3D overlapped spectrum of both drugs which has good linearity. The system suitability test parameters for the developed method are shown in Table 1.

3.2. Validation of the Proposed Methods

3.2.1. Linearity. Figure 6 shows that linear correlation was obtained between peak areas and concentrations of EPE and PCM in the range of 200–700 ng/spot for EPE with $R^2 = 0.994$ and 1300–4550 ng/spot for PCM with $R^2 = 0.996$, respectively, and data are shown in Table 2.

3.2.2. Accuracy. The recovery experiments were performed by the standard addition method. The HPTLC method was found to be accurate with % recovery of 98.61–100.94% for EPE and 99.18–100.57% for PCM, respectively, (Table 3). The high values indicate that the method is accurate.

TABLE 4: Summary of validation parameters of HPTLC.

Parameters	EPE	PCM
Recovery %	99.59–100.94	99.18–100.57
Repeatability (CV, $n = 6$)	0.69	0.35
Precision (CV)		
Intraday ($n = 3$)	0.69–0.76	0.35–0.59
Interday ($n = 3$)	0.77–1.34	0.72–1.65
Limit of detection (ng/spot)	8.05	87.64
Limit of quantitation (ng/spot)	24.41	265.57
Robustness	Robust	Robust
Solvent suitability	Suitable for 24 hr	Suitable for 24 hr

TABLE 5: Assay result of marketed formulation.

Formulation	Drug	Amount taken (ng/spot)	Amount found (ng/spot) ($n = 3$)	Labeled claim (mg)	Amount found per tablet (mg)	% Label claim ± SD
MYOSONE PLUS	EPE	400	395.17	50	49.39	98.79 ± 1.26
(tablet)	PCM	2600	2544.47	325	318.06	97.86 ± 0.84

3.2.3. Repeatability. The CV values for EPE and PCM were found to be 0.69 and 0.35, respectively. The CV values were found to be <1%, which indicates that the proposed methods are repeatable.

3.2.4. Precision. The CV values were found to be <2%, which indicates that the proposed method is precise.

3.2.5. Limit of Detection (LOD) and Limit of Quantification (LOQ). LOD values for EPE and PCM were found to be 8.05 ng/spot and 87.64 ng/spot, respectively. LOQ values for EPE and PCM were found to be 24.41 ng/spot and 265.57 ng/spot, respectively. These data show that nanogram quantity of both drugs can be accurately determined (Table 4).

3.2.6. Specificity. Excipients (Starch) used in the specificity studies did not interfere with the estimation of either of the drugs by the proposed methods. Hence, the methods were found to be specific for estimation of EPE and PCM.

3.2.7. Robustness. Peak area and retention time variation were found to be <1%. Also, no significant change in peak area was observed during 24 hr. No decomposition was observed in either the first or second direction of the 2-dimensional analysis for both drugs on the HPTLC plate. Hence, the method was found to be robust for estimation of EPE and PCM.

3.2.8. Assay of the Tablet Dosage Form (EPE 50 mg and PCM 325 mg per Tablet). The proposed validated method was successfully applied to determine EPE and PCM in their tablet dosage form (MYOSONE PLUS). The results obtained for EPE and PCM was comparable with the corresponding labelled amounts (Table 5).

4. Conclusion

Thus, the objective of project work was development and comparison of analytical method of EPE and PCM in their combined dosage form. The developed and validated HPTLC method for EPE and PCM was found to be simple, specific, and cost effective and can be routinely applied for analysis of EPE and PCM in their combined dosage form. We can say that HPTLC method is more sensitive giving precise results (interday, intraday) for both drugs, and also HPTLC method is more sensitive in terms of LOD and LOQ. It also requires least solvents for analysis. The proposed method has the advantages of simplicity and convenience for the separation and quantitation of EPE and PCM in combination and can be used for the assay of their dosage form. Also, the low solvent consumption and short analytical run time lead to environmentally friendly chromatographic procedures. The additives usually present in the pharmaceutical formulations of the assayed analytes did not interfere with determination of EPE and PCM. The method can be used for the routine simultaneous analysis of EPE and PCM in pharmaceutical preparations.

Disclosure

The usage of this trade mark symbol or company name is for proving the genuinity of the work and not for any another purpose. The authors of the paper, do not have any financial relation with the commercial identity mentioned in the paper.

Conflict of Interests

The authors have no conflict of interests or no financial gains in mentioning the company names or trade marks.

Acknowledgments

The authors are thankful to Macleods Pharmaceuticals Ltd. (Mumbai, India) and Elysium Pharmaceutical Ltd. (Vadodara, India) for providing gratis sample with the great pleasure. The authors are also thankful to Principal and Staff of Indukaka Ipcowala College of Pharmacy (New Vallabh Vidyanagar, India) and Director of SICART (New Vallabh Vidyanagar, India) for providing the necessary facilities for research work.

References

[1] *Japanese Pharmacopoeia*, Government of Japan, The Ministry of Health, Labour and Welfare, 16th edition.

[2] The Merck index, *An Encyclopedia of Chemicals, Drugs and Biological*, Merck Research Laboratories, 13th edition, 2006.

[3] A. Beltrame, S. Grangiè, and L. Guerra, "Clinical experience with eperisone in the treatment of acute low back pain," *Minerva Medica*, vol. 99, no. 4, pp. 347–352, 2008.

[4] K. D. Tripathi, *Essentials of Medical Pharmacology*, vol. 1, Jaypee Brothers, 5th edition, 2008.

[5] *British Pharmacopoeia*, vol. 2, Stationery Office on Behalf of the Medicines and Healthcare Products Regulatory Agency (MHRA), London, UK, 2011.

[6] *United States Pharmacopoeia and National Formulary*, The United States Pharmacopoeia Convention, 24th Asian edition.

[7] *Indian Pharmacopoeia*, vol. 3, Government of India, Ministry of Health & Family Welfare, the Controller & Publication, Delhi, India, 2010.

[8] P. U. Patel, S. K. Patel, and U. J. Patel, "Spectrophotometric method for simultaneous estimation of eperisone HCl and diclofenac sodium in synthetic mixture," *International Research Journal of Pharmacy*, vol. 3, no. 9, pp. 203–206, 2012.

[9] Y. Zhang, L. Ding, X. Wei, S. Zhang, and J. Sheng, "Rapid and sensitive liquid chromatography-electrospray ionization-mass spectrometry method for the determination of eperisone in human plasma: method and clinical applications," *Journal of Chromatographic Science*, vol. 42, no. 5, pp. 254–258, 2004.

[10] M. S. Kondawar, R. R. Shah, J. J. Waghmare, N. D. Shah, and M. K. Malusare, "UV spectrophotometric estimation of paracetamol and lornoxicam in bulk drug and tablet dosage form using multiwavelength method," *International Journal of PharmTech Research*, vol. 3, no. 3, pp. 1603–1608, 2011.

[11] R. Sawant, L. Bhangale, R. Joshi, and P. Lanke, "Validated spectrophotometric methods for simultaneous estimation of Paracetamol, Domperidone and Tramadol HCl in pure and tablet dosage form," *Journal of Chemical Metrology*, vol. 4, no. 1, pp. 21–27, 2010.

[12] K. Kalra, S. Naik, G. Jarmal, and N. Mishra, "Spectrophotometric method for simultaneous estimation of paracetamol and domperidone in tablet formulation," *Asian Journal of Research in Chemistry*, vol. 2, no. 2, pp. 112–114, 2009.

[13] C. Narajji, H. R. Patel, M. D. Karvekar, and A. R. S. Babu, "Simultaneous estimation of aceclofenac, paracetamol and tizanidine in their combined dosage forms by spectrophotometric and RP- HPLC method," *Journal of Analytical and Bioanalytical Techniques*, vol. 2, article 123, 2011.

[14] S. Mahaparale, R. S. Telekone, R. P. Raut, S. S. Damle, and P. V. Kasture, "Simultaneous spectrophotometric determination of drotaverine hydrochloride and paracetamol in tablet," *Indian Journal of Pharmaceutical Sciences*, vol. 72, no. 1, pp. 133–136, 2010.

[15] R. B. Prasanna and M. S. Reddy, "RP-HPLC method for simultaneous estimation of paracetamol and ibuprofen in tablets," *Asian Journal of Research in Chemistry*, vol. 2, no. 1, pp. 70–72, 2009.

[16] L. Monser and F. Darghouth, "Simultaneous LC determination of paracetamol and related compounds in pharmaceutical formulations using a carbon-based column," *Journal of Pharmaceutical and Biomedical Analysis*, vol. 27, no. 6, pp. 851–860, 2002.

[17] N. P. Dudhane, M. J. Umekar, and R. T. Lohiya, "Validated RP-HPLC method for estimation of metoclopramide hydrochloride and paracetamol in solid dosage form," *Journal of Pharmaceutical Research*, vol. 3, no. 12, pp. 3064–3066, 2010.

[18] A. Yadav, R. Singh, S. Mathur, P. Saini, and G. Singh, "A simple and sensitive HPTLC method for simultaneous analysis of domperidone and paracetamol in tablet dosage forms," *Journal of Planar Chromatography*, vol. 22, no. 6, pp. 421–424, 2009.

[19] P. B. Deshpanday, V. S. Gandhi, and S. Y. Bhangale, "HPTLC method for simultaneous estimation of etoricoxib and paracetamol in combined tablet dosage," *Journal of Pharmaceutical and Biomedical Analysis*, vol. 3, pp. 1–5, 2010.

[20] ICH Harmonized Tripartite Guidelines, *Validation of Analytical Procedures: Text and Methodology, Q2 (R1)*, Geneva, Switzerland, 2005.

18

Biological Fingerprinting of Herbal Samples by Means of Liquid Chromatography

Łukasz Cieśla[1, 2]

[1] Department of Inorganic Chemistry, Medical University of Lublin, Chodźki 4a, 20-093 Lublin, Poland
[2] Department of Biochemistry, Institute of Soil Science and Plant Cultivation, State Research Institute, Czartoryskich 8, 24-100 Puławy, Poland

Correspondence should be addressed to Łukasz Cieśla, lukecarpenter@poczta.onet.pl

Academic Editor: Teresa Kowalska

Biological chromatographic fingerprinting is a relatively new concept in the quality control of herbal samples. Originally it has been developed with the application of HPLC, and recently herbal samples' biological profiles have been obtained by means of thin-layer chromatography (TLC). This paper summarizes the application of liquid chromatographic techniques for the purpose of biological fingerprint analysis (BFA) of complex herbal samples. In case of biological TLC fingerprint, which is a relatively novel solution, perspectives of its further development are outlined in more detail. Apart from already published data, some novel results are also shown and briefly discussed. The paper aims at drawing scientists' attention to the unique solutions offered by biological fingerprint construction.

1. Introduction

Fingerprint construction has become an important quality control tool of herbal samples in the light of constantly growing interest in natural origin medicines. Fingerprint analysis has been accepted by WHO as a methodology for the quality control of herbal samples [1, 2]. It is applied to identify closely related plant species, to detect adulterations, to control the extraction process or to study the quality of a finished product. Herbal sample fingerprint can be defined as a set of characteristic chromatographic or spectroscopic signals, whose comparison leads to an unambiguous sample recognition. Several chromatographic methods have been applied for fingerprint construction, namely, high-performance liquid chromatography (HPLC), thin-layer chromatography (TLC), gas chromatography (GC), or high-speed counter current chromatography (HSCCC). However, it is difficult to indicate with 100% certainty which signals (peaks, bands, etc.) should be present in the obtained fingerprint to confirm sample identity. For that purpose the analyzed sample can be compared with a defined Botanical Reference Material (BRM) or a set of standard compounds. Defining and obtaining BRM for every

plant species is a difficult task, therefore, new solutions are sought for fingerprint comparison. More recently a concept of multiple fingerprints construction and multidimensional fingerprinting have gained much attention, as large amount of chromatographic and/or spectroscopic signals enable more comprehensive data analysis [3]. Multiple chromatographic fingerprint consists of more than one chromatographic profile [3], while in multidimensional fingerprinting hyphenated detectors are used (e.g., DAD and MS) that both record eluting compounds [1]. Usually the set of obtained data is further processed by means of similarity measure or chemometrics approach (e.g., principal component analysis, hierarchical clustering, multivariate calibration, etc.) [1, 3]. In case of multiple fingerprints, a data fusion-based method is proposed [3]. It is also a common approach to construct the reference/standard fingerprints that are further used for data comparison. The similarity of a tested sample's fingerprint and a reference chromatographic profile is performed usually using the correlation coefficient measure [3].

As far as herbal medicines are considered their biological activity is an important issue. However, traditional chromatographic fingerprint analysis provides the researchers only with qualitative and quantitative information. A critical issue is the fact, that compounds present in low concentration may exert more potent biological activity than those present in greater amounts. It is, therefore, important to introduce screening of the biological activity into the chromatographic fingerprint analysis. To screen the natural samples for the presence of most active compounds biological fingerprinting analysis has been introduced. It was originally developed with the use of high-performance liquid chromatography. Apart from qualitative and quantitative data, it gives also the possibility to fish out the active compounds from a myriad of compounds present in herbal samples. HPLC biological fingerprinting techniques have been well reviewed by Su et al. [4] and Yu et al. [5].

2. Biological Fingerprinting by Means of HPLC

The concept of biological fingerprinting has been first developed for the purposes of the quality control of complex traditional Chinese medicines [4]. By analogy with the definition of traditional chromatographic fingerprint, biological fingerprint (biofingerprint) can be defined as a set of chromatographic and/or spectroscopic signals that enable the identification of the active compounds present in a complex herbal sample. Biological fingerprints have been also coupled with traditional chromatographic profiles (binary fingerprints) for obtaining more comprehensive information on the complex sample. However, the main goal of biofingerprint construction is to pinpoint individual active compounds present in a complex matrix, that can, for example, constitute good candidates for potential drugs. Biofingerprint analysis can also be used to simulate and evaluate the actions of active compounds *in vivo* (e.g., interaction with cell membranes, receptors, enzymes, serum proteins, etc.) [6]. Therefore, biofingerprints combine the data obtained in traditional chromatographic fingerprint (qualitative and quantitative) with biological activity. The majority of biofingerprints have been obtained basing on small molecule-biomacromolecule interactions [4]. The compounds present in herbal formulations were checked for their possible interactions with DNA, serum proteins, liver homogenate, or liposomes [4]. Usually RP-HPLC column has been coupled with a column containing immobilized proteins or other macromolecules (affinity chromatography). The most popular techniques have been those using DNA, as it is a molecular target of many drugs, for example, anticancer, antiviral, or antibacterial [4]. To construct biological profiles, two different approaches have been used: affinity chromatography with immobilization of DNA onto silica gel [7] and microdialysis after the interaction with DNA [8]. In the former case, the authors used immobilized DNA column with silica monolithic ODS column. With the application of such chromatographic conditions, the authors constructed binary fingerprints for *Coptis chinensis Franch* and *Rheum palmatum* (see Figure 1). The major drawback of such solution is the column efficiency decrease due to DNA degradation; therefore, the column should be stored at 4°C [7]. In case of the method with microdialysis step,

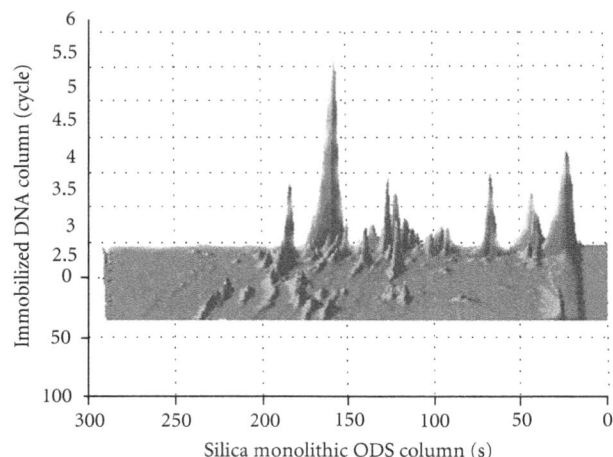

FIGURE 1: Binary chromatographic fingerprint obtained for *Rheum palmatum* L. For experimental details, see publication by Su et al. [7] (originally published in [7]).

the authors compared chromatographic profiles obtained before and after the interaction with DNA [8]. Apart from data multiplication, which eases the comparative studies, DNA-binding fingerprint gives an opportunity to indicate compounds with a potential to be used as DNA-target drugs. Biofingerprinting chromatogram analysis constitutes also an alternative for a conventional procedure for discovery of bioactive compounds present in complex samples.

Apart from DNA-interaction profiles, biological fingerprints have been developed with the use of human serum albumin (HSA) immobilized on a surface of a column packing [9]. This protein is one of the most important drug-binding macromolecules in human plasma. Wang et al. applied silica-bonded human serum albumin column in the first direction and monolithic ODS column in the second one to screen traditional Chinese medicine prescription consisting of ten medicinal materials [9]. The authors underline the importance of applying HSA-immobilized column in the first dimension, because ODS as second-dimensional column is characterized with greater peak capacity and it is better suited for MS detection [9]. With the application of this multidimensional liquid chromatography system, 100 compounds interacting with HSA were separated and 19 of them identified (see Figure 2). The use of monolithic column, characterized with high permeability and excellent mass-transfer properties, enabled fast separation. The authors underline the need for coupling biochromatography fingerprints with traditional chromatographic profiles (2D system), as in the case of the former ones, low column efficiency and peak capacity limit its application as single fingerprint technique. 2D biochromatography was found to be advantageous for biological fingerprint analysis of complex samples, for example, traditional Chinese medicines [9].

The microdialysis step coupled with HPLC has been also used to screen the interaction of herbal samples' components with human plasma proteins [10] and cells [6]. One of the

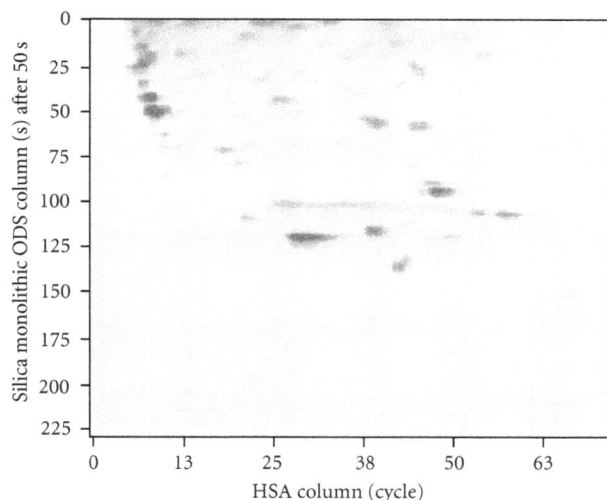

FIGURE 2: Binary chromatographic fingerprint obtained for Longdan Xiegan Decoction (originally published in [9]).

advantages of the proposed solutions is the possibility of direct sample's injection into the HPLC system after the microdialyisis [10]. In case of analyte-protein interactions, the authors studied the binding degree of *Rhizoma Chuanxiong* components to human serum albumin (HSA) and other human blood serum proteins [10]. It was shown that the binding degree of the analytes was influenced by pH, what has been explained as the result of the changing ionization degree of the active herbal compounds and changes in the structure of the HSA binding site. The importance of biological fingerprinting based on analyte-blood protein interactions is also explained, as molecules with lower and higher binding degrees would not be probably active after oral administration [10]. In case of the other paper, mentioned in this paragraph, biological fingerprinting analysis was used to reveal the character of anticancer activities of bioactive compounds in *Cortex Pseudolarix* and *Radix Stephaniae* [6]. The compounds that interacted with drug sensitive human breast cancer MCF-7 cells and multidrug-resistant MCF-7/ADR cells were confined in the semipermeable membrane. The peaks' areas of bioactive compounds decreased when compared with chromatographic fingerprints obtained for the non-microdialised samples. The authors report on the possibility of obtaining false-positive results; however, a solution for excluding such results is also proposed [6]. It is also indicated that low recoveries of some compounds can be a limitation of microdialysis that is performed prior to HPLC analysis. The construction of biological fingerprints, basing on analyte-cancer cells interactions, can be a valuable tool to identify the potential anticancer drugs in complex natural samples [6].

Biological profile of a herbal sample can also be obtained with the application of immobilized liposome chromatography (ILC), as originally reported by Mao et al. [11]. ILC can also be a valuable tool to study drug-membrane interactions. With the application of this technique, it is possible to identify compounds that can penetrate through

biological membranes, as immobilized liposomes resemble the bilayer structure of phospholipid cellular membrane. Using this technique, the authors have tested *Angelica sinensis* sample to evaluate the permeability of its compounds. Due to a complexity of drug-biological membrane interactions in ILC (combination of hydrophobic, ion pairing and hydrogen bonding) it is assumed to be a better model for compounds' permeability testing when compared with models based on interactions with ODS column surface [11]. Limitations concerning the use of some organic solvents and mobile phase additives, which may cause liposome bilayer destruction, are the major drawbacks of ILC. As the ILC could not be directly coupled with MS, a complementary RP-HPLC method was developed to identify the permeable compounds. Coupling ILC column with RP column has also been reported for the screening of a complex traditional Chinese medicine, Longdan Xiegan Decoction (see Figure 3) [12]. The authors identified eight flavonoids and two iridoids that could penetrate biomembranes, basing on their interactions with ILC column. As concluded by the authors, this two-dimensional system shows high suitability for biological fingerprinting analysis of complex natural samples. Reports on the use of cell membranes immobilized on chromatographic columns' packing and on their application for biological fingerprinting have also been published [4].

An interesting approach for biological fingerprint analysis has been proposed by Zhang et al., who reported on the coupling of chromatographic and metabolic fingerprint for the quality control of popular traditional Chinese prescription [13]. The chromatographic profile was obtained with the application of HPLC-UV equipment. The metabolic HPLC fingerprint was obtained after intravenous injection of the analyzed formulation in rats. After proper preparation steps, plasma samples were analyzed by means of HPLC-UV and HPLC-MS. The authors concluded that combination of chemical and metabolic fingerprints is a useful tool for the quality control and revealing possible mechanism of action of herbal samples [13]. The idea of metabolic fingerprint can be based on the concept that herbal-active compounds should be present in blood (but also in urine) after its administration [8].

Another group of biofingerprinting techniques are focused on combining traditional chromatographic fingerprint with an antioxidant profile of a sample. In this case, the sample undergoes separation only on one column; however, the separated compounds are derivatized postcolumn to check their possible antioxidant potential [14]. There are three main categories of antioxidant activity assays, namely, (1) tests that use true reactive oxygen species, (2) assays involving a relatively stable single oxidizing regents, and (3) methods applying electrochemical detection. Coupling high resolution screening techniques (HRS) with (bio)chemical detection has been reviewed by Niederländer et al. [14]. Antioxidant biofingerprinting interactions between compounds and applied derivatizing agents are of chemical rather than biological nature. Therefore, a question arises whether such fingerprints should also be termed "biological". Due to the fact that antioxidant mechanism of natural compounds *in vivo* seems to be analogical with that observed

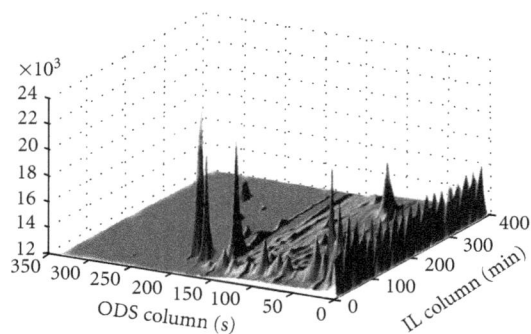

FIGURE 3: 3D chromatogram for Longdan Xiegan Decoction obtained by coupling immobilized liposome chromatography column in the first direction and ODS column in the second one (originally published in [12]).

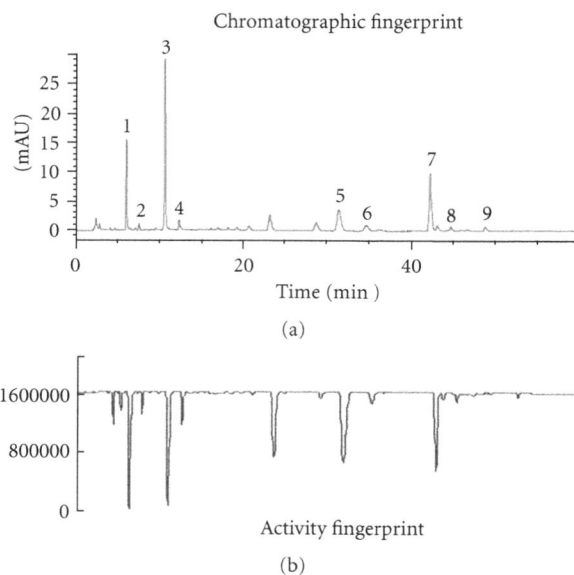

FIGURE 4: Comparison of chromatographic and antioxidant activity fingerprints of Danshen injection (originally published in [16]).

in vitro (e.g., electron or hydrogen transfer to a radical [15]), these methods have also been classified, for the needs of this paper, as belonging to biological fingerprints.

Chang et al. reported on the coupling high-performance liquid chromatography with chemiluminescence detection for antioxidant activity fingerprint construction of Danshen injections (see Figure 4) [16]. Phenolic compounds able to scavenge hydrogen peroxide produced negative peaks in antioxidant-activity fingerprint.

Antioxidant activity fingerprint has been proposed for this preparation, as its protective effect on reperfusion injuries has been linked with antioxidant properties. A data fusion-based method, which combined information encoded both in antioxidant and chemical fingerprints, was applied to evaluate the investigated samples. Significant difference was observed between chromatographic profiles and activity fingerprints. Such results indicate the need to revise the common belief, that samples with similar chromatographic (chemical) profiles likely have similar properties. The authors concluded that quality control of complex herbal samples, by means of simultaneous construction of chemical and biological fingerprints, is more comprehensive when compared with traditional approach. A predominance of antioxidant-activity-integrated fingerprint over traditional chromatographic profiles has been proved for the quality control of Danshen samples.

HPLC online assays for antioxidants have been often performed with the use of postcolumn derivatization with relatively stable free radicals DPPH• or ABTS•+. With the use of this technique, two profiles are always obtained: chemical (termed also "normal") with positive peaks and antioxidant profile with "negative" peaks [17]. Several solutions for the screening of complex natural samples have been proposed. They have been recently reviewed by Niederländer et al. [14] and van Beek et al. [17].

Antioxidant activity of a sample can also be predicted from its chromatographic fingerprint with the use of multivariate calibration techniques, what has been proved in a series of papers [18–23]. The authors have shown that combining the information from chromatographic fingerprints

with the results obtained in spectrophotometric antioxidant assay allows good quality control of herbal samples.

There have been also trials to introduce methods aimed at identification of inhibitors of selected enzymes by means of RP-HPLC. However, their major drawbacks include: relatively large amount of enzymes needed, long reaction time as well as the unsuitability of the majority of organic solvents, used as mobile phase components, for studying analyte-enzyme interactions [17]. Further optimization steps are required to adjust these methods for routine laboratory use.

3. Biological Fingerprinting by Means of TLC

Thin-layer chromatography is considered to be an ideal method for fingerprint construction of herbal samples [24]. The advantages of this technique are well known and have been characterized in numerous publications. However, only a few research groups, working in the field of phytochemical analysis, are aware of its potential for biological detection. The concept of biological fingerprint development in TLC has been introduced by Cieśla et al. [25], who constructed a so-called "binary chromatographic fingerprint" combining chemical and biological detection systems. In the former case, the plates were sprayed with the use of vanillin reagent, while in the case of biological fingerprint methanolic solution of a stable free radical, DPPH•, was applied (see Figure 5). Apart from videoscans, documented for fingerprint comparison, real chromatograms were obtained by means of freely available image processing program—ImageJ. The application of this software gives the possibility to process the fingerprints, documented in the form of videoscans without the need to use densitometers. In fact, the application of densitometers is a difficult task in case of

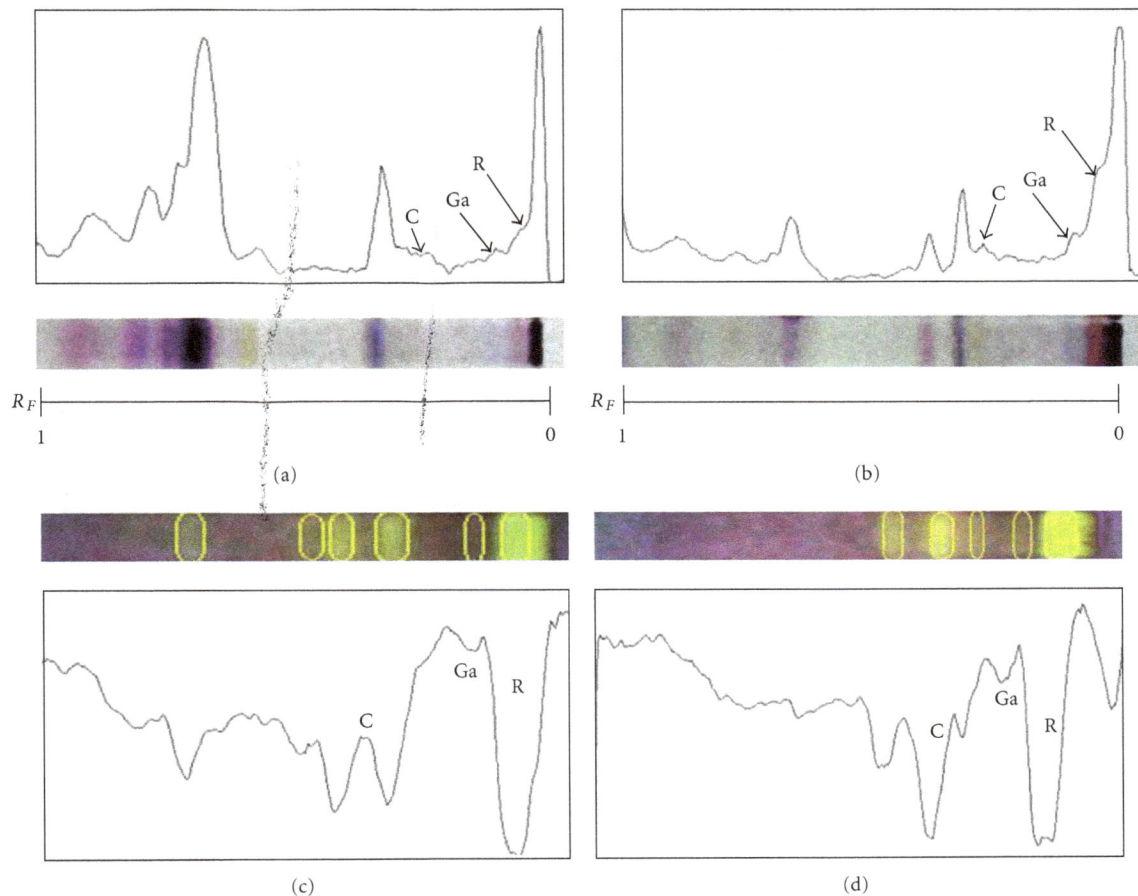

FIGURE 5: Comparison of chemical and free radical scavenging TLC fingerprints obtained with use of ImageJ program, for the extracts prepared from two *Salvia* species: (a) *S. lavandulaefolia* and (b) *S. atropatana*. Symbols' meaning: C:caffeic acid, Ga: gallic acid, R: rosmarinic acid (originally published in [25]).

the results, that are changing in time, as the one obtained with DPPH• as derivatizing agent [26]. With the application of the technique, four *Salvia* species were characterized as a rich source of free radical scavengers, active *in vitro*, namely, *S. officinalis*, *S. triloba*, *S. canariensis*, and *S. lavandulaefolia* [25]. The comparison of both chemical and free radical scavenging fingerprints led to a conclusion, that *S. triloba* can be further investigated as a possible equivalent of the pharmacopoeial *S. officinalis*. The authors underline also that coupling chemical and biological fingerprint enables more comprehensive investigation of the analyzed samples, as some features barely seen in chemical profile may be more distinct in the biological one.

The analogous procedure has been applied for the quality control of pharmaceutical preparations containing *Salvia officinalis* extract [27]. Both chemical and biological chromatographic fingerprints of properly processed chromatographic formulations were compared with chromatographic profiles obtained for a botanical reference material (BRM). It was concluded that the proposed technique may be successfully applied for the comprehensive quality control of the finished products containing sage extract.

The idea of binary chromatographic fingerprint construction by means of thin-layer chromatography is not a new one, as it had been already described by Chen et al. [28]. However, the concept of binary chromatographic fingerprints in the aforementioned papers is somewhat different. In case of the paper by Chen et al., the term "binary" relates to fingerprints obtained separately for glycoside and aglycone fractions, which are bounded by a peak that appears in both profiles. In the paper by Cieśla et al., the focus is on joining the biological and chemical chromatographic fingerprints in order to obtain more information needed for species differentiation and identification of bioactive compounds.

Effect directed analysis, aimed at the isolation of compounds characterized with desired activity, has been the primary application for biological detection in TLC. However, as already shown, it can also be used for the purposes of the quality control of different herbal samples. The potential of thin-layer chromatography for performing simple benchtop bioassays has been recently reviewed, and perspectives of its further development were outlined by several authors [29–31]. Apart from the aforementioned application of TLC for assessing free radical activity, it

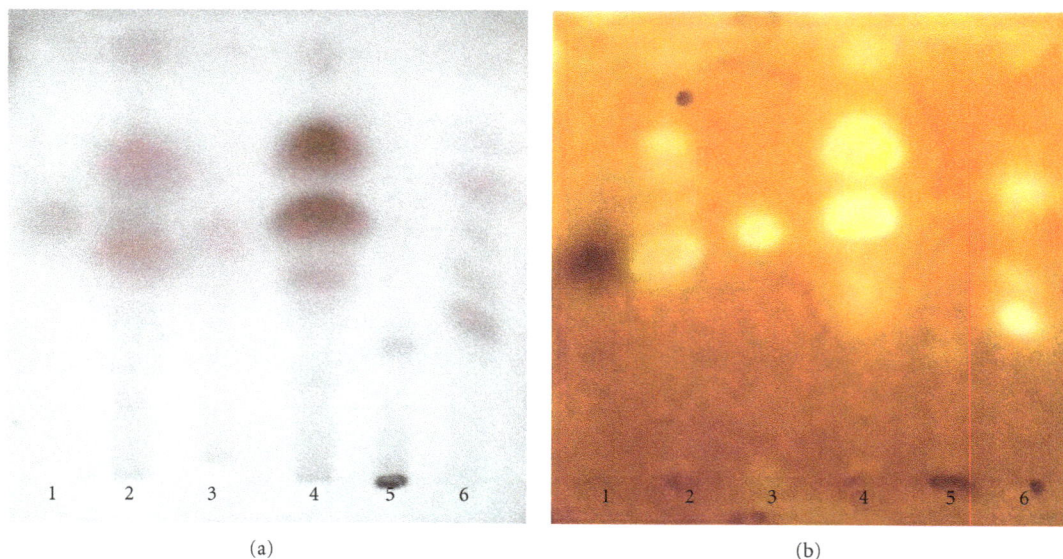

(a)   (b)

FIGURE 6: Comparison of (a) chemical and (b) biological profiles (AchE inhibitory activity) obtained for selected *Lamiaceae* plant samples. In case of chemical detection the plate was sprayed with 20% methanolic solution of sulfuric acid; in case of biological profile, the plate was derivatized according to diazotization method [30]. Symbols' meaning: 1: *Thymus vulgaris* essential oil, 2: *Rosmarinus officinalis* essential oil, 3: *Mentha piperita* essential oil, 4: *Lavandula officinalis* essential oil, 5: *Salvia officinalis* (ethanolic tincture), and 6: *Melissa officinalis* essential oil [34].

can also be used to screen the natural samples for the presence of the inhibitors of selected enzymes or to detect compounds with antibacterial or antifungal properties [29–31]. Due to the large number of people suffering from neurodegenerative ailments (e.g., Alzheimer's or Parkinson's disease) and limited amount of approved drugs, there is a growing need for finding new medicines. Screening natural samples for the presence of acetylcholinesterase (AChE) inhibitors, by means of simple TLC benchtop bioassays, has recently become very popular among scientists, whose researches are focused on the discovery of new potential drugs to treat Alzheimer's disease. New solutions, for better TLC tests' performance, are still being sought and published [32]. The recent results obtained at the Department of Inorganic Chemistry have shown, it is possible to apply low-temperature TLC for screening volatile samples for the presence of AChE inhibitors [33]. An example of using TLC-AChE inhibitory test for selected volatile samples has been shown in Figure 6. Low-temperature TLC bioassays can be a method of choice for effect directed analysis and biological fingerprint construction of volatiles. In case of essential oils, the analytical method of choice is usually GC-MS; however, the use of this technique excludes the possibility of applying effect-directed analysis (detection of direct antioxidants or AChE inhibitors). Thus it may be concluded that TLC biological fingerprinting can be a valuable tool in case of volatile samples, whose content is variable, which may result in changes of its efficacy, when used for medical purposes.

Attention should be paid to rational use of thin-layer chromatographic fingerprint analysis. TLC biofingerprints should only be constructed in case when the analyzed extract or formulation is intended to be used due to its properties

screened in the study. For example, there is no need to perform AChE inhibitory tests for samples that will never be used for treating dementia of Alzheimer's type. Recently, the abuse of different DPPH$^\bullet$ (including TLC-DPPH$^\bullet$ tests) screening techniques can be observed. In many papers, a direct link between the results of DPPH$^\bullet$ studies and pharmacological activity is claimed, which is not supported by any scientific data. The easiness of TLC-DPPH$^\bullet$ test performance can be one of the reasons of its popularity. The most common abuses of simple *in vitro* tests have been described by Houghton et al. [35]. An important issue, that should also be taken into account in case of biological fingerprints, is false-positive results. They may appear for example, due to interactions of analyzed compounds with the active sites of the used adsorbents. Such interactions have been observed while performing tests aimed at the detection of acetylcholinesterase inhibitors or free radical scavengers in plant extracts [33, 36, 37].

4. Perspectives of BFA Development

As it was already stated, the idea of biofingerprint is a new concept in the analysis of multicomponent herbal samples; therefore, its potential for further development is great. In the future, HPLC-based biofingerprint analysis can be extended to screen herbal samples for the presence of bioactive compounds that can interact with particular receptors. New solutions may also be proposed to overcome the common problems encountered in studying analyte-enzyme interactions. Novel detection techniques as well as hyphenations may also be proposed to detect bioactive compounds present in plant extracts. Yu et al. describe the potential of coupling BFA with omics technologies in the discovery of bioactive compounds in traditional Chinese medicines [5]. Thus BFA can be regarded as a powerful and still developing tool in drug discovery.

In case of planar chromatography, special modes of chromatogram development (e.g., multidimensional and multimodal separations, two-dimensional TLC) can be combined with chemical and biological detection for the purposes of effect directed analysis (EDA) as well as fingerprint construction. This approach may be beneficial especially in the case of very complex samples, for example, polyherbal formulations. Special modes of chromatogram development have been well described in the literature [38, 39] with examples of their application for fingerprint construction [40, 41]. Supervised and unsupervised chemometric analysis of two-dimensional images has also been recently described that turned out to be beneficial for recognizing the differences between closely related plant species [42]. Apart from chemometric pretreatment, a concept of an average fingerprint construction by means of freely available program ImageJ (Wayne Rasband, Natonal Institutes of Health, USA; http://rsbweb.nih.gov/ij/) has also been proposed [43]. This idea has been introduced to overcome the problem of the R_F values shift encountered in TLC. The average fingerprint is constructed with the use of "Calculator" function of the ImageJ program. The detailed description of the procedure can be found elsewhere [43]. Planar chromatography is ideally suited for different hyphenations, and even superhyphenations (hypernations) [44]. Coupling low-temperature TLC with MS detection and gas chromatography has recently been shown as suitable for fingerprint construction of essential oils from different *Salvia* species [45, 46]. All the aforementioned solutions may also be further applied in biological fingerprint analysis. the first report that describes the combination of two-dimensional TLC with DPPH• staining, in the analysis of secondary plant metabolites, has already been published [47]. The authors show also an interesting way of data presentation with the use of ImageJ program processing (Figure 7).

Coupling biological detection with fingerprint construction in TLC may in future even outperform some biological fingerprint solutions offered by HPLC.

5. Conclusions

Biological detection in liquid chromatography gives an opportunity to comprehensively analyze herbal samples. Apart from greater amount of data, biofingerprints enable screening plant samples for the presence of new active compounds (effect-directed analysis). It is possible to distinguish the bioactive compounds from among the set of chromatographic and spectroscopic signals. HPLC and TLC can be regarded as complementary techniques, as some biofingerprint solutions are easier to be realized by means of HPLC, in other cases TLC may outperform HPLC. For example, affinity chromatography (DNA-target analysis, HSA binding, or liposome-binding chromatography) can

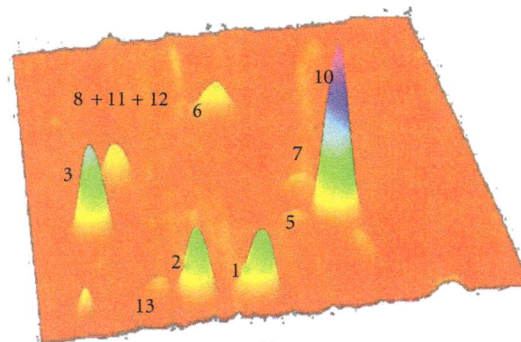

FIGURE 7: Three dimensional plot obtained for a set of standard compounds separated by means of two-dimensional thin-layer chromatography. For symbols see the original publication [47].

be performed by means of HPLC. Construction of volatile samples' biofingerprints is possible to be realized by means of low-temperature TLC. TLC-based screening techniques, for potential plant-derived enzyme inhibitors still outperform those based on HPLC separations. What is more libraries of active compounds may be created with the use of liquid chromatography biofingerprints. HPLC and TLC can be both used in dereplication step, aimed at avoiding isolation of compounds with proven activity. However, the real potential of biological fingerprinting is yet to be explored.

Acknowledgment

Financial support from the 7FP EC project "Proficiency" is gratefully acknowledged.

References

[1] C. Tistaert, B. Dejaegher, and Y. V. Heyden, "Chromatographic separation techniques and data handling methods for herbal fingerprints: a review," *Analytica Chimica Acta*, vol. 690, no. 2, pp. 148–161, 2011.

[2] WHO, *General Guidelines for Methodologies on Research and Evaluation of Traditional Medicine*, World Health Organization, Geneva, Switzerland, 2000.

[3] X. H. Fan, Y. Y. Cheng, Z. L. Ye, R. C. Lin, and Z. Z. Qian, "Multiple chromatographic fingerprinting and its application to the quality control of herbal medicines," *Analytica Chimica Acta*, vol. 555, no. 2, pp. 217–224, 2006.

[4] X. Su, L. Kong, X. Lei, L. Hu, M. Ye, and H. Zou, "Biological fingerprinting analysis of traditional Chinese medicines with targeting ADME/Tox property for screening of bioactive compounds by chromatographic and MS methods," *Mini-Reviews in Medicinal Chemistry*, vol. 7, no. 1, pp. 87–98, 2007.

[5] F. Yu, L. Kong, H. Zou, and X. Lei, "Progress on the screening and analysis of bioactive compounds in traditional Chinese medicines by biological fingerprinting analysis," *Combinatorial Chemistry and High Throughput Screening*, vol. 13, no. 10, pp. 855–868, 2010.

[6] X. Lei, L. Kong, H. Zou, H. Ma, and L. Yang, "Evaluation of the interaction of bioactive compounds in Cortex Pseudolarix and Radix Stephaniae by the microdialysis probe coupled with

high performance liquid chromatography-mass spectrometry," *Journal of Chromatography A*, vol. 1216, no. 11, pp. 2179–2184, 2009.

[7] X. Su, L. Hu, L. Kong, X. Lei, and H. Zou, "Affinity chromatography with immobilized DNA stationary phase for biological fingerprinting analysis of traditional Chinese medicines," *Journal of Chromatography A*, vol. 1154, no. 1-2, pp. 132–137, 2007.

[8] X. Su, L. Kong, X. Li, X. Chen, M. Guo, and H. Zou, "Screening and analysis of bioactive compounds with biofingerprinting chromatogram analysis of traditional Chinese medicines targeting DNA by microdialysis/HPLC," *Journal of Chromatography A*, vol. 1076, no. 1-2, pp. 118–126, 2005.

[9] Y. Wang, L. Kong, L. Hu et al., "Biological fingerprinting analysis of the traditional Chinese prescription Longdan Xiegan Decoction by on/off-line comprehensive two-dimensional biochromatography," *Journal of Chromatography B*, vol. 860, no. 2, pp. 185–194, 2007.

[10] M. Guo, X. Su, L. Kong, X. Li, and H. Zou, "Characterization of interaction property of multicomponents in Chinese Herb with protein by microdialysis combined with HPLC," *Analytica Chimica Acta*, vol. 556, no. 1, pp. 183–188, 2006.

[11] X. Mao, L. Kong, Q. Luo, X. Li, and H. Zou, "Screening and analysis of permeable compounds in *Radix Angelica Sinensis* with immobilized liposome chromatography," *Journal of Chromatography B*, vol. 779, no. 2, pp. 331–339, 2002.

[12] Y. Wang, L. Kong, X. Lei et al., "Comprehensive two-dimensional high-performance liquid chromatography system with immobilized liposome chromatography column and reversed-phase column for separation of complex traditional Chinese medicine Longdan Xiegan Decoction," *Journal of Chromatography A*, vol. 1216, no. 11, pp. 2185–2191, 2009.

[13] J. L. Zhang, M. Cui, Y. He, H. L. Yu, and D. A. Guo, "Chemical fingerprint and metabolic fingerprint analysis of Danshen injection by HPLC-UV and HPLC-MS methods," *Journal of Pharmaceutical and Biomedical Analysis*, vol. 36, no. 5, pp. 1029–1035, 2005.

[14] H. A. G. Niederländer, T. A. van Beek, A. Bartasiute, and I. I. Koleva, "Antioxidant activity assays on-line with liquid chromatography," *Journal of Chromatography A*, vol. 1210, no. 2, pp. 121–134, 2008.

[15] M. Musialik, R. Kuzmicz, T. S. Pawlowski, and G. Litwinienko, "Acidity of hydroxyl groups: an overlooked influence on antiradical properties of flavonoids," *Journal of Organic Chemistry*, vol. 74, no. 7, pp. 2699–2709, 2009.

[16] Y. X. Chang, X. P. Ding, J. Qi et al., "The antioxidant-activity-integrated fingerprint: an advantageous tool for the evaluation of quality of herbal medicines," *Journal of Chromatography A*, vol. 1208, no. 1-2, pp. 76–82, 2008.

[17] T. A. van Beek, K. K. R. Tetala, I. I. Koleva et al., "Recent developments in the rapid analysis of plants and tracking their bioactive constituents," *Phytochemistry Reviews*, vol. 8, no. 2, pp. 387–399, 2009.

[18] A. M. van Nederkassel, M. Daszykowski, D. L. Massart, and Y. Vander Heyden, "Prediction of total green tea antioxidant capacity from chromatograms by multivariate modeling," *Journal of Chromatography A*, vol. 1096, no. 1-2, pp. 177–186, 2005.

[19] M. Daszykowski, Y. Vander Heyden, and B. Walczak, "Robust partial least squares model for prediction of green tea antioxidant capacity from chromatograms," *Journal of Chromatography A*, vol. 1176, no. 1-2, pp. 12–18, 2007.

[20] M. Dumarey, A. M. van Nederkassel, E. Deconinck, and Y. Vander Heyden, "Exploration of linear multivariate calibration techniques to predict the total antioxidant capacity of green tea from chromatographic fingerprints," *Journal of Chromatography A*, vol. 1192, no. 1, pp. 81–88, 2008.

[21] N. Nguyen Hoai, B. Dejaegher, C. Tistaert et al., "Development of HPLC fingerprints for *Mallotus* species extracts and evaluation of the peaks responsible for their antioxidant activity," *Journal of Pharmaceutical and Biomedical Analysis*, vol. 50, no. 5, pp. 753–763, 2009.

[22] C. Tistaert, B. Dejaegher, N. Nguyen Hoai et al., "Potential antioxidant compounds in *Mallotus* species fingerprints. Part I: indication, using linear multivariate calibration techniques," *Analytica Chimica Acta*, vol. 649, no. 1, pp. 24–32, 2009.

[23] M. Dumarey, I. Smets, and Y. Vander Heyden, "Prediction and interpretation of the antioxidant capacity of green tea from dissimilar chromatographic fingerprints," *Journal of Chromatography B*, vol. 878, no. 28, pp. 2733–2740, 2010.

[24] E. Reich and A. Schibli, *High-Performance Thin-Layer Chromatography for the Analysis of Medicinal Plants*, Thieme, New York, NY, USA, 2006.

[25] Ł. Cieśla, D. Staszek, M. Hajnos, T. Kowalska, and M. Waksmundzka-Hajnos, "Development of chromatographic and free radical scavenging activity fingerprints by thin-layer chromatography for selected *Salvia* species," *Phytochemical Analysis*, vol. 22, no. 1, pp. 59–65, 2011.

[26] T. Yrjönen, L. Peiwu, J. Summanen, A. Hopia, and H. Vuorela, "Free radical-scavenging activity of phenolics by reversed-phase TLC," *Journal of the American Oil Chemists' Society*, vol. 80, no. 1, pp. 9–14, 2003.

[27] Ł. Cieśla and M. Waksmundzka-Hajnos, "Application of thin-layer chromatography for the quality control and screening the free radical scavenging activity of selected pharmacuetical preparations containing *Salvia officinalis* L. extract," *Acta Poloniae Pharmaceutica—Drug Research*, vol. 67, no. 5, pp. 481–485, 2010.

[28] S. B. Chen, H. P. Liu, R. T. Tian et al., "High-performance thin-layer chromatographic fingerprints of isoflavonoids for distinguishing between *Radix Puerariae Lobate* and *Radix Puerariae Thomsonii*," *Journal of Chromatography A*, vol. 1121, no. 1, pp. 114–119, 2006.

[29] Ł. Cieśla, "Thin-layer chromatography with biodetection for the search of new potential drugs to treat neurodegenerative diseases—state of the art and future perspectives," *Medicinal Chemistry*. In press.

[30] A. Marston, "Thin-layer chromatography with biological detection in phytochemistry," *Journal of Chromatography A*, vol. 1218, no. 19, pp. 2676–2683, 2011.

[31] I. M. Choma and E. M. Grzelak, "Bioautography detection in thin-layer chromatography," *Journal of Chromatography A*, vol. 1218, no. 19, pp. 2684–2691, 2011.

[32] Z.-D. Yang, Z.-W. Song, J. Ren, M.-J. Yang, and S. Li, "Improved thin-layer chromatography bioautographic assay for the detection of acetylcholinesterase inhibitors in plants," *Phytochemical Analysis*, vol. 22, no. 6, pp. 509–515, 2011.

[33] Ł. Cieśla, J. Kryszeń, A. Stochmal, and M. Waksmundzka-Hajnos, "TLC-DPPH test revisited," in *Proceedings of the 34th Symposium: Chromatographic Methods of Investigating the Organic Compounds*, Katowice-Szczyrk, Poland, June 2011.

[34] Ł. Cieśla, J. Kryszeń, and M. Waksmundzka-Hajnos, "unpublished results".

[35] P. J. Houghton, M. J. Howes, C. C. Lee, and G. Steventon, "Uses and abuses of in vitro tests in ethnopharmacology: visualizing an elephant," *Journal of Ethnopharmacology*, vol. 110, no. 3, pp. 391–400, 2007.

[36] I. K. Rhee, M. van de Meent, K. Ingkaninan, and R. Verpoorte, "Screening for acetylcholinesterase inhibitors from Amaryllidaceae using silica gel thin-layer chromatography in combination with bioactivity staining," *Journal of Chromatography A*, vol. 915, no. 1-2, pp. 217–223, 2001.

[37] S. Di Giovanni, A. Borloz, A. Urbain et al., "In vitro screening assays to identify natural or synthetic acetylcholinesterase inhibitors: thin layer chromatography versus microplate methods," *European Journal of Pharmaceutical Sciences*, vol. 33, no. 2, pp. 109–119, 2008.

[38] Ł. Cieśla and M. Waksmundzka-Hajnos, "Two-dimensional thin-layer chromatography in the analysis of secondary plant metabolites," *Journal of Chromatography A*, vol. 1216, no. 7, pp. 1035–1052, 2009.

[39] Ł. Cieśla and M. Waksmundzka-Hajnos, "Multidimensional and multimodal separations by HPTLC in phytochemistry," in *High-Performance Thin-Layer Chromatography (HPTLC)*, M. M. Srivastava, Ed., Springer, Berlin, Germany, 2011.

[40] Ł. Cieśla, A. Bogucka-Kocka, M. Hajnos, A. Petruczynik, and M. Waksmundzka-Hajnos, "Two-dimensional thin-layer chromatography with adsorbent gradient as a method of chromatographic fingerprinting of furanocoumarins for distinguishing selected varieties and forms of *Heracleum spp*," *Journal of Chromatography A*, vol. 1207, no. 1-2, pp. 160–168, 2008.

[41] L. Cieśla, K. Skalicka-Woźniak, M. Hajnos, M. Hawryl, and M. Waksmundzka-Hajnos, "Multidimensional TLC procedure for separation of complex natural mixtures spanning a wide polarity range; Application for fingerprint construction and for investigation of systematic relationships within the *Peucedanum* genus," *Acta Chromatographica*, vol. 21, no. 4, pp. 641–657, 2009.

[42] Ł. Komsta, T. Cieśla, A. Bogucka-Kocka, A. Józefczyk, J. Kryszeń, and M. Waksmundzka-Hajnos, "The start-to-end chemometric image processing of 2D thin-layer videoscans," *Journal of Chromatography A*, vol. 1218, no. 19, pp. 2820–2825, 2011.

[43] L. Cieśla, M. Hajnos, and M. Waksmundzka-Hajnos, "Application of hydrophilic interaction TLC systems for separation of highly polar glycosidic compounds from the flowers of selected Verbascum species," *Journal of Planar Chromatography—Modern TLC*, vol. 24, no. 4, pp. 295–300, 2011.

[44] G. Morlock and W. Schwack, "Hyphenations in planar chromatography," *Journal of Chromatography A*, vol. 1217, no. 43, pp. 6600–6609, 2010.

[45] M. Sajewicz, Ł. Wojtal, M. Hajnos, M. Waksmundzka-Hajnos, and T. Kowalska, "Low-temperature TLC-MS of essential oils from five different sage (*Salvia*) species," *Journal of Planar Chromatography—Modern TLC*, vol. 23, no. 4, pp. 270–276, 2010.

[46] M. Sajewicz, L. Wojtal, D. Staszek, M. Hajnos, M. Waksmundzka-Hajnos, and T. Kowalska, "Low temperature planar chromatography-densitometry and gas chromatography of essential oils from different sage (*Salvia*) species," *Journal of Liquid Chromatography and Related Technologies*, vol. 33, no. 7-8, pp. 936–947, 2010.

[47] M. A. Hawryl and M. Waksmundzka-Hajnos, "Two-dimensional thin-layer chromatography of selected *Polygonum* sp. extracts on polar-bonded stationary phases," *Journal of Chromatography A*, vol. 1218, no. 19, pp. 2812–2819, 2011.

19

Micellar LC Separation of Sesquiterpenic Acids and Their Determination in *Valeriana officinalis* L. Root and Extracts

Artem U. Kulikov

Laboratory of Pharmacopoeial Analysis, Scientific and Expert Pharmacopoeial Centre, Astronomicheskaya Street 33, Kharkov 61085, Ukraine

Correspondence should be addressed to Artem U. Kulikov, kulikov@phukr.kharkov.ua

Academic Editor: Samuel Carda-Broch

A simple micellar liquid chromatography (MLC) method was developed and validated according to ICH Guidelines for the determination of sesquiterpenic acids (valerenic, hydroxyvalerenic, and acetoxyvalerenic acids) in root and rhizome extract from *Valeriana officinalis* L. and valerian dry hydroalcoholic extract. Samples were analyzed on Nucleosil C18 column (150 mm×4.6 mm, 5 μm) using an isocratic mobile phase which consisted of Brij 35 (5% (w/v) aqueous solution; pH 2.3 $\pm$ 0.1 by phosphoric acid) and 1-butanol (6% (v/v)); UV detection was at 220 nm. Micellar mobile phase using allows to fully separate valerenic acids within 25 minutes. Linearity for hydroxyvalerenic, acetoxyvalerenic, and valerenic acids was 1.9–27.9, 4.2–63.0, and 6.1–91·3 μg·mL^{-1}, and limit of detection was 0.14, 0.037, and 0.09 μg·mL^{-1}, respectively. Intraday and interday precisions were not less than 2% for all investigated compounds. The proposed method was found to be reproducible and convenient for quantitative analysis of sesquiterpenic acids in valerian root and related preparations.

1. Introduction

Valeriana officinalis L. is the most common species of the genus *Valeriana* that is used for its medicinal properties [1]. The root and rhizome of the *valerian* plant is used medicinally for its sedative properties with indications including nervous tension, insomnia, anxiety, and stress. *Valerian* is also considered to have antispasmodic, anticonvulsant, and antidepressant effects.

The roots and rhizomes of *Valeriana officinalis* contain two main groups of constituents: sesquiterpenes of the volatile oil (valerenic acid and its derivatives, valeranone, valeranal, and kessyl esters) and valepotriates (valtrate, didrovaltrate, acevaltrate, and isovaleroxyhydroxyvaltrate), in addition to other constituents such as flavonoids, triterpenes, lignans, and alkaloids [2, 3].

Valerian is available in a variety of formulations, including tablets, capsules, liquid, teas, and tinctures. Products may contain whole herb and/or a proprietary blend or may also be combined with other herbal supplements (e.g., lemon balm, hops, kava, St. John's worth, etc.). *Valerian* root and some commercial products (drugs) are standardized according to

the content of valerenic acid, but concentrations vary among products.

Valerenic acids are sesquiterpenes based on the dual ring valerane structure with the main representatives being valerenic acid ((2E)-3-[(4S,7R,7aR)-3,7-dimethyl-2, 4,5,6,7,7a-hexahydro-1H-inden-4-yl]-2-methylacrylic acid; $C_{15}H_{22}O_2$), acetoxyvalerenic acid, and hydroxyvalerenic acid. Chemical structures of these acids are presented in Figure 1.

Several methods, adopting different analytical techniques, have been reported to analyze sesquiterpenic acids: TLC [4–6], SFC [7], and HPLC-UV, which is the most commonly used method.

Bos et al. [8, 9] simultaneously analyzed valepotriates (valtrate, isovaltrate, acevaltrate, and didrovaltrate), valerenic acids (valerenic, hydroxyvalerenic, and acetoxyvalerenic), baldrinal and homobaldrinal by HPLC-DAD using LiChrospher 100 RP-18 column, and gradient elution of acetonitrile-water mobile phases. Total analysis time was 30 min.

Gobbeto and Lolla [10] proposed the HPLC method for the determination of valerenic acids in *valerian* extracts using Hypersil ODS column and gradient elution with acetonitrile

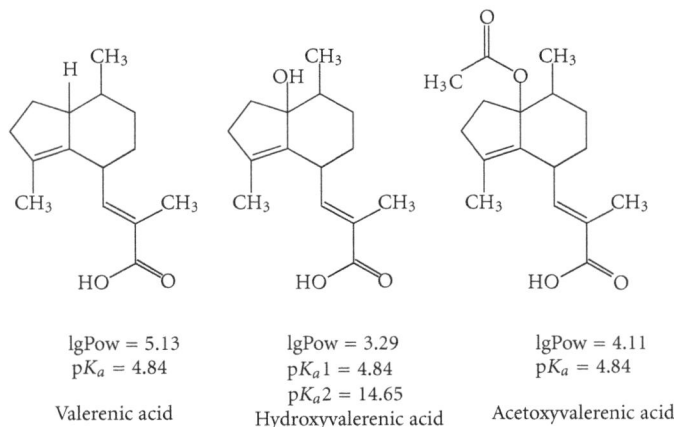

lgPow = 5.13
pK_a = 4.84

Valerenic acid

lgPow = 3.29
pK_a1 = 4.84
pK_a2 = 14.65

Hydroxyvalerenic acid

lgPow = 4.11
pK_a = 4.84

Acetoxyvalerenic acid

FIGURE 1: Sesquiterpenic acids chemical structures, calculated values of logarithm of partition constant in 1-octanol-water system lgPow, and protonization constants (pK_a).

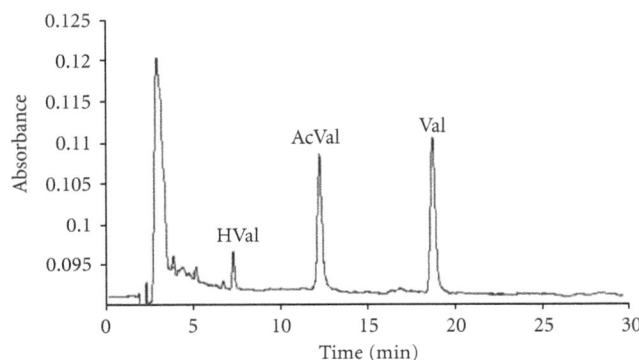

FIGURE 2: Chromatogram of sesquiterpenic acids separation obtained by approved RP HPLC method. Kromasil C18 chromatographic column; linear gradient elution with 5 g·L^{-1} aqueous H_3PO_4-acetonitrile mobile phases; flow rate 1.5 mL·min^{-1}; UV 220 nm. HVal: hydroxyvalerenic acid, AcVal: acetoxyvalerenic acid, and Val: valerenic acid.

containing 0.1% phosphoric acid-acetonitrile mobile phase; UV detection at 220 nm; analysis time was 40 min.

Bicchi et al. [7] developed supercritical fluid chromatography method with UV detection for the determination of valerenic acids and valepotriates in *valerian* extracts (analysis time is 20 min) and compared results, obtained during *valerian* extracts analysis that were determined by SFC and HPLC methods.

European Pharmacopoeia [11] for the analysis of valerenic acids in *valerian* root end extracts proposed HPLC-UV method with C18 column using an acetonitrile-phosphoric acid aqueous solution gradient elution. Detection at 220 nm and total analysis time is about 30 min (Figure 2).

As we can see, all LC methods required gradient elution techniques for separation and assay of sesquiterpenic acids in raw plant materials and drugs.

It was already described that micellar liquid chromatography (MLC) is an alternative for RP HPLC method [12, 13]. MLC is successfully applied to the analysis of drugs [14, 15] and biological fluids [16], and nowadays this method is widely used for the analysis of raw plant materials and plant containing drugs [17, 18]. One of MLC advantages that made this method preferable to HPLC is the simultaneous separa-

tion of both ionic and nonionic compounds, substances with different hydrophobicity without needing gradient elution.

The aim of this work is an investigation of possibility to separate valerenic acids using MLC method, to validate method according ICH validation requirements (specificity, accuracy, precision, and linearity), and an application of the developed method for the rapid sesquiterpenic acid analysis in raw materials (roots and rhizomes of *Valeriana officinalis*) and *valerian* dry extracts.

2. Experimental

2.1. Reagents and Materials. Acetoxyvalerenic acid (AcVal), hydroxyvalerenic acid (HVal), and valerenic acid (Val) were isolated from *Valeriana officinalis* L. in Plant Material Department of the Scientific and Expert Pharmacopoeial Centre, Ukraine, as described in [19]. Chemical structure was confirmed using IR, ^{1}H, and ^{13}C NMR spectra (in the State Scientific Institution "Institute for Single Crystals" of the National Academy of Sciences of Ukraine). Chromatographic purity was determined by RP HPLC method [11]; those were more than 98.5% for all cases.

Valerian standardized dry extract EP CRS (0.38% $C_{15}H_{22}O_2$ (valerenic acid) was from European Pharmacopoeia (Strasbourg, France).

1-butanol (1-BuOH), concentrated phosphoric acid, polyethylene glycol dodecyl ether (Brij 35), methanol (HPLC grade), and other chemicals were purchased from Fluka Chemie (Buchs, Switzerland). Double-distilled water was used in all experiments.

Fragmented underground parts of *Valeriana officinalis* L. were from "Sumyfitofarmacia Ltd," Ukraine; *valerian* dry extracts were from "AIM Ltd," Ukraine.

2.2. Apparatus. The chromatographic measurements were carried out with Hewlett Packard equipment (Agilent Technologies, Waldbronn, Germany) consisting of a Series 1050 pump, a Series 1050 spectrophotometric detector with the variable wavelength, and a Series 3395 integrator. The analytical column was the reversed-phase Nucleosil C18 (150 mm × 4.6 mm, 5 μm, Macherey-Nagel, Germany). The pH values were determined with a Beckman Φ-200 pH meter (Beckman Instruments, Fullerton, CA, USA) and Paratrode electrode (Metrohm AG, Herisau, Switzerland).

For the determination of peak purity, the Waters 2695 Separation Module (Waters, Milford, MA, USA) with Waters 996 Photodiode Array Detector (Waters) was used.

2.3. Standard Preparation. A stock standard solution was prepared by weighting an amount of hydroxyvalerenic acid, acetoxyvalerenic acid, and valerenic acid in a 100 mL volumetric flask, dissolving it in methanol to obtain concentrations of investigated compounds 0.0465 mg·mL^{-1}, 0.105 mg·mL^{-1}, and 0.152 mg·mL^{-1}, respectively. The solution was stored in a dark place at the temperature 5°C during 7 days.

Working standard solutions (used for linearity evaluation, evaluation of precision, repeatability, and accuracy of the MLC method) were prepared by dilution of the stock standard solution with a micellar mobile phase to obtain solutions with different concentrations within the range of interest.

2.4. Sample Preparation

Valerian Root. Extract 1.50 g of powdered underground parts of *Valeriana officinalis* (1 mm) with 20 mL of methanol and heat on a water bath under a reflux condenser for 30 min. Allow to cool and filter through a paper filter into a 50 mL volumetric flask. Add another 20 mL of methanol, heat on a water bath under a reflux condenser for 15 min, cool, filter into the same 50 mL volumetric flask, and dilute to 50.0 mL with methanol [11].

Valerian Dry Extract. Place about 0.5 g of *valerian* dry extract in a 25 mL volumetric flask, add 15 mL methanol, mix, and sonicate during 10 min. Then the solution is diluted to 25.0 mL with methanol and mix [11].

All samples must be filtered through a 45 μm HPLC filter (DynaCard HPLC filter, Microgon, USA) before analysis.

2.5. Chromatographic Separation. A 5% (w/v) aqueous solution of Brij 35 with 6% (v/v) of 1-butanol, adjusted to pH 2.3 ± 0.1 by phosphoric acid, was prepared and used as mobile phase. The flow rate was set at 1.0 mL·min^{-1}, and the injection volume was 10 μL. The chromatographic runs were carried out at 40.0 ± 0.1°C. Detection wavelength was chosen at 220 nm.

2.6. Data Sources, Software, and Processing. The statistical analyses were performed with Microsoft Excel (2007, Microsoft Corporation, http://office.microsoft.com/). The values of lgPow and pK_a were calculated using ACDLabs 10.0 (2010, http://www.ACDLabs.com/) and ChemBioDraw 11.0 (2010, Cambridgesoft, http://www.cambridgesoft.com/).

3. Results and Discussion

Nonionic surfactants, unfortunately, are not widely used for micellar liquid chromatographic separation of compounds with different hydrophobicity. It was not found any recommendation in the literature about an approximate nonionic surfactant concentration, which is required for different hydrophobicity substances separation, as described for sodium dodecyl sulfate as micellar-forming agent [12]. In [20–22] it was found that concentration of nonionic surfactants to be micellar eluents should be sufficiently large—5–10% (w/v).

Figure 1 provides the chemical structures and some physical-chemical characteristics of investigated compounds: hydrophobicity as lgPow (logarithms of partition constant in 1-octanol-water system) and pKa values. Hydrophobicity of compounds is not quite different and ranges from 3 for hydroxyvalerenic acid to 5 for valerenic acid. Wang et al. [20] separated quinazoline derivatives with hydrophobicity 2–4 using 0.06 M Brij 35 (6% w/v) as a micellar mobile phase.

Preliminary experiment for valerenic acids separation showed that its retention in approximately 0.06 M Brij 35 mobile phase was about 75 min, and peaks asymmetries of investigated compounds were high.

According to pK_a values, at pH value below 4, the predominant form of compounds in solution is molecular forms—all ionizable groups are fully protonated, and at pH, above 6 compounds are negatively charged. A mobile phase for acid separation should be acidified. Preliminary experiments have shown that when acidified 0.06 M Brij 35 containing micellar mobile phase used for separation of the investigated compounds, peak symmetry is lower than 2; it is appropriate to the assay determination.

It was investigated retention time, peak symmetry, and resolution for different micellar mobile phases to choose an optimal micellar mobile phase composition. Brij 35 concentrations in mobile phases were varying from 2% to 8% (w/v) with a step of 1%, and micellar mobile phase organic modifier (1-bithanol) was varying from 1% to 10% (v/v) with the same step. The mobile phase pH in all cases was about 2.3 ± 0.1.

The optimization procedure [23, 24] that was used for choosing the optimal mobile phase composition showed that

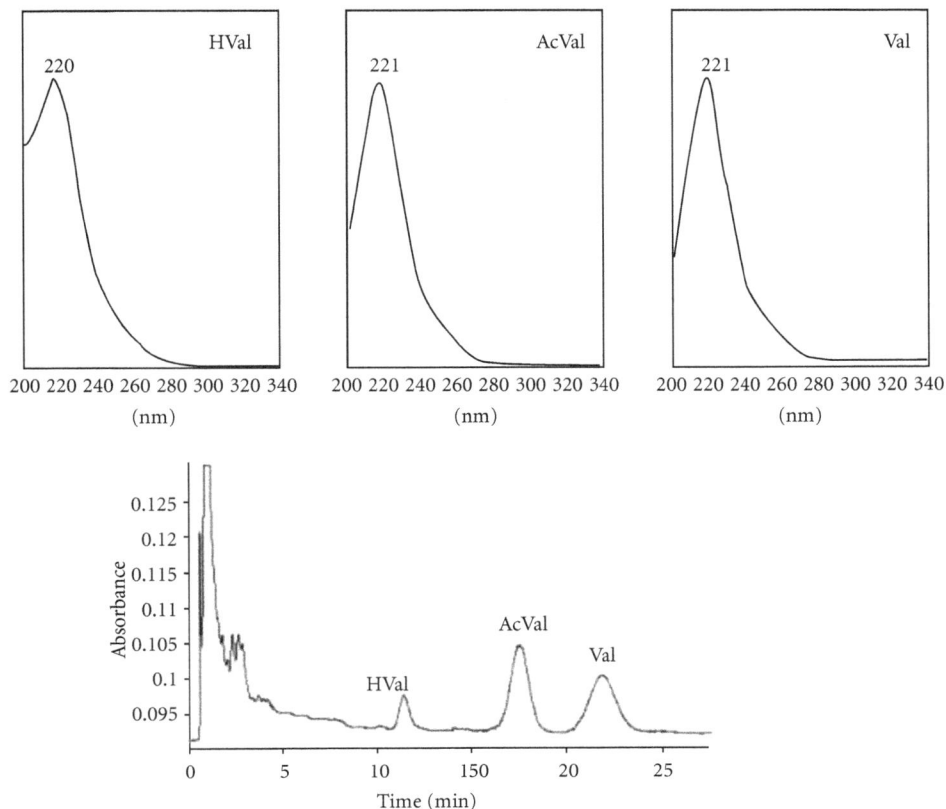

FIGURE 3: Micellar LC separation of sesquiterpenic acids during selectivity determination. HVal: hydroxyvalerenic acid, AcVal: acetoxyvalerenic acid, and Val: valerenic acid.

for all micellar mobile phases were obtained good chromatographic characteristics: sesquiterpenic acid resolution was more than 1.5, and peak symmetry was 1.5–2. Lowest retention time (12 min) was obtained when 8% (w/v) of Brij 35 and 10% (v/v) of 1-BuOH as micellar mobile phase components were used. Unfortunately, this micellar mobile phase has high viscosity, and column input pressure was unacceptable: 350–370 bar whereas chromatograph pump upper pressure limit is 400 bar. When a mobile phase with low viscosity (2% (w/v) of Brij 35 and 1% (v/v) of 1-BuOH) was used for sesquiterpenic acids separation, total analysis time was about 90 min.

Optimum between micellar mobile phase viscosity (relatively low viscosity) and analysis time (till 25 min) was found, when Brij 35 concentration of 5% (w/v) and 1-BuOH volume fraction of 6% (v/v) were chosen.

In addition, the efficiency of the separation and peak asymmetry can be improved by using high temperatures due to faster mass transfer of solutes between mobile and stationary phases [15], and for reducing of mobile phase viscosity. A column temperature 40°C was chosen, because higher temperature (50 and 60°C) had not exerted significant influence on the chromatographic behavior of investigated compounds.

3.1. Validation of the MLC Method.
The procedures and parameters used for the validation of the chromatographic

method developed in this study are those described in [25, 26].

3.1.1. Specificity (Selectivity).
The specificity of the method was determined by analyzing *valerian*-standardized dry extract EP CRS and methanolic extract of *valerian* root. Methanolic extract was analyzed without and with addition of investigated compounds; increasing peak areas indicate the presence of valerenic acids in the chromatogram. Selectivity of the MLC method was assessed by the comparison of the spectrum extracted from the chromatogram of *valerian*-standardized dry extract and methanolic extract obtained using photodiode array detector; peak purity was also determined.

Figure 3 shows a chromatogram and UV spectra obtained for *valerian*-standardized dry extract, which contains (see Figure 2) hydroxyvalerenic, acetoxyvalerenic, and valerenic acids. The same spectra were obtained for *valerian* methanolic extract. The peaks purity was more than 98.5% in all cases.

3.1.2. Linearity.
A linear plot was obtained from six different concentrations of working standard solutions using four replicate injections [25, 26]. The regression line was calculated as $Y = a + bC$, where C was the analyte concentration

TABLE 1: Characteristic parameters of the calibration equations for the proposed MLC method for sesquiterpenic acids determination.

Parameters	Hydroxyvalerenic acid	Acetoxyvalerenic acid	Valerenic acid
Calibration range (μg·mL^{-1})	1.9–27.9	4.2–63.0	6.1–91.2
Detection limit (μg·mL^{-1})	0.037	0.09	0.14
Quantitation limit (μg·mL^{-1})	0.11	0.27	0.42
Parameters of regression equation $Y = a + b \cdot C$, where C is the concentration of compound in μg·mL^{-1}, Y is the peak area			
Intercept (a)	$-5.5 \cdot 10^3$	$-7.8 \cdot 10^4$	$-1.6 \cdot 10^4$
SD of the intercept (S_a)	$0.3 \cdot 10^3$	$0.6 \cdot 10^4$	$0.1 \cdot 10^4$
Slope (b)	$27.1 \cdot 10^3$	$21.3 \cdot 10^4$	$23.8 \cdot 10^4$
SD of the slope (S_b)	$0.2 \cdot 10^3$	$0.2 \cdot 10^4$	$0.2 \cdot 10^4$
RSD of the slope, %	0.74	0.94	0.84
Correlation coefficient (r)	0.9995	0.9994	0.9995

(μg mL^{-1}), and y was the peak area. The calibration plot was obtained by linear least-squares regression.

Calibration parameters, such as slope, intercept, and correlation coefficient are provided in Table 1. The correlation coefficient (r) was close to unity; hence, there was a linear relationship between amount of analyte and detector signal. The systematic deviations between experimental and predicted responses were not observed.

Since the coefficient of correlation is not suitable as a general acceptance criterion to the linearity performance of an analytical procedure [27], the relative standard error of slope was used as a parameter with respect to precision of the regression. This parameter should be comparable to the relative standard deviation obtained in precision studies within the given concentrations range. Relative standard error of slopes was obtained to be less than RSD of precision.

3.1.3. Limit of Detection and Quantitation.

The limits of detection (LOD) and quantitation (LOQ) were calculated in accordance with the 3.3 s/m and 10 s/m criteria, respectively, where s is the standard deviation of the peak area for the sample, and m is the slope of the calibration curve, determined from linearity [25]. Both values were calculated and presented in Table 1.

3.1.4. Accuracy.

The accuracy of the proposed method was tested in some ways. First, samples of known concentration (working standard solutions) were analyzed, and measured values were compared with the true value. The results obtained from determination of accuracy, expressed as percentage recovery, are summarized in Table 2. Accuracy criterion for the assay method is that the mean recovery will be $100 \pm 2\%$ for each concentration over the investigation range [28]. The recovery of the proposed method was good.

Second, this study was performed by addition of known amounts of studied compounds to a solution of *valerian* dry extract (standard addition method). The resulting mixtures were analyzed by proposed MLC method, and the obtained results were compared with the expected results. The excellent recoveries of standard addition method (Table 2) suggested good accuracy of the proposed method.

Third, accuracy of MLC method was determined by comparing results obtained by proposed method with results from the validated method using reversed-phase HPLC [11]. MLC and RP-HPLC results for sesquiterpenic acids assay were compared. Statistically analysis of the results obtained for both methods using variance ratio F-test shows that there is no significant difference between these results. The calculated F values were less than those of the theoretical values at 95% confidence level (see data from Table 3).

3.1.5. Precision.

The precision as intraday and interday reproducibility, expressed as RSD %, was characterized by the spread of data from replicate determinations.

For the intraday reproducibility, that is, repeatability, we performed nine determinations covering the specified range of the method. Working standard solutions were analyzed (four replicates each). Interday precision of the method was checked on three different days by preparing and analyzing working standard solutions (four replicate injections) under the same conditions.

The results obtained from these analyses are listed in Table 4 as mean recovery. The table shows that there is no significant difference between assay results either within days or between days, implying that the reproducibility of MLC method was good. In all instances, the % R.S.D. values were less than 2%—criterion for an intra-assay precision [28].

3.2. Analysis of Plant Material and Drug.

Sesquiterpenic acids (valerenic, acetoxyvalerenic, and hydroxyvalerenic acids) were quantified using developed MLC method in the *valerian* root and rhizomes and *valerian* dry extract (chromatograms of investigated compounds are presented in Figure 4). Fragmented underground parts of *Valeriana officinalis* L. cultivated in 2008–2011 years were from "Sumyfitofarmacia Ltd," Ukraine; *valerian* dry hydroalcoholic extracts obtained from above-mentioned *valerian* root were from "AIM Ltd," Ukraine.

The percentage of valerenic acid, acetoxyvalerenic acid, and hydroxyvalerenic acid, in root and rhizomes was, according to [11] not less than 0.17% of sesquiterpenic acids expressed as valerenic acid, calculated with the reference to

TABLE 2: Accuracy results of MLC method: analysis of sesquiterpenic acid mixtures and standard addition technique.

Compound	Added (μg·mL^{-1})	As found* (μg·mL^{-1})	% Recovery	Δ	Added, μg·mL^{-1}	As found*, μg·mL^{-1} with addition	As found*, μg·mL^{-1} without addition	Recovery, %
	1.90	1.88	99.0	−1.0	0	—	2.10	
Hydroxyvalerenic acid	9.30	9.25	99.5	−0.5	1.90	4.20	2.20	104.8
	27.9	27.8	99.6	−0.4	5.60	7.74	2.14	102.1
					9.3	11.4	2.13	101.7
Mean, %R.S.D.			99.4, 0.8				Mean, %	102.9
	4.20	4.19	99.8	−0.2	0	—	33.4	
Acetoxyvalerenic acid	21.0	20.85	99.3	−0.7	12.6	46.3	33.7	102.1
	63.0	63.1	100.1	+0.1	21.0	54.2	33.2	99.4
					29.4	63.5	34.1	102.1
Mean, %R.S.D.			99.7, 1.1				Mean, %	101.2
	6.10	6.07	99.5	−0.5	0	—	50.7	
Valerenic acid	30.4	30.5	100.3	+0.3	18.2	69.4	51.2	101.0
	91.2	91.4	100.2	+0.2	30.4	80.7	50.3	99.2
					42.6	93.6	51.0	100.6
Mean, %R.S.D.			100.0, 1.1				Mean, %	100.3

* Mean value of the four determinations.

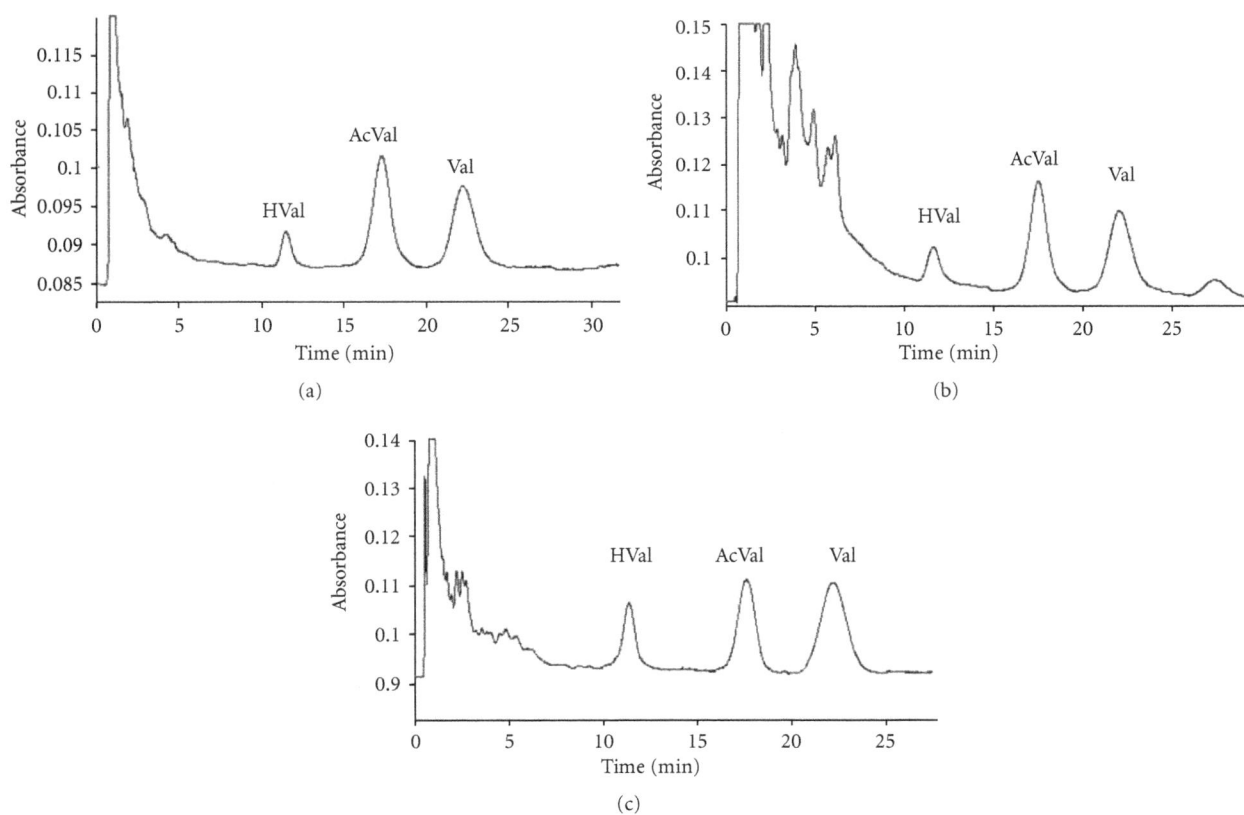

FIGURE 4: Chromatograms obtained during *Valeriana officinalis* L. root and extract analysis. HVal: hydroxyvalerenic acid, AcVal: acetoxyvalerenic acid, and Val: valerenic acid. (a) valerenic acids standard solution; (b) methanolic extract of *Valeriana officinalis* L. root and rhizome; (c) chromatogram of *valerian* hydroalcoholic dry extract.

TABLE 3: Results from sesquiterpenic acids assay in various samples using RP-HPLC and MLC methods and verification of the variance of significance between two methods.

MLC*				HPLC*				Δ				Calculated F values**		
HVal, %	AcVal, %	Val, %	Sum, %	HVal, %	AcVal, %	Val, %	Sum, %	HVal, %	AcVal, %	Val, %	Sum, %	HVal	AcVal	Val
Valerian root and rhizome (3 samples)														
0.005	0.073	0.088	0.166	0.006	0.070	0.087	0.163	−0.001	−0.003	+0.001	+0.003	3.152	4.261	2.154
0.001	0.071	0.084	0.156	0.001	0.069	0.082	0.152	0	+0.002	+0.002	+0.004	1.555	3.644	2.233
0.002	0.082	0.098	0.182	0.001	0.085	0.099	0.185	+0.001	−0.003	−0.001	−0.003	4.176	3.954	3.771
Valerian dry extract (3 samples)														
0.031	0.142	0.178	0.351	0.033	0.145	0.177	0.355	−0.002	−0.003	+0.001	−0.004	3.244	1.547	3.334
0.045	0.133	0.201	0.379	0.047	0.132	0.204	0.383	+0.002	+0.001	−0.003	−0.004	3.672	3.674	1.780
0.028	0.186	0.214	0.428	0.030	0.188	0.213	0.431	−0.002	−0.002	+0.001	−0.003	2.966	2.425	2.655

HVal: hydroxyvalerenic acid, AcVal: acetoxyvalerenic acid, Val: valerenic acid, and Sum: total contents of valerenic acids.
*Mean value of the four determinations
**Tabulated F value is 6.38.

TABLE 4: Intraday and inter-day precision data.

Analyte	Target concentration (100%), μg/mL	Inter-day variation*(% ± R.S.D)	Inter-day variation (% ± R.S.D)
Hydroxyvalerenic acid	1.90	101.3 ± 1.1	101.7 ± 1.8
	9.30	100.9 ± 0.9	100.3 ± 1.1
	27.9	99.9 ± 0.7	100.2 ± 0.9
Acetoxyvalerenic acid	4.20	100.7 ± 0.5	100.2 ± 1.1
	21.0	100.1 ± 0.2	100.5 ± 0.8
	63.0	99.8 ± 0.3	100.1 ± 0.5
Valerenic acid	6.10	99.5 ± 0.9	100.2 ± 1.4
	30.4	100.8 ± 1.0	100.5 ± 1.8
	91.2	100.3 ± 0.5	100.9 ± 1.1

*Mean value of the four determinations.

the dried drugs. Ukrainian Pharmacopoeia [29] required plants that grown in Ukraine and Russia regions, sum of sesquiterpenic acids not less than 0.10% (as valerenic acid, calculated with the reference to the dried drugs).

Valerian dry hydroalcoholic extract should content minimum 0.25% of sesquiterpenic acids, expressed as valerenic acid (dry extract) [11].

Table 3 provides the results of the determination of sesquiterpenic acids in various analytical subjects, which was obtained using micellar liquid chromatographic method. As can be seen from Table 3, not all raw materials satisfied the European Pharmacopoeia [11] requirements for *valerian* root, but all results fulfilled requirements of Ukrainian Pharmacopoeia. All results for *valerian* dry extract analysis comply with requirements of European Pharmacopoeia [11].

4. Conclusions

A simple and reliable MLC method for the simultaneous determination of sesquiterpenic acids (valerenic, acetoxy-valerenic, and hydroxyvalerenic acids) in *valerian* root and rhizomes and *valerian* dry extract was developed. Chromatographic method with using micellar mobile phase does not require gradient elution which is widely used in reversed-phase HPLC methods. That is the main advantage

of MLC that allows separating compounds with different hydrophobicity in a single run without the gradient elution.

The method was completely validated showing satisfactory data for all the parameters tested. This method is also ecofriendly for its low concentration of organic solvent, as compared to other analytical techniques.

References

[1] S. L. Plushner, "Valerian: *Valeriana officinalis*," *American Journal of Health-System Pharmacy*, vol. 57, no. 4, pp. 328–335, 2000.

[2] M. Goppel and G. Franz, "Stability control of valerian ground material and extracts: a new HPLC-method for the routine quantification of valerenic acids and lignans," *Pharmazie*, vol. 59, no. 6, pp. 446–452, 2004.

[3] D. Shohet, R. B. H. Wills, and D. L. Stuart, "Valepotriates and valerenic acids in commercial preparations of valerian available in Australia," *Pharmazie*, vol. 56, no. 11, pp. 860–863, 2001.

[4] N. Singh, A. P. Gupta, B. Singh, and V. K. Kaul, "Quantification of valerenic acid in *Valeriana jatamansi* and *Valeriana officinalis* by HPTLC," *Chromatographia*, vol. 63, no. 3-4, pp. 209–213, 2006.

[5] A. P. Gupta, M. M. Gupta, and S. Kumar, "Simultaneous determination of curcuminoids in Curcuma samples using high

performance thin layer chromatography," *Journal of Liquid Chromatography and Related Technologies*, vol. 22, no. 10, pp. 1561–1569, 1999.

[6] N. Singh, A. P. Gupta, B. Singh, and V. K. Kaul, "Quantification of picroside-I and picroside-II in Picrorhiza kurroa by HPTLC," *Journal of Liquid Chromatography and Related Technologies*, vol. 28, no. 11, pp. 1679–1691, 2005.

[7] C. Bicchi, A. Binello, and P. Rubiolo, "Packed column SFC/UV versus HPLC/UV analysis of valerenic acids and valepotriates in extracts of *Valeriana officinalis* L.," *Phytochemical Analysis*, vol. 11, no. 3, pp. 179–183, 2000.

[8] R. Bos, H. J. Woerdenbag, H. Hendriks et al., "Analytical aspects of phytotherapeutic valerian preparations," *Phytochemical Analysis*, vol. 7, no. 3, pp. 143–151, 1996.

[9] R. Bos, H. J. Woerdenbag, F. M. S. Van Putten, H. Hendriks, and J. J. C. Scheffer, "Seasonal variation of the essential oil, valerenic acid and derivatives, and valepotriates in *Valeriana officinalis* roots and rhizomes, and the selection of plants suitable for phytomedicines," *Planta Medica*, vol. 64, no. 2, pp. 143–147, 1998.

[10] S. Gobbeto and E. Lolla, "A new HPLC method for the analysis of valerenic acids in *Valeriana officinalis* extracts," *Fitoterapia*, vol. 67, no. 2, pp. 159–162, 1996.

[11] *European Pharmacopoeia*, Council of Europe, Strasbourg, France, 7th edition, 2011.

[12] A. Berthod and M. C. Garcia-Alvarez-Coque, *Micellar Liquid Chromatography*, Marcel Dekker, New York, NY, USA, 2000.

[13] A. U. Kulikov, A. G. Verushkin, and L. P. Loginova, "Comparison of micellar and reversed-phase liquid chromatography for determination of sulfamethoxazole and trimethoprim," *Chromatographia*, vol. 61, no. 9-10, pp. 455–463, 2005.

[14] I. Rapado-Martínez, M. C. García-Alvarez-Coque, and R. M. Villanueva-Camañas, "Performance of micellar mobile phases in reversed-phase chromatography for the analysis of pharmaceuticals containing β-blockers and other antihypertensive drugs," *Analyst*, vol. 121, no. 11, pp. 1677–1682, 1996.

[15] B. L. Kolte, B. B. Raut, A. A. Deo, M. A. Bagool, and D. B. Shinde, "Simultaneous determination of metformin in its multicomponent dosage forms with glipizide and gliclazide using micellar liquid chromatography," *Journal of Liquid Chromatography and Related Technologies*, vol. 26, no. 7, pp. 1117–1133, 2003.

[16] D. Bose, A. Durgbanshi, S. Carda-Broch, M. Gil-Agustí, M. E. Capella-Peiró, and J. Esteve-Romero, "Direct injection analysis of epinephrine, norepinephrine, and their naturally occurring derivatives in serum by micellar liquid chromatography with electrochemical detection," *Journal of Liquid Chromatography and Related Technologies*, vol. 28, no. 20, pp. 3265–3281, 2005.

[17] A. U. Kulikov, M. N. Galat, and A. P. Boichenko, "Optimization of micellar LC conditions for the flavonoid separation," *Chromatographia*, vol. 70, no. 3-4, pp. 371–379, 2009.

[18] J. Gu, X. Zeng, B. Kong, Y. Mao, W. Liu, and W. Wei, "Rapid determination of polyphenols in tobacco by MLC," *Chromatographia*, vol. 71, no. 9-10, pp. 769–774, 2010.

[19] R. Bos, H. Hendriks, A. P. Bruins, J. Kloosterman, and G. Sipma, "Isolation and identification of valerenane sesquiterpenoids from *Valeriana officinalis*," *Phytochemistry*, vol. 25, no. 1, pp. 133–135, 1985.

[20] S. Wang, G. Yang, Z. Li, L. Haiyan, J. Bai, and Y. Zhang, "Micellar liquid chromatography study of quantitative retention-activity relationships for antihypertensive drugs," *Chromatographia*, vol. 64, no. 1-2, pp. 23–29, 2006.

[21] M. F. Borgerding, W. L. Hinze, L. D. Stafford, G. W. Fulp, and W. C. Hamlin, "Investigations of stationary phase modification by the mobile phase surfactant in micellar liquid chromatography," *Analytical Chemistry*, vol. 61, no. 13, pp. 1353–1358, 1989.

[22] M. F. Borgerding and W. L. Hinze, "Characterization and evaluation of the use of nonionic polyoxyethylene(23)dodecanol micellar mobile phases in reversed-phase high-performance liquid chromatography," *Analytical Chemistry*, vol. 57, no. 12, pp. 2183–2190, 1985.

[23] L. P. Loginova, L. V. Samokhina, A. P. Boichenko, and A. U. Kulikov, "Micellar liquid chromatography retention model based on mass-action concept of micelle formation," *Journal of Chromatography A*, vol. 1104, no. 1-2, pp. 190–197, 2006.

[24] A. U. Kulikov, A. P. Boichenko, and A. G. Verushkin, "Optimization of micellar LC conditions for separation of opium alkaloids and their determination in pharmaceutical preparations," *Analytical Methods*, vol. 3, no. 12, pp. 2749–2757, 2011.

[25] "Validation of Analytical procedures: Text and Methodology Q2(R1)," http://www.ich.org/.

[26] J. A. Adamovics, *Chromatographic Analysis of Pharmaceuticals*, Marcel Dekker, New York, NY, USA, 1997.

[27] J. Ermer and H. -J. Ploss, "Validation in pharmaceutical analysis—part II: central importance of precision to establish acceptance criteria and for verifying and improving the quality of analytical data," *Journal of Pharmaceutical and Biomedical Analysis*, vol. 37, no. 5, pp. 859–870, 2005.

[28] G. A. Shabir, "Validation of high-performance liquid chromatography methods for pharmaceutical analysis: Understanding the differences and similarities between validation requirements of the US Food and Drug Administration, the US Pharmacopeia and the International Conference on Harmonization," *Journal of Chromatography A*, vol. 987, no. 1-2, pp. 57–66, 2003.

[29] *Ukrainian State Pharmacopoeia*, REREIG, Kharkov, Ukraine, 1st edition, 2001.

20

A Comparison of Methodical Approaches to Fingerprinting of the Volatile Fraction from Winter Savory (*Satureja montana*)

Józef Rzepa,[1] Mieczysław Sajewicz,[1] Tomasz Baj,[2] Patrycja Gorczyca,[1] Magdalena Włodarek,[1] Kazimierz Głowniak,[2] Monika Waksmundzka-Hajnos,[3] and Teresa Kowalska[1]

[1] Institute of Chemistry, University of Silesia, 9 Szkolna Street, 40-006 Katowice, Poland
[2] Department of Pharmacognosy with Medicinal Plant Unit, Faculty of Pharmacy, Medical University of Lublin,
 1 Chodźki Street, 20-093 Lublin, Poland
[3] Department of Inorganic Chemistry, Faculty of Pharmacy, Medical University of Lublin, Collegium Pharmaceuticum,
 4a Chodźki Street, 20-093 Lublin, Poland

Correspondence should be addressed to Teresa Kowalska, teresa.kowalska@us.edu.pl

Academic Editor: Yvan Vander Heyden

It was the aim of this study to compare the efficiency of the different essential oil extraction methods upon the two winter savory (*Satureja montana*) samples of different origin. The compared techniques were the headspace gas chromatography with mass spectrometric detection (HS-GC/MS) run at the two different headspace temperatures (i.e., at 80 and 100°C) and the three different steam distillation techniques preceding the GC/MS analysis. HS-GC/MS is considered as the technique of the first choice, and the compared steam distillation techniques are recommended, respectively, by Polish Pharmacopoeia, European Pharmacopoeia, and the Polish Patent. Adequate conclusions were drawn as to the advantage of HS-GC/MS (not having the pharmacopoeial recommendation) over the different steam distillation techniques and the drawbacks of each individual analytical procedure were discussed.

1. Introduction

The genus *Satureja* L. contains over 30 species. Winter savory (*Satureja montana*) is a perennial plant belonging to the family Lamiaceae, growing mainly in the regions of South Europe. It is a semievergreen subshrub growing to about 50 cm tall with the oval-lanceolate leaves and white flowers. *Satureja montana* contains numerous subspecies, and there is much variability in morphologic characteristics of the species *Satureja montana* L. [1]. It is similar in use and flavor to the annual summer savory (*Satureja hortensis*) and it is cultivated as a culinary herb having spicy flavor. Both summer and winter savory have a long history of use in traditional medicine as tonics, carminatives, astringents, and expectorants, and for the treatment of intestinal problems such as diarrhea and nausea. However, the scientific literature primarily documents *Satureja hortensis* (and not *Satureja montana* L.) as a folk remedy in treating various ailments such as cramps, muscle pains, nausea, indigestion, diarrhea, and infectious diseases [1–3].

Winter savory contains ca. 1.6% volatile oil, whereas summer savory only ca. 1.0%. Some authors document the dominant components of the volatile oil as caryophyllene and geraniol, or as carvacrol. The relative composition of the volatile oil varies with the location of cultivation, the species, and the strain [3–6]. The essential oil of the *Satureja* sp. has a broad spectrum of antimicrobial activity [7–10]. *Satureja montana* L. also has a potent anti-HIV-1 activity [11].

With this study on fingerprinting of the volatile fraction contained in *Satureja montana*, we continue our earlier commenced methodical approach to fingerprinting of the volatile fractions derived from the other medicinal and culinary plants belonging to the same family Lamiaceae. In our earlier studies (e.g., [12–15]), we have focused our attention on the selected representatives of the *Salvia* genus, the largest one in the Lamiaceae family. In spite of great popularity of many plants belonging to the Lamiaceae family both in traditional European medicine and the Mediterranean cuisine, an insufficient attention has been paid so far to fingerprinting of this particular botanical material. Fingerprints, that is, chromatographic profiles derived from the plant extracts, can provide a sufficient basis for differentiation of plants, their chemotaxonomic comparison, identification within a given family, and so forth. Moreover, they can be used for rapid screening of plant material to prevent the forgery in the bulk trading on the medicinal and culinary herbs. Due to the variation in the consistency of plant material both in the fresh and the dried form, the efficiency of the different extraction methods can also differ and hence, the best performing extraction technique has to be experimentally confirmed for each individual genus.

The main goal of this study was to make a methodical comparison of the performance of the basic fingerprinting strategies applied to the volatile fraction contained in the dried samples of *Satureja montana*. To this effect, we compared the headspace-GC/MS fingerprints of the volatile fraction (HS-GC/MS) and with the fingerprints originating from the three different steam distillation techniques preceding the GC/MS analysis (Figure 2). HS-GC/MS is generally considered as the technique of the first choice, and the compared steam distillation techniques are recommended, respectively, by Polish Pharmacopoeia [16], European Pharmacopoeia [17], and the Polish Patent [18]. Partial identification of the volatile fraction components was also performed.

2. Experimental

2.1. Herbal Material and Reagents. Two different samples of winter savory (*Satureja montana*) were investigated in this study. Lot 1 was harvested in Pharmacognosy Garden of the Medical University, Lublin, Poland. The plant material comprised all parts of the plant (i.e., roots and the aerial parts) and it was dried for 40 h in an oven with a forced air flow at 35 to 40°C. Then the obtained dry material was stored in the deep-freeze compartment of refrigerator until the commencement of the analysis. Finally, plant material from lot 1 was weighed and ground with a mechanical blender. Lot 2 originated from the farm market in Belgrade (Serbia) as the dried herbs and it was not pretreated in our laboratory in any way, except for identical grinding as with lot 1.

The following solvents were used in this study: methanol (HPLC purity grade, P.O.CH, Gliwice, Poland) and water (double distilled and deionized with use of the Elix Advantage model Millipore system, Molsheim, France).

2.2. Headspace Gas Chromatography-Mass Spectrometry (HS-GC/MS) of the Volatile Fraction. The headspace gas chromatography-mass spectrometry (HS-GC/MS) analyses were carried out with use of a TRACE 2000 model GC with an MS TRACE model mass detector (ThermoQuest, Waltham, MA, USA), equipped with a CTC Analytics model autosampler (Combi PAL, Basel, Switzerland), used in the headspace mode. Temperatures and time of the headspace desorption were, respectively, 80 and 100°C, and 15 min. The 0.5 mL volume of the headspace phase was introduced on to the DB-5 capillary column (30 m ↔ 0.25 mm i.d., 0.25-μm film thickness; Agilent Technologies, Palo Alto, CA, USA).

Helium (p = 100 kPa) was used as carrier gas. Gradient analysis was run using the following temperature program: 40°C (1 min); 40–180°C (12°C/min); and 180°C (20 min). The temperature of the injector was kept constant at 180°C. Mass spectrometer was fitted with an EI source operated at 70 eV. Identification of individual compounds was based on a comparison of the obtained mass spectra of the individual chromatographic peaks with those valid for the standards and available from the National Institute of Standards and Technology software library (Gaithersburg, MD, USA). A comparison was also carried out of the retention times valid for individual peaks from the *Satureja montana* samples with those of the known essential oils components. To this effect, we used pine oil, peppermint oil, eucalyptus oil, juniper oil, thyme oil, and lavender oil as the sets of the volatile fraction standards (Apotheca Pacis, Rybnik, Poland). The identified compounds originating from lots 1 and 2 are listed in Table 1.

2.3. Steam Distillation Modes. In this study, we compared the performance of the three steam distillation modes, as given below:

(a) mode 1, applied to the lot 1 and lot 2 samples (50 g) with use of the Deryng apparatus recommended by Polish Pharmacopoeia VI [16];

(b) mode 2, applied to the lot 1 sample (50 g) with use of the Clevenger apparatus recommended by European Pharmacopoeia [17];

(c) mode 3(a), applied to the lot 1 sample (20 g) with use of the Clevenger apparatus (like in Mode 2);

(d) mode 3(b), applied to the lot 1 sample (20 g) with use of the Clevenger apparatus, yet with the herbal/aqueous mixture additionally ultrasonicated in order to enhance maceration of herbal material, as recommended by the Polish Patent [18].

In the experiments valid for modes 1 and 2, the dried plant material (50 g) was placed in the round-bottomed flask, 400 mL water was added, and the steam distillation was performed for 3 h. In the experiments valid for modes 3(a) and 3(b), aimed to compare the efficiency of the Clevenger apparatus without and with ultrasonication, the 20 g amounts of the plant material with 400 mL water was placed in the round-bottomed flask, 400 mL water was added, and the steam distillation was performed for 3 h.

From the distillates obtained in each individual experiment, the 5% (*v/v*) solutions in methanol were prepared and the 1-μL aliquots of the respective solutions were analyzed by means of GC/MS. The applied chromatographic conditions were the same as those given in Section 2.2.

3. Results and Discussion

This study was focused on a comparison of the efficiency of the methods used for derivation and fingerprinting of the volatile fraction from winter savory (*Satureja montana*). To this effect, we compared the results originating from the method of the first choice (which is HS-GC/MS)

TABLE 1: Volatile compounds in *Satureja montana* (lot 1 and lot 2), their respective retention times (t_R), and semi-quantitative evaluation of their relative contributions (%) to the overall volatile fraction* depending on the applied extraction method (HS-80: headspace at 80°C; HS-100: headspace at 100°C; Deryng: mode 1; Clevenger: mode 2).

Volatile compound	α-Thujene	α-Pinene	Camphene	β-Myrcene	1-Octen-3-ol	Phelandrene	α-Terpinene	D-Limonene	p-Cymene	γ-Terpinene	β-Terpineol	β-Linalool	Borneol	p-Mentha-3,6-diene-2,5-dione	Carvacrol	Caryophyllene
No.	1	2	3	4	5	6	7	8	9	10	11	12	13	14	15	16
Retention time (min)	6.61	6.81	7.26	7.77	7.99	8.24	8.38	8.64	8.82	9.13	9.55	9.85	11.50	13.06	13.49	14.62
Peak Area (%)																
L1 HS-80	1.61	0.87	—	0.98	1.47	—	0.94	—	14.51	8.01	1.29	—	++	3.12	68.48	—
L1 HS-100	1.31	0.86	0.35	1.71	4.30	++	1.79	—	21.92	9.69	3.07	1.37	++	11.49	44.82	1.18
L1 Deryng	0.65	0.35	0.18	0.62	1.57	—	0.89	—	12.51	6.03	0.52	0.53	++	—	76.16	—
L1 Clevenger	—	+	—	—	0.39	—	—	—	17.77	1.42	—	+	++	—	79.90	—
L2 HS-80	—	11.79	5.10	—	—	—	—	12.94	32.37	—	5.10	6.92	++	—	19.63	3.60
L2 HS-100	—	4.77	2.09	0.19	—	—	0.56	7.64	20.46	3.10	3.50	6.06	++	1.62	31.73	2.88
L2 Deryng	—	+	—	—	—	—	—	9.70	48.27	1.57	—	7.89	++	++	2.73	—

* values estimated from the respective peak heights.
+ data falling between the limit of detection (LOD) and the limit of quantification (LOQ).
++ lack of an adequate (i.e., enabling quantification) peak separation.
— compound not detected.

with those of the steam distillation of the volatile fraction (recommended by many national and international pharmacopoeias), followed by the GC/MS analysis. Our results originate from the HS-GC/MS experiment with the two different headspace temperatures and from the three different steam distillation modes. All experiments presented in this study were performed twice, upon the two different batches of plant material from lot 1 and lot 2. Due to a limited amount of the available plant material and hence, due to a limited number of repetitions, the quantitative results presented in this study were taken as the mean values, but not statistically evaluated. Although these results bear a semi-quantitative importance only, the estimated relative percent error of quantification of the individual volatile compounds (based on the two repetitions) never surpassed the ±3% level. In the forthcoming paragraphs, the obtained results are going to be presented and discussed.

In Table 1, all volatile compounds identified and semi-quantitatively assessed in the analyzed *Satureja montana* samples (lots 1 and 2) are listed, depending on the applied extraction mode. The most frequently appearing volatile compounds were α-pinene, p-cymene, γ-terpinene, terpineol, β-linalool, borneol, and carvacrol, as shown in Table 2 (together with the respective structural formulas and mass spectra). These results remain in good agreement with those known from the literature (e.g., [19, 20]).

From the general phytochemical knowledge and also from the obtained results it is evident that the origin of the plant material (and also its preprocessing like, e.g., the drying technique) plays a meaningful role in chemical composition of the volatile fraction. Thus, in lot 1 (originating from Pharmacognosy Garden of the Medical University in Lublin, Poland), α-thujene and 1-octen-3-ol were identified, which were absent from lot 2 (originating from the farm market in Belgrade, Serbia). To the contrary, D-limonene was identified exclusively in lot 2 and it was not detected in lot 1. These differences are evidently due to the different origin (in terms of the different cultivation regions, the climatic and strain differences, the different drying techniques, etc.) of the two examined specimens.

A comparison of analytical results dealing with the headspace extraction of the volatile fraction carried out at 80 and 100°C allowed drawing very practical conclusions. The chromatograms addressing this particular issue are given in Figure 1 (chromatograms (a) versus (b) valid for lot 1 and chromatograms (e) versus (f) valid for lot 2). A comparison of the aforementioned chromatograms showed that at the higher headspace temperature (100°C) a higher number of the volatile compounds were isolated than at the lower one. Thus applying the headspace temperature of 100°C, in lot 1 camphene, phellandrene, β-linalool, and caryophyllene were identified, which were not extracted from the same plant sample at 80°C. When applying the headspace temperature of 100°C to lot 2, β-myrcene, α-terpinene, γ-terpinene, p-mentha-3,6-diene-2,5-dione, and caryophyllene were isolated, which were absent from the volatile fraction extracted from the same plant sample at the lower headspace temperature.

A consecutive comparison was made of the analytical results valid for the headspace extraction of the volatile fraction carried out at 100°C with that obtained for the steam distillation in the Deryng apparatus (mode 1). The conclusions can be drawn from a comparison of the chromatograms shown in Figures 1(b) and 1(c) (lot 1, HS-100 versus mode 1), and of those shown in Figures 1(f) and 1(g) (lot 2, HS-100 versus mode 1). This comparison allows a conclusion that the headspace extraction carried out at 100°C is more effective particularly with the more volatile compounds than the steam distillation in the Deryng apparatus. From a comparison of the data contained in Table 1 and valid for the two investigated *S. montana* samples (lot 1 and lot 2), it is evident that the percent contributions of the more volatile compounds (i.e., those with the relatively low retention times, t_R) are considerably higher, when the headspace extraction is carried out at 100°C than with the steam distillation in mode 1.

Then the two steam distillation techniques with use of the Deryng and the Clevenger apparatus (modes 1 and 2) are compared for lot 1 (Figure 1, chromatograms (c) versus (d)). Essential oil derived with use of the Deryng apparatus (mode 1, Figure 1(c)) proved richer in terms of the isolated and identified chemical species than that derived with use of the Clevenger apparatus (mode 2, Figure 1(d)). Namely, in the essential oil extracted with use of the Clevenger apparatus, only four compounds were identified (i.e., 1-octen-3-ol, p-cymene, γ-terpinene, and carvacrol), while in the essential oil isolated with use of the Deryng apparatus, six additional compounds (i.e., α-thujene, camphene, β-myrcene, α-terpinene, β-terpineol, and p-mentha-3,6-diene-2,5-dione) were found. This experimental outcome can be due to the construction differences of these two steam distillation apparatuses, and specifically to the less efficient reflux condensation system in the Clevenger apparatus, allowing an easier loss of the most volatile compounds than the Deryng apparatus.

The last comparison refers to the contents of the volatile compounds in the samples belonging to lot 1 and originating from the steam distillation in the Clevenger apparatus without and with ultrasonication (derivation modes 3(a) and 3(b), resp.). It was shown that the steam distillation combined with ultrasonication resulted in a lower concentration yield with one essential oil component (p-cymene) and in an absence of the two components (1-octen-3-ol and γ-terpinene; see Table 3), as compared with the distillation without ultrasonication. The only compound which showed slightly higher extraction yield in the case of steam distillation supported with ultrasonication was carvacrol, one of the least volatile compounds on the list (98.54% contribution to the overall volatile fraction obtained in mode 3(b) versus 92.97% obtained in mode 3(a)). This result is probably due to the too long ultrasonication period (equal to 180 min), although from the literature it comes out that the steam distillation yields remain practically constant even with the two or more hours lasting ultrasonication (although the shortest recommended ultrasonication period is 20 minutes; [21]). Thus with *Satureja montana*, maceration of plant material with use of ultrasonication cannot

TABLE 2: Seven volatile compounds most frequently appearing in *Satureja montana* with their respective structural formulas and mass spectra.

Volatile compound	Structural formula and mass spectrum
α-Pinene	
p-Cymene	
γ-Terpinene	
β-Terpineol	
β-Linalool	
Borneol	
Carvacrol	

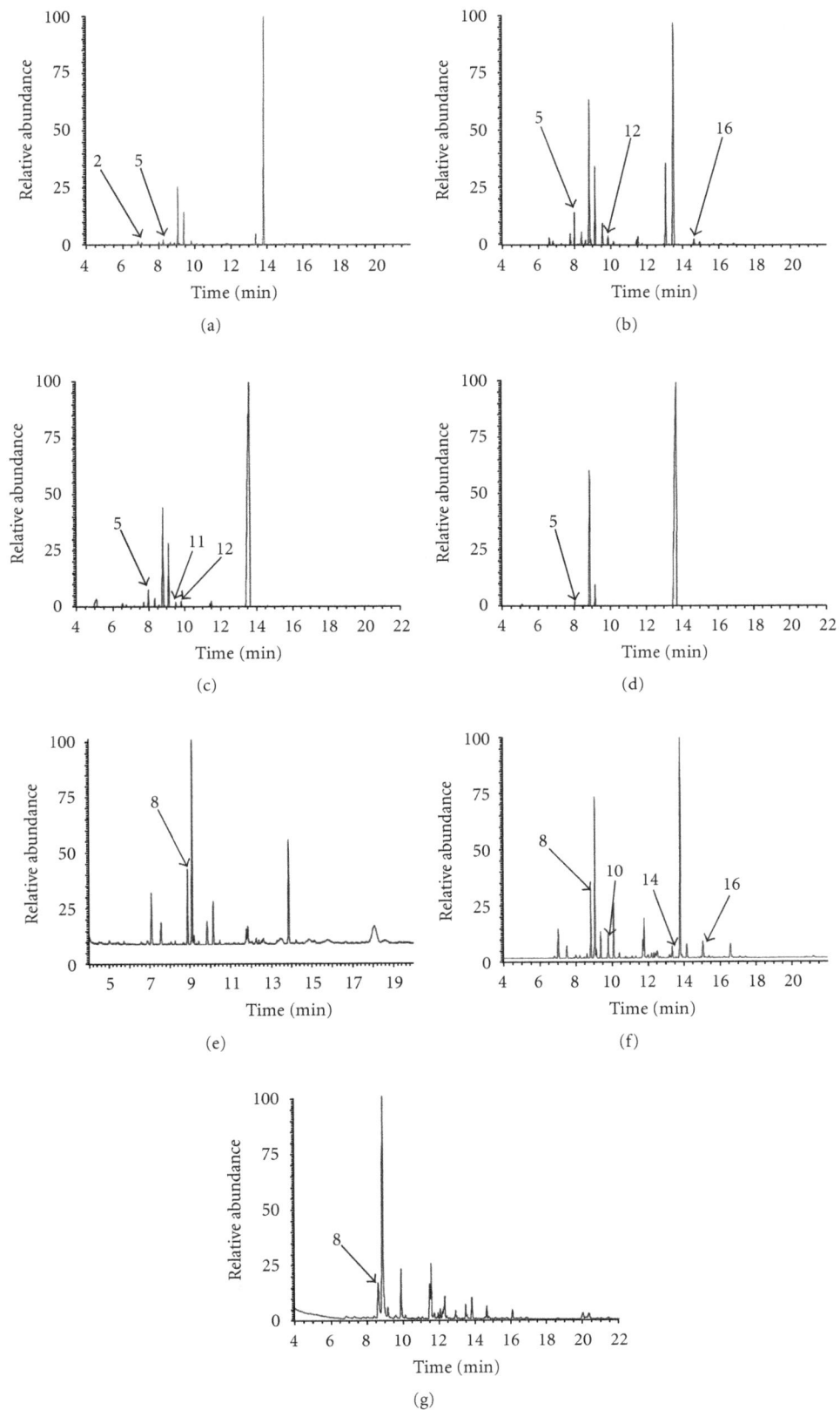

FIGURE 1: The GC/MS fingerprints of the volatile *Satureja montana* fraction for the following samples: (a): Lot 1, HS-80; (b): Lot 1, HS-100; (c): Lot 1, Deryng; (d): Lot 1, Clevenger; (e): Lot 2, HS-80; (f): Lot 2, HS-100; (g): Lot 2, Deryng. Extraction modes and peak numbers remain in conformity with those given in Tables 1 and 2.

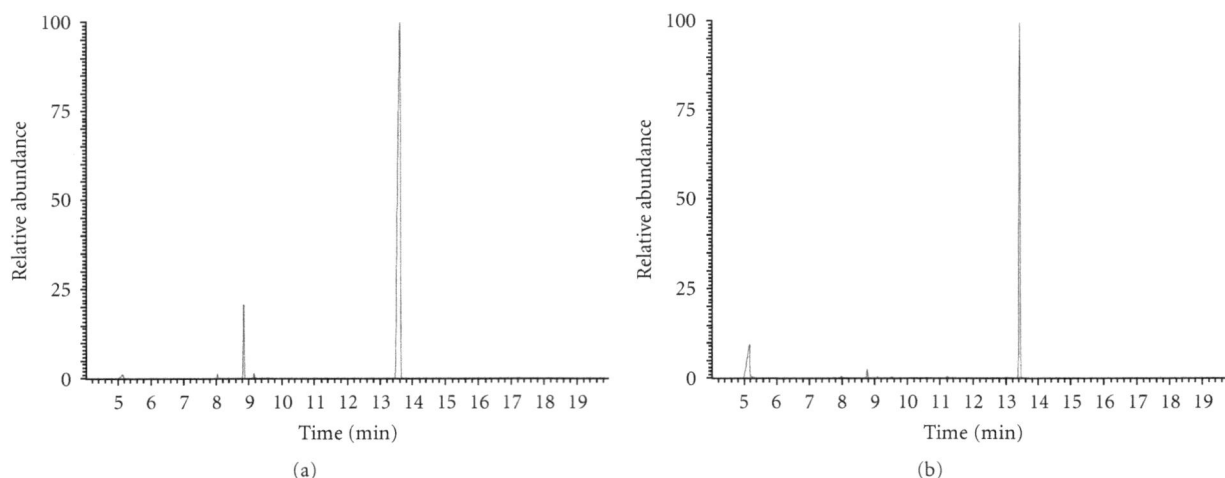

FIGURE 2: The GC/MS fingerprints of the volatile *Satureja montana* fraction (lot 1) vapor distilled with use of the Clevenger apparatus. (a) without ultrasonication; (mode 3(a) and 3(b)) with ultrasonication (mode 3(b)).

TABLE 3: Volatile compounds in *Satureja montana* (lot 1), their respective retention times (t_R), and semiquantitative evaluation of their relative contributions (%) to the overall volatile fraction* depending on the applied extraction method (mode 3(a) and mode 3(b)).

Volatile compound	1-Octen-3-ol	*p*-Cymene	*γ*-Terpinene	Carvacrol
No.**	5	9	10	15
Retention time (min)	8.01	8.81	9.12	13.56
Peak Area (%)				
Mode 3(a)	0.32	0.31	0.40	92.97
Mode 3(b)	+	0.19	+	98.54

*values estimated from the respective peak heights.
** numbering of the volatile compounds in conformity with Table 1.
+ data falling between the limit of detection (LOD) and the limit of quantification (LOQ).

be recommended, because the loss of the more volatile essential oil components is in this case higher than without ultrasonication.

4. Conclusions

The methodical aspects of the analysis of the volatile fraction contained in winter savory (*Satureja montana*) were discussed in this study and they can be summarized in the following way.

(i) It was demonstrated that the headspace desorption of the volatile fraction run at 100°C outperforms the steam distillation modes with use of the Deryng apparatus, the Clevenger apparatus, and the Clevenger apparatus in combination with ultrasonication of the hydrodistilled material.

(ii) This result is most probably due to an uncontrolled loss of the most volatile components both from the Deryng apparatus and the Clevenger apparatus (as an evident partial "leaking" of the two reflux condensation systems).

(iii) Furthermore, it was shown that this uncontrolled loss (both in qualitative and quantitative terms) is somewhat higher with the Clevenger than the Deryng apparatus.

(iv) Additionally, it was shown that the ultrasonication-supported maceration of plant material belonging to the *S. montana* species in the Clevenger apparatus results in an even greater overall loss of the volatile fraction components than without using this supplementary technique.

(v) In order to make phytochemical analysis of the volatile fractions contained in medicinal plants more up to the date and in the first instance considerably more accurate, it seems recommendable (or even urgent) to replace pharmacopoeial hydrodistillation techniques with the headspace-GC/MS technique (as a standard separation and quantification tool in modern analytical chemistry laboratories).

References

[1] P. Schauenberg and F. Paris, *Guide to Medicinal Plants*, Keats Publishing, New Canaan, Conn, USA, 1990.

[2] C. Uslu, R. M. Karasen, F. Sahin, S. Taysi, and F. Akcay, "Effects of aqueous extracts of *Satureja hortensis* L. on rhinosinusitis treatment in rabbit," *Journal of Ethnopharmacology*, vol. 88, no. 2-3, pp. 225–228, 2003.

[3] J. A. Duke, *CRC Handbook of Medicinal Herbs*, CRC Press, Boca Raton, Fla, USA, 1985.

[4] V. Slavkovska, R. Jancic, S. Bojovic, S. Milosavljevic, and D. Djokovic, "Variability of essential oils of *Satureja montana* L. and *Satureja kitaibelii* Wierzb. ex Heuff. from the central part of the Balkan peninsula," *Phytochemistry*, vol. 57, no. 1, pp. 71–76, 2001.

[5] M. De Vincenzi, A. Stammati, A. De Vincenzi, and M. Silano, "Constituents of aromatic plants: carvacrol," *Fitoterapia*, vol. 75, no. 7-8, pp. 801–804, 2004.

[6] A. Radonic and M. Milos, "Chemical composition and in vitro evaluation of antioxidant effect of free volatile compounds from *Satureja montana* L.," *Free Radical Research*, vol. 37, no. 6, pp. 673–679, 2003.

[7] M. Güllüce, M. Sökmen, D. Daferera et al., "In vitro antibacterial, antifungal, and antioxidant activities of the essential oil and methanol extracts of herbal parts and callus cultures of *Satureja hortensis* L.," *Journal of Agricultural and Food Chemistry*, vol. 51, no. 14, pp. 3958–3965, 2003.

[8] F. Şahin, I. Karaman, M. Güllüce et al., "Evaluation of antimicrobial activities of *Satureja hortensis* L.," *Journal of Ethnopharmacology*, vol. 87, no. 1, pp. 61–65, 2003.

[9] M. Ciani, L. Menghini, F. Mariani, R. Pagiotti, A. Menghini, and F. Fatichenti, "Antimicrobial properties of essential oil of *Satureja montana* L. on pathogenic and spoilage yeasts," *Biotechnology Letters*, vol. 22, no. 12, pp. 1007–1010, 2000.

[10] K. Oberg, L. Rolling, and C. Oberg, "Selection of essential oil components to inhibit *Candida* without affecting normal microbiota," *The Journal of the Utah Academy of Sciences, Arts, and Letters*, vol. 82, pp. 60–72, 2005.

[11] K. Yamasaki, M. Nakano, T. Kawahata et al., "Anti-HIV-1 activity of herbs in *Labiatae*," *Biological and Pharmaceutical Bulletin*, vol. 21, no. 8, pp. 829–833, 1998.

[12] J. Rzepa, Ł. Wojtal, D. Staszek et al., "Fingerprint of selected salvia species by HS-GC-MS analysis of their volatile fraction," *Journal of Chromatographic Science*, vol. 47, no. 7, pp. 575–580, 2009.

[13] M. Sajewicz, J. Rzepa, M. Hajnos et al., "GC-MS study of the performance of different techniques for isolating the volatile fraction from sage (*Salvia* L.) species, and comparison of seasonal differences in the composition of this fraction," *Acta Chromatographica*, vol. 21, no. 3, pp. 453–471, 2009.

[14] M. Sajewicz, L. Wojtal, D. Staszek, M. Hajnos, M. Waksmundzka-Hajnos, and T. Kowalska, "Low temperature planar chromatography-densitometry and gas chromatography of essential oils from different sage (*Salvia*) species," *Journal of Liquid Chromatography and Related Technologies*, vol. 33, no. 7-8, pp. 936–947, 2010.

[15] M. Sajewicz, Ł. Wojtal, M. Natić, D. Staszek, M. Waksmundzka-Hajnos, and T. Kowalska, "TLC-MS versus TLC-LC-MS fingerprints of herbal extracts. Part I. essential oils," *Journal of Liquid Chromatography and Related Technologies*, vol. 34, no. 10-11, pp. 848–863, 2011.

[16] Polish Pharmacopoeia VI, Polish Pharmaceutical Society, Warsaw, Poland, 2002.

[17] European Pharmacopoeia, Vol. 3, pp.68, Maisonneuve SA, Sainte Ruffine, France, 1975.

[18] Polish Patent No 208058, 2008.

[19] G. S. Ćetković, A. I. Mandić, J. M. Čanadanović-Brunet, S. M. Djilas, and V. T. Tumbas, "HPLC screening of phenolic compounds in winter savory (*Satureja montana* L.) extracts," *Journal of Liquid Chromatography and Related Technologies*, vol. 30, no. 2, pp. 293–306, 2007.

[20] J. Masteli and I. Jerkovi, "Gas chromatography-mass spectrometry analysis of free and glycoconjugated aroma compounds of seasonally collected *Satureja montana* L.," *Food Chemistry*, vol. 80, no. 1, pp. 135–140, 2003.

[21] R. Kowalski and J. Wawrzykowski, "Effect of ultrasound-assisted maceration on the quality of oil from the leaves of thyme *Thymus vulgaris* L.," *Flavour and Fragrance Journal*, vol. 24, no. 2, pp. 69–74, 2009.

21

Development and Validation of Dissolution Test for Fluconazole Capsules by HPLC and Derivative UV Spectrophotometry

Josilene Chaves Ruela Corrêa,[1] Cristina Duarte Vianna-Soares,[2] and Hérida Regina Nunes Salgado[1]

[1] *School of Pharmaceutical Sciences, Universidade Estadual Paulista, Rodovia Araraquara-Jaú km 1, 14801-902 Araraquara, SP, Brazil*
[2] *Department of Pharmaceutical Products, Federal University of Minas Gerais, Avenida Antônio Carlos 6627, 31270-901 Belo Horizonte, MG, Brazil*

Correspondence should be addressed to Josilene Chaves Ruela Corrêa, josilenechavescorrea@gmail.com

Academic Editor: Bengi Uslu

The purpose of this study is to develop and validate a dissolution test for fluconazole, an antifungal used for the treatment of superficial, cutaneous, and cutaneomucous infections caused by *Candida* species, in capsules dosage form. Techniques by HPLC and UV first derivative spectrophotometry (UV-FDS) were selected for quantitative evaluation. In the development of release profile, several conditions were evaluated. Dissolution test parameters were considered appropriate when a most discriminative release profile for fluconazole capsules was yielded. Dissolution test conditions for fluconazole capsules were 900 mL of HCl 0.1 M, $37 \pm 0.5\,°C$ using baskets with 50 rpm for 30 min of test. The developed HPLC and UV-FDS methods for the antifungal evaluation were selective and met requirements for an appropriate and validated method, according to ICH and USP requirements. Both methods can be useful in the registration process of new drugs or their renewal. For routine analysis application cost, simplicity, equipment, solvents, speed, and application to large or small workloads should be observed.

1. Introduction

The dissolution can be defined in a narrow sense as the process by which a solid substance is incorporated into the solvent to form a solution. However, in a broad sense, it is more than a simple measurement of solubility rate and can be better described as physical test to predict the drug release from a dosage form, for a given area for some precise time. Fundamentally, this process is controlled by the affinity between the solvent and the solid substance and the way by which the pharmaceutical system releases the drug [1, 2]. According to Mehta and coworkers [3] dissolution test provides an indication of bioavailability of a drug and, thus, pharmaceutical equivalence from batch to batch. The dissolution test is an important tool in quality control of drugs and it becomes more important for drugs with relatively low water solubility, including broad spectrum antifungal fluconazole. Some characteristics of this

drug are not well defined, for instance, its classification as high or low soluble in water or its classification in the biopharmaceutical classification system [4–6]. Well-defined drug features, together with the dissolution studies, can be an important tool to justify the use of in vitro methods instead of in vivo methods when they are requested. The aim of this dissolution study is to contribute to define fluconazole dissolution conditions, what can be the focus for further studies.

A dissolution method should be discriminatory, and it should allow evaluating the performance of the product and possible changes it may suffer from during the stability study. According to Marcolongo [7], many variables can influence the results of a dissolution test. Among these variables we can find solubility, chemical nature of the drug, the dosage form, excipients and manufacturing technology employed, the apparatus used, the stirring speed, the use of devices for floating dosage forms (sinkers), the volume of media used, pH and temperature of the media, the filtration, and analytical method employed.

To develop a dissolution method the characteristics of the drug and its behavior in the chosen test media should be taken in account. Moreover, the dissolution conditions must follow the sink conditions (final concentration equivalent to 10% of saturation concentration) [7], and the quantitation method should be sensitive, selective, accurate, and precise.

Although there are many published analytical methods for fluconazole, there are few dissolution studies for fluconazole, as it has been well compiled in a review by Corrêa and Salgado [8]. Fluconazole dissolution has not been recommended in any pharmacopeia until December 2010 when it has been incorporated in the capsule monograph in the Brazilian Pharmacopeia, 5th edition [9]. However, FDA [10] has recommended dissolution methods since 2004 for fluconazole suspension and since 2006 for tablets; the FDA-recommended method for tablets was tested in this study without good results. In the Brazilian Pharmacopeia [9], the dissolution method recommended for fluconazole capsule uses 900 mL of 0.1 M HCl at 37°C and baskets with 100 rpm for 30 min with quantitation by spectrophotometry at $\lambda = 261$ nm.

The aim of this work is to develop and validate a discriminative dissolution method for fluconazole capsules employing two analytical methods to determine fluconazole by high-performance liquid chromatography (HPLC) and by first-order derivative UV spectrophotometry (FDS). Problems encountered by the UV spectrophotometric method, recommended by the Brazilian Pharmacopeia 5th edition [9] are discussed.

2. Experimental

2.1. Material and Equipment. Fluconazole chemical reference (assigned purity 100%) was purchased from Sigma Aldrich. Bulk drug was kindly donated (EMS, Hortolândia, SP, Brazil) and was standardized against fluconazole chemical reference. Capsules were purchased from local market with 150 mg drug label claim. A Hanson SR 8 *Plus* dissolution system containing six vessels was used for dissolution tests.

LC grade methanol was purchased from Tedia (Fairfield, USA). Purified water was prepared in-house by using Direct-Q water system (Millipore, Billerica, MA, USA). Prior to use, mobile phase solvents were degassed in an ultrasonic bath for 30 min. Purified water (>18 MOhm cm) was used to prepare the mobile phase. Solvents were filtered through a 0.45 μm membrane filter.

An HP 8453 UV-Visible spectrophotometer (Agilent Technologies, Inc., Santa Clara, CA, USA) with photodiode array (PDA) and HP ChemStation software with automatic differentiation was used. A liquid chromatograph (Waters Corporation, Milford, MA, USA) equipped with a Waters 1525 binary pump, a Rheodyne Breeze 7725i manual injector, and a Waters 2487 UV-VIS wavelength detector was used. HPLC analysis was conducted in an RP C18 column (Symmetry, 5 μm, 4.6 mm × 250 mm, Waters, Milford, MA, USA).

2.2. Preparation of Standard and Sample Solutions

2.2.1. Dissolution Performance. Fluconazole 150 mg capsules were dissolved in 900 mL of HCl 0.1 M at 37°C for 30 min, using baskets. The collected samples were filtered through quantitative filter paper, when evaluated by FDS, or through 0.45 μm regenerated cellulose membranes, when evaluated by HPLC.

2.2.2. LC and UV Determination Method. A stock solution was prepared by dissolving 50.0 mg of fluconazole in a 100 mL volumetric flask using HCl 0.1 M with 30 min sonication. Five standard solutions were prepared from stock solution in different concentrations (135, 150, 165, 180, and 195 μg/mL) by dilution with the same solvent. The range of concentrations chosen has the used concentration (150 mg of fluconazole in 900 mL of dissolution media) as the central concentration. The samples were filtered using quantitative filter paper, when they were evaluated by FDS, and using membranes of regenerated cellulose, 0.45 μm, when they were evaluated by HPLC. Sample, standard, and placebo-enriched solutions were prepared in the selected central concentration in the same way.

3. Results and Discussion

3.1. Analytical Development

3.1.1. UV Determination Method. The selectivity of the Brazilian-Pharmacopoeia-recommended method was evaluated using the compounding fluconazole and its excipients (capsules and placebo). The calculations of interference were performed in accordance with the recommendations of USP 32 [11], since the Brazilian Pharmacopoeia does not cite this calculation, using the following equation. The interference must not exceed 2%:

$$100C \times \left(\frac{Ap}{As}\right) \times \left(\frac{V}{L}\right), \tag{1}$$

where C is the concentration (mg/mL) of standard solution, Ap and As are the absorbances of placebo and standard solutions, respectively, V is the volume of medium (mL), and L is the dosage of the product (mg).

Spectrum results showed a strong interference, caused by both capsule shells and placebo in the analytical result. Using the analytical method recommended by the Brazilian Pharmacopoeia the mean percentage of dissolution for six fluconazole capsules was 115.21% and the mean percentage of response for placebo and capsule was 3.94% and 10.70%, respectively, a total of 14.64%. These results are in accordance with those obtained by Oliveira and coworkers [12], who reported fluconazole capsules dissolution. They have shown a huge interference of the excipients in fluconazole determination by UV.

In order to eliminate the interference of the excipients (placebo and capsule) UV-FDS was tested. In general, the spectral derivation provides simultaneous drugs determinations in association, as well as, increased selectivity. In addition, there are often an increased sensitivity and

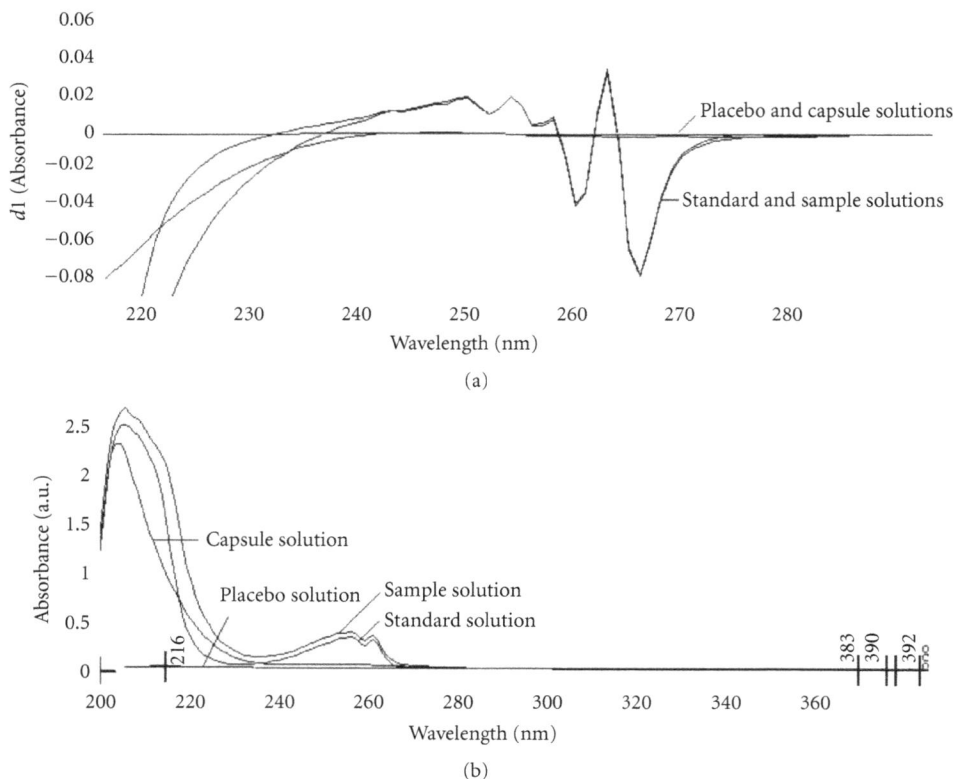

FIGURE 1: (a) First-order derivative spectra overlay of fluconazole standard, dissolution sample, and placebo and capsule shells solutions. (b) Zero-order derivative spectra overlay of standard solution, dissolution of fluconazole sample, and capsule shells and placebo solutions.

improved detection limits. The increased sensitivity observed in the derivative spectroscopy is based on the observation that the amplitude of the absorbance derivative relative to the wavelength is inversely proportional to the bandwidth of the ordinary spectrum. The order of the derivative must be carefully selected since there is usually an increase in noise level with increasing order of differentiation [13, 14].

Figure 1 shows the first-order derivative of absorbance of fluconazole that has an intense and well-defined valley at 268 nm. It was the wavelength chosen. Fluconazole capsule samples were tested according to the method recommended by the Brazilian Pharmacopeia [9]. The same samples were employed to evaluate the derivative spectrophotometric method. The average interference from placebo and capsules was again calculated in the same way but using the derivative spectrophotometric method. The interference was equal to 0.1% due to placebo and 1.98% due to capsules shells. These values have low analytical significance and are satisfactorily acceptable. Thus, the interference of excipients (placebo and capsule) was considerably reduced by using the differentiation.

3.1.2. HPLC Determination.
HPLC is a widely used method of separation with high precision and accuracy. It allows the separation between the drug and excipients, as well as degradation products, which is useful as an indicative stability method.

The analytical development must be developed to obtain a simple and optimal method. Good results were obtained using reversed phase C18 (250×4.6 mm, 5 mm) Water Symmetry endcapped column, 1.0 mL/min, water, and methanol ($60 : 40$, v/v), injection of 20 μL monitored at $\lambda = 261$ nm.

The samples used to test the method recommended by the Brazilian Pharmacopeia [9] were now employed to evaluate the HPLC method. The results showed that the HPLC method is selective, and it was able to separate the drug from the placebo and capsule (Figure 2).

Figure 2 shows that capsule shells and placebo absorb energy in UV region (peaks at 5.15 min); however, they could be well separated from fluconazole (peak at 3.65 min). The capsule peak was observed in capsule shells and sample solutions. The peak at 2.45 min, present in all samples, refers to the dissolution media (HCl 0.1 M), including the blank solvent.

3.1.3. Dissolution Performance.
Fluconazole three different pK_a [15] values, 11.01 ± 0.29, 2.94 ± 0.10, and 2.56 ± 0.12, correspond to the groups alcohol (proton donor) and two nitrogens (proton acceptors), respectively. Thus, the aqueous solubility of fluconazole is greater in solutions with extreme pH values (above 11.01 and below 2.94), situations in which the drug would be fully ionized.

However, dissolution of fluconazole capsules was evaluated in deionized water, as recommended by FDA [10], and

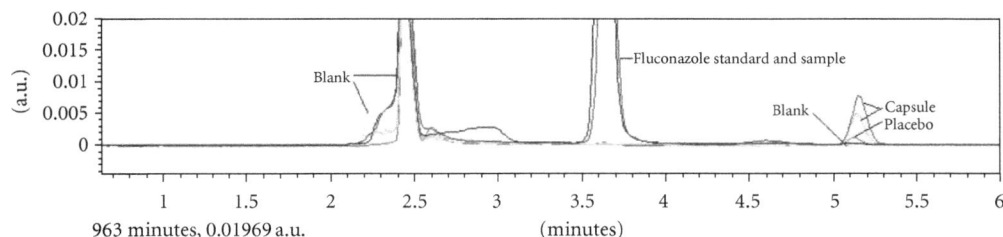

FIGURE 2: Chromatograms overlay of fluconazole dissolution samples, placebo and capsule shells, at $\lambda = 261$ nm.

in HCl 0.1 M at 37°C. Both baskets and paddles (with and without sinkers) apparatuses were used in a total medium volume of 900 mL in all tests to evaluate dissolution profile. The dissolution media were degassed by sonication for 30 min at 37°C before initiating the dissolution test. The final concentration of fluconazole in 900 mL was nearly 165 mg/mL as capsule products contained 150 mg of the drug. This is in agreement with required sink conditions, desirable to prevent saturation. Thus, the concentration 165 mg/mL was included to be the central point of the range used to validate the method.

In the dissolution profile, samples were collected after 5, 10, 15, 20, 30, 45, 60, and 65 min, filtered through quantitative filter paper, and fluconazole was quantified by UV-FDS. The dissolution medium was used as blank. The replacement of media was performed after each collection using the dissolution media at 37°C. Both mass withdrawn and the dilution made for each replacement of media were taken into account in the calculations for the construction of the dissolution profiles.

Rotation speeds 75 and 100 rpm were used in each test with baskets and 50 and 75 rpm with paddles. During the last 5 min, the speed of rotation was changed to 150 rpm in all tests. The speed can be increased at this point to verify if the drug contained in the capsule was entirely released during the test. The results of dissolution profiles using the described conditions are shown in Table 1 and Figure 3.

Water was used as dissolution medium in Tests 1 and 2, as recommended by FDA [10]. It shows not to be an appropriate medium for fluconazole capsules even after 65 min test at the highest speed because the drug percentage release was below 57%. The HCl 0.1 M dissolution media tested showed better results.

The paddle apparatus was tested at 50 and 75 rpm, with and without sinker. The use of sinker made the dissolution slower, and it is shown comparing Tests 4 and 6 to Tests 3 and 5. Tests 3 and 4, when 75 rpm was employed, showed fast dissolution in the beginning and low discriminatory power. Tests 5 and 6 showed slow dissolution in the beginning; however, at the end the drug was not released from the dosage form to the dissolution media. It could be realized after employing 150 rpm for 5 minutes in the end of test that the concentration of drug increased rapidly.

Baskets were employed in Tests 7 and 8 using HCl 0.1 M as dissolution media and 100 and 75 rpm, respectively. Test

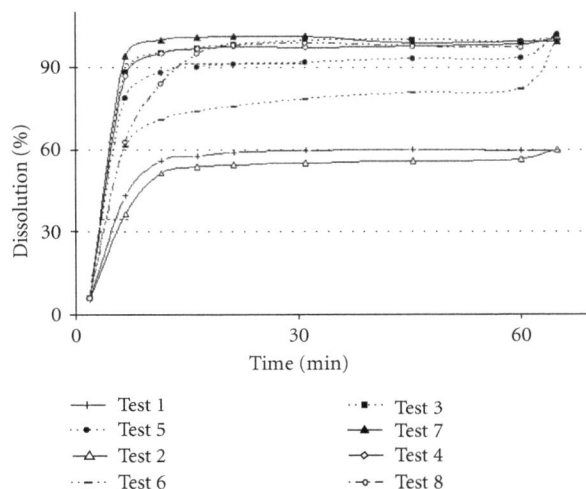

FIGURE 3: Dissolution profile obtained for testing fluconazole 150 mg capsules by UV-FDS. Conditions: as in tests 1–8 specified in Table 1.

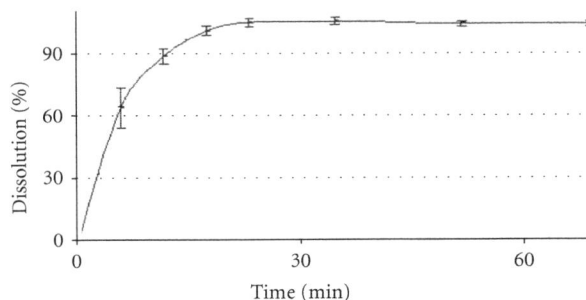

FIGURE 4: Dissolution profile for fluconazole capsules. Samples analyzed using 0.1 M HCl at 37°C as dissolution media, basket, and 75 rpm. UV method was employed.

7 has shown fast initial dissolution with more than 90% of drug release after 5 min and therefore a low discriminatory power. In Test 8, there were small release increases in fluconazole amount along the test, which has shown a discriminative profile (Figure 4).

Therefore, basket apparatus, 75 rpm, 900 mL of 0.1 M HCl at 37°C as dissolution media, and 30 min of testing

TABLE 1: Dissolution profiles for fluconazole determination: tested parameters and results.

Test	Time (min)	Medium	Medium volume (mL)	Apparatus	Rotation (rpm)	Average % dissolution	% R.S.D.
1	5	water	900	Basket	100	39.05	10.58
	10					52.79	3.26
	15					55.73	3.14
	20					56.73	1.16
	30					56.65	0.38
	45					56.82	0.31
	60					56.74	0.60
	65				150	56.69	1.11
2	5	water	900	Paddle	75	32.29	0.07
	10					48.09	4.78
	15					50.31	3.21
	20					51.01	2.02
	30					51.77	1.57
	45					52.48	1.25
	60					53.52	1.26
	65				150	56.99	0.89
3	5	0.1 M HCl	900	Paddle	75	86.79	6.23
	10					94.14	3.47
	15					95.89	2.72
	20					97.32	3.43
	30					98.64	2.40
	45					99.19	1.58
	60					98.24	1.68
	65				150	100.24	2.37
4	5	0.1 M HCl	900	Paddle + sinker	75	85.45	9.69
	10					93.92	1.13
	15					95.70	0.26
	20					96.63	0.73
	30					96.34	1.31
	45					97.06	1.71
	60					97.57	1.37
	65				150	100.00	1.43
5	5	0.1 M HCl	900	Paddle	50	76.67	5.84
	10					86.41	1.35
	15					88.70	2.58
	20					89.78	3.47
	30					90.44	3.65
	45					91.79	2.94
	60					92.36	3.81
	65				150	101.24	2.77
6	5	0.1 M HCl	900	Paddle + sinker	50	58.10	4.62
	10					68.09	6.98
	15					71.54	7.02
	20					73.34	5.75
	30					76.28	4.53
	45					78.76	4.70
	60					80.19	4.78
	65				150	99.05	4.55

TABLE 1: Continued.

Test	Time (min)	Medium	Medium volume (mL)	Apparatus	Rotation (rpm)	Average % dissolution	% R.S.D.
	5					92.96	6.34
	10					98.96	0.32
	15					100.07	0.20
7	20	0.1 M HCl	900	Basket	100	100.56	0.39
	30					100.47	0.38
	45					98.09	0.15
	60					98.66	0.23
	65				150	98.92	0.87
	5					59.50	15.07
	10					82.07	4.45
	15					93.61	2.19
8	20	0.1 M HCl	900	Basket	75	97.00	1.93
	30					97.64	1.36
	45					96.43	1.35
	60					96.38	1.23

were the conditions chosen for the dissolution of fluconazole capsules. This method was validated by FDS and HPLC.

3.2. Method Validation. The quantitative methods were validated according to the ICH [16] and USP [11] guidelines for development and validation of dissolution methods. Because nonuniform drug distribution may affect the dissolution test in the single-dose units, fluconazole capsule samples were evaluated regarding uniformity content (UC). All fluconazol capsules UC results obtained by HPLC (between 91.42–100.2% of labeled value) were within accepted specifications.

The calibration curves were obtained at five fluconazole concentration levels from 135 to 195 μg/mL for HPLC (λ = 261 nm) and UV-FDS (λ = 268 nm). Lambert-Beer law was observed at this concentration range. Linearity was evaluated by the least square method with determinations in triplicate at each concentration level. Both methods were linear with this model. Regression equations were $y = 2.5 \times 10^6 X - 1.1 \times 10^4$ ($r^2 = 0.996$) for HPLC and $y = 0.51439X - 2.9 \times 10^{-3}$, ($r^2 = 0.998$) for UV-FDS. The standard deviations of the regression were 0.83% for HPLC and 0.62% for UV-FDS. The validity of the assay was verified by means of ANOVA. According to ANOVA there is a statistically significant linear regression ($F_{calculated} > F_{critical}$; $P = 0.05$) and there is no deviation from linearity ($F_{calculated} < F_{critical}$; $P = 0.05$) for both methods.

The precision of the methods was determined by repeatability (intraday) and intermediate precision (interday). For repeatability test three curves were constructed with the established five concentration levels using standard solutions in the same day; for intermediate precision three curves were constructed with three concentration levels (high, intermediate, and low) using standard solutions in a different day. An interval of two days between repeatability and intermediate precision was observed. The results were expressed as percentage of relative standard deviation (R.S.D.). The

intraday precision tests showed R.S.D. of 0.81%, HPLC and 0.75%, FDS and interday precision tests showed R.S.D. of 2.29%, HPLC and 2.55%, FDS. These results indicate good precision.

The accuracy was performed in triplicate using the standard addition method (enriched placebo). Known amounts of standard of fluconazole were added to placebo in order to reach five established concentration levels. The mean percentage recovery of fluconazole standard found was 98.56 ± 0.82% for HPLC and 98.35 ± 0.88% for UV-FDS (Table 2). These results indicate an agreement between the true values and found values.

This paper compares the methods to determine fluconazole after dissolution test regarding their precision accuracy, and repeatability (Table 3). Both methods showed to be specific, precise, accurate, and linear in the range of concentration tested.

3.2.1. Dissolution Performance Validation. After several conditions tested for the dissolution test development, appropriate parameters were considered optimized whether provided most discriminatory dissolution profile for fluconazole capsules. That means test conditions must be able to show differences in drug release from batch to batch products, as well as to distinguish possible changes that may occur during stability studies or shelf-life of the product. The optimal parameters for fluconazole capsules dissolution are 900 mL of HCl 0.1 M, 37 ± 0.5°C using baskets with 50 rpm during 30 min.

Validation of dissolution performance was carried out by the two methods of quantification, HPLC and FDS. The concentration of 150 mg of fluconazole in 900 mL of dissolution media is nearly equal to the central concentration (165 mg/mL) at the range established.

The precision was determined by repeatability (intraday) and intermediate precision (interday). The repeatability

TABLE 2: Recovery data for fluconazole standard solutions added to the placebo by using the proposed HPLC and UV-FDS.

Method	Added amount (μg/mL)	Found[a] amount (μg/mL)	Bias (%)	Recovery[a] (%) ± R.S.D.
	135	133.92	0.80	99.20 ± 0.77
	150	148.00	1.33	98.67 ± 1.36
HPLC	165	161.28	2.25	97.75 ± 0.30
	180	177.30	1.50	98.50 ± 0.20
	195	192.41	1.33	98.67 ± 0.65
	135	134.54	0.34	99.66 ± 1.02
	150	147.51	1.66	98.34 ± 0.17
FDS	165	161.93	1.86	98.14 ± 0.08
	180	175.68	2.40	97.60 ± 0.22
	195	190.96	2.07	97.93 ± 0.15

[a] Average of three replicates.

TABLE 3: Validation parameters for different analytical methods, UV-FDS and HPLC, to determine fluconazole in capsules.

Parameters	FDS	HPLC
Analytical curve	$0.51439X - 2.9 \times 10^{-3}$	$2.5 \times 10^6 X - 1.1 \times 10^4$
Intercept values	-2.9×10^{-3}	-1.1×10^4
Standard error of slope	8.8346×10^{-7}	19.63
Correlation coefficient (r^2)	0.998	0.996
R.S.D. of repeatability (%)	0.75	0.81
R.S.D. intermediate (%)	2.55	2.29
Accuracy (%)	98.35	98.56
R.S.D. of accuracy (%)	0.88	0.82
LOQ	4.9×10^{-3}	1.28×10^{-8}
LOD	1.4×10^{-3}	3.84×10^{-9}

was tested by dissolution of five fluconazole capsules containing 150 mg of drug in duplicate, in the same day; and intermediate test was evaluated by the same way in a different day. An interval of two days between repeatability and intermediate test was observed. The results were expressed as percentage of dissolution and the R.S.D. The repeatability results were 97.96% ± 2.01% for FDS and 97.10% ± 2.44% for HPLC, and the interday results were 96.94% ± 2.48% for FDS and 95.02% ± 2.80% for HPLC. These results indicate good precision.

The accuracy of the dissolution performance was determined using enriched placebos. Known amounts of fluconazole standard were added to placebos in order to obtain three concentrations (high, intermediate, and low) of the range established. The accuracy test was performed in triplicate. The mean recovery was found to be 98.04% ± 0.89% for HPLC and 98.86% ± 1.20% for FDS (Table 4) indicating an agreement between the true values and the values found.

TABLE 4: Recovery data of dissolution performance obtained by HPLC and UV-FDS methods.

Method	Added amount (μg/mL)	Found[a] amount (μg/mL)	Bias (%)	Recovery[a] (%) ± R.S.D.
	135	133.55	1.07	98.93 ± 0.23
HPLC	165	160.33	2.83	97.17 ± 0.55
	195	191.14	1.98	98.02 ± 0.63
	135	135.11	0.08	100.08 ± 0.42
FDS	165	163.40	0.97	99.03 ± 0.32
	195	190.05	2.54	97.46 ± 0.34

[a] Average of three replicates.

4. Conclusions

A discriminative dissolution test for fluconazole capsules determination was presented in this study. Selective, sensitive, precise, and accurate analytical methods were used for quantitation. The results showed that the determination of fluconazole capsules using direct UV spectrophotometry, recommended in the Brazilian Pharmacopeia, is not enough selective; however, the developed first-order UV derivative spectrophotometry and the HPLC showed to be selective and meet requirements for an appropriate validated method. Both methods are useful for the registration of new drugs or their renewal. The application of each method, as a routine analysis, should be observed considering cost, simplicity, equipment, solvents, speed, and application to large or small workloads.

Acknowledgments

The authors wish to thank the EMS Pharmaceutical Company (Hortolândia, Brazil) for the kind supply of the raw material and thank the Fapesp, CNPq, FUNDUNESP, and PADC-UNESP for financial support.

References

[1] P. Costa and J. M. S. Lobo, "Formas farmacêuticas de libertação modificada," Revista Portuguesa de Farmacia, vol. 49, no. 4, pp. 181–190, 1999.

[2] P. Costa and J. M. S. Lobo, "Influence of dissolution medium agitation on release profiles of sustained-release tablets," Drug Development and Industrial Pharmacy, vol. 27, no. 8, pp. 811–817, 2001.

[3] J. Mehta, K. Patidar, V. Patel, N. Kshatri, and N. Vyas, "Development & validation of an in vitro dissolution method with HPLC analysis for misoprostol in formulated dosage form," Analytical Methods, vol. 2, no. 1, pp. 72–75, 2010.

[4] J. de Souza, Z. M. F. Freitas, and S. Storpirtis, "Modelos in vitro para determinação da absorção de fármacos e previsão da relação dissolução/absorção," Brazilian Journal of Pharmaceutical Sciences, vol. 43, no. 4, pp. 515–527, 2007.

[5] N. A. Kasim, M. Whitehouse, C. Ramachandran et al., "Molecular properties of WHO essential drugs and provisional biopharmaceutical classification," Mol Pharm, vol. 1, no. 1, pp. 85–96, 2004.

[6] M. Lindenberg, S. Kopp, and J. B. Dressman, "Classification of orally administered drugs on the World Health Organization Model list of Essential Medicines according to the biopharmaceutics classification system," *European Journal of Pharmaceutics and Biopharmaceutics*, vol. 58, no. 2, pp. 265–278, 2004.

[7] R. Marcolongo, *Dissolução de medicamentos: fundamentos, aplicações, aspectos regulatórios e perspectivas na área farmacêutica*, Thesis-Drugs and Medicines Post Graduation Program, University of São Paulo, São Paulo, Brazil, June 2011.

[8] J. C.R. Corrêa and H. R. N. Salgado, "Review of fluconazole properties and analytical methods for its determination," *Critical Reviews in Analytical Chemistry*, vol. 41, no. 2, pp. 124–132, 2011.

[9] Farmacopeia Brasileira. 5 ed; Rio de Janeiro,Fiocruz/ANVISA, 2010.

[10] U.S. Department of Health and Human Services, Food and Drug Administration, FDA-Recommended Dissolution Methods, 2011, http://www.accessdata.fda.gov/scripts/cder/dissolution/dsp_SearchResults_Dissolutions.cfm.

[11] United States Pharmacopeial Convention Inc., Maryland, The United States pharmacopeia (USP32), 2009.

[12] D. M. Oliveira, B. E. O. Markman, O. Uessugui, E. M. Wu, R. F. Magnelli, and E. M. Wu, "Ensaio de dissolução de cápsulas de fluconazol: problemas encontrados na determinação por espectrofotometria na região do UV," *Revista de Ciencias Farmaceuticas Basica e Aplicada*, vol. 31, no. 2, pp. 211–213, 2010.

[13] F. R. P. Rocha and L. S. G. Teixeira, "Estratégias para aumento de sensibilidade em espectrofotometria UV-VIS," *Química Nova*, vol. 27, no. 5, pp. 807–812, 2004.

[14] F. S. Rojas, C. B. Ojeda, and J. M. C. Pavon, "Derivative ultraviolet-visible region absorption spectrophotometry and its analytical applications," *Talanta*, vol. 35, no. 10, pp. 753–761, 1988.

[15] Predicting ACD/pKa DB. Advanced Chemistry Development Inc. Copyright 1994–2002.

[16] International Conference of Harmonization—ICH, "Validation of Analytical Procedures: Text and Methodology, Q2 (R1), 2005," Harmonised Tripartite Guideline, 2005, http://www.ich.org/fileadmin/Public_Web_Site/ICH_Products/Quality/Q2_R1/Step4/Q2_R1__Guideline.pdf.

22

Implementation of QbD Approach to the Analytical Method Development and Validation for the Estimation of Propafenone Hydrochloride in Tablet Dosage Form

Monika L. Jadhav and Santosh R. Tambe

Pharmaceutical Chemistry Department, M.G.V's Pharmacy College, Panchavati, Nashik 422003, India

Correspondence should be addressed to Monika L. Jadhav; monika.jadhav@yahoo.co.in

Academic Editor: Irene Panderi

Chromatographic and spectrophotometric methods were developed according to Quality by Design (QbD) approach as per ICH Q8(R2) guidelines for estimation of propafenone hydrochloride in tablet dosage form. QbD approach was carried out by varying various parameters and these variable parameters were designed into Ishikawa diagram. The critical parameters were determined by using principal component analysis as well as by observation. Estimated critical parameters in HPTLC method include solvent methanol, mode of detection absorbance, precoated aluminium backed TLC plate (10 cm × 10 cm), wavelength: 250 nm, saturation time: 20 min, band length: 8 mm, solvent front: 70 mm, volume of mobile phase: 5 mL, type of chamber: 10 cm × 10 cm, scanning time: 10 min, and mobile phase methanol : ethyl acetate : triethylamine (1.5 : 3.5 : 0.4 v/v/v). Estimated critical parameters in zero order spectrophotometric method were solvent methanol, sample preparation tablet, wavelength: 247.4 nm, slit width: 1.0, scan speed medium, and sampling interval: 0.2, and for first order derivative spectrophotometric method it was scaling factor: 5 and delta lambda 4. The above methods were validated according to ICH Q2(R1) guidelines. Proposed methods can be used for routine analysis of propafenone hydrochloride in tablet dosage form as they were found to be robust and specific.

1. Introduction

Quality by Design approach suggests looking into the quality of analytical process during the development stage itself. It says that quality should be built into the process design rather than testing into final results of analytical process [1]. QbD is defined as "a systematic approach to development that begins with predefined objectives and emphasizes product and process understanding based on sound science and quality risk management" [2]. In alignment with the approach proposed in the draft FDA guidance for process validation, a three-stage approach [3] can be applied to method validation.

Stage 1. Method Design. Define method requirements and conditions and identify critical controls.

Stage 2. Method Qualification. Confirm that the method is capable of meeting its design intent.

Stage 3. Continued Method Verification. Gain ongoing assurance to ensure that the method remains in a state of control during routine use.

A critical function of Stage 1 is the design of an Analytical Target Profile (ATP) for the method. To design the ATP, it is necessary to determine the characteristics that will be indicators of method performance for its intended use. These are selected from the performance characteristics described in ICH Q2 as per the traditional approach [4]. Instead of being applied in a tick box manner, they are investigated by a risk assessment exercise as described in ICH Q9 [5] in combination with carefully designed development studies to identify the critical method and sources of variation [6]. Variables are then investigated by robustness and ruggedness experiments to understand the functional relationship between method input variables and each of the method performance characteristics and the results are compared to the desired outcome defined in the ATP. From this, one

Implementation of QbD Approach to the Analytical Method Development and Validation for the Estimation of Propafenone
Hydrochloride in Tablet Dosage Form

149

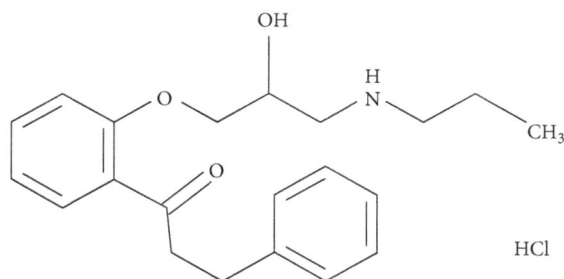

FIGURE 1: Chemical structure of propafenone hydrochloride.

can identify a set of operational method controls. Also, having evaluated the critical method parameters and gained a better understanding of the method through structured experimentation, a control strategy can be built into the method to ensure a consistent performance throughout its life cycle [7]. A key advantage of the QbD approach for all of the above situations is the flexibility to perform a qualification against the specific ATP defined for the intended use of the method [8].

Propafenone hydrochloride (PFH) (Figure 1) is Class 1C antiarrhythmic drug with local anesthetic effects and a direct stabilizing action on myocardial membranes [9]. Literature survey revealed that few chromatographic methods including enantiomeric separation using chiral stationary phases [10, 11] and liquid chromatography mass spectrometry [12] have been reported for determination of PFH from pharmaceutical formulation and biological fluids. Reported methods included use of chiral columns such as chiral AGP chiral pack AD [13] and involves various sophisticated techniques such as selected reaction monitoring mode via electrospray ionization [14]. Hence, it was planned to develop simple, economical, and less time consuming methods including High Performance Thin Layer Chromatographic method (method 1), zero order spectrophotometric method (method 2), and first order spectrophotometric method (method 3) for estimation of PFH using QbD approach. Applying the principles of QbD to analytical methods could result in more robust methods which produce consistent, reliable, and quality data throughout the life cycle and in turn will lead to less method incidents when used in the routine environment. This would mean less time spent on investigations and ultimately save time and money.

2. Implementation of QbD Approach

According to ICH Q8 (R2) guidelines, an experimental work was planned and QbD approach was implemented as follows.

2.1. Method Design. The method design stage includes establishing the method performance requirements, developing a method that will meet these requirements and then performing appropriate studies to understand the critical method variables that must be controlled to assure the method is robust and rugged.

2.2. Method Performance Requirements. Utilizing a QbD approach, it is essential at this stage that sufficient thought be given to the intended use of the method and that the objectives or performance requirements of the method be fully documented. This represents the Analytical Target Profile (ATP) [15] for the method.

ATP is the estimation of propafenone hydrochloride in tablet dosage form using spectrophotometric and chromatographic methods.

2.3. Method Development. Once the ATP has been defined, an appropriate technique and method conditions must be selected in order to meet the requirements of the ATP.

2.3.1. Method Understanding. Based on an assessment of risk (i.e., the method complexity and the potential for robustness and ruggedness issues) one can perform an exercise focused on understanding the method to better understand what impact key input variables might have on the method's performance characteristics. From this, one can identify a set of operational method controls.

2.3.2. Risk Assessment. Experiments can be run to understand the functional relationship between method input variables and each of the method performance characteristics. Knowledge accumulated during the development and initial use of the method provides input into a risk assessment (using tools such as the Fishbone diagram and FMEA) which may be used to determine which variables need studying and which require controls.

2.3.3. Design of Experiments. Robustness experiments are typically performed on parametric variables using Design of Experiments (DoE) to ensure that maximum understanding is gained while minimizing the total number of experiments. Depending on the type of method, surrogate measures of characteristics such as accuracy or precision may be evaluated.

2.4. Method Design Output. A set of method conditions will have been developed and defined which are expected to meet the ATP. Those conditions will have been optimized based on understanding of their impact on method performance.

QbD-based treatment of the robustness of an analytical method requires the assessment of all parameters (factors) which most strongly influence selectivity (results) alone and in combination. The experimental verification of many factors simultaneously is impractical and associated with extreme technical difficulties and expense. Some authors, have employed statistical studies, such as Plackett-Burman or fractional factorial designs and risk-based approaches [16–20] to overcome the challenge and reduce the experimental workload. Other procedures include running automated robustness experiments [21–24]. The present paper, however, employs statistical analysis that is principal component analysis which exhibits factor extraction of variable parameters to evaluate robustness.

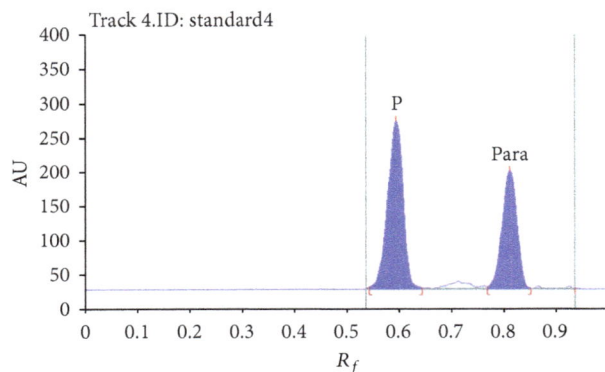

FIGURE 2: Densitogram of PFH and paracetamol (internal standard) for method 1.

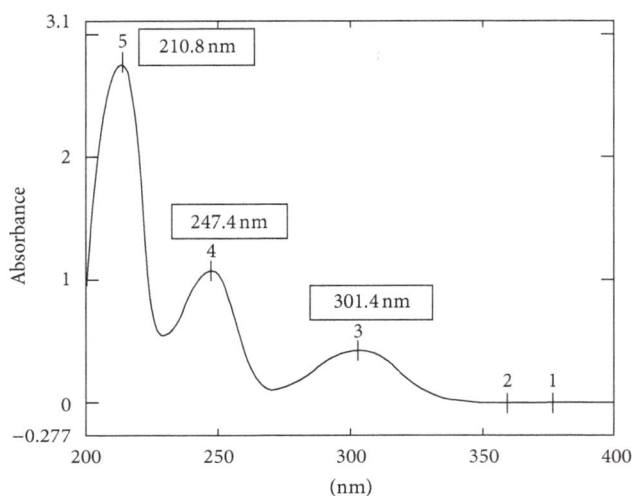

FIGURE 3: Spectrum of PFH in Methanol for method 2.

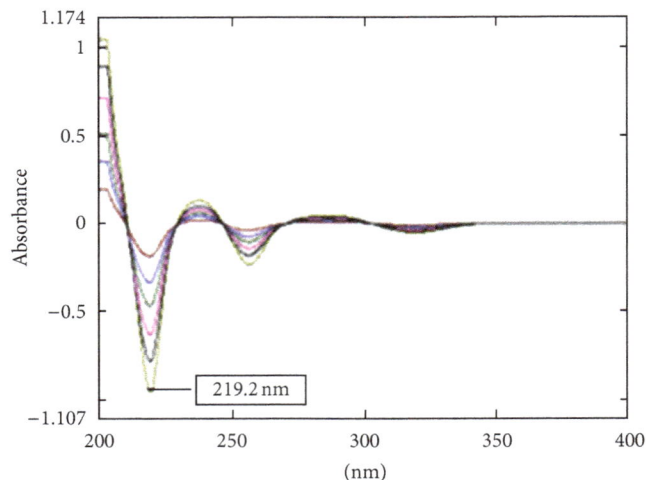

FIGURE 4: Linearity spectra of PFH in Methanol for method 3.

3. Materials and Methods

3.1. Chemicals and Reagents. PFH was procured from Procos and RLD Ltd., India. All chemicals and reagents used were of analytical grade and purchased from Merck chemicals, India. As the marketed formulation was not available, the internal formulation was prepared as tablet dosage form.

3.2. Instrumentation. Chromatographic measurements were carried out using Camag HPTLC, Wincats 1.4.2 software, Linomat 5 applicator, twin trough chamber, 100 μL syringe, and TLC scanner 3. Spectrophotometric measurements were carried out using a double beam UV visible spectrophotometer of Shimadzu UV-2450 PC (Japan), UV Probe 2.21 software, and 10 mm path length with 1 cm quartz cells.

3.3. Chromatographic Conditions. Standard stock solution (400 ng–2400 ng) and internal standard paracetamol (200 μg/mL) were applied in the form of bandwidth 8 mm on precoated silica gel 60 F_{254} (10 cm × 10 cm × 0.2 mm thickness) plates using Camag Linomat 5 applicator. A

constant application rate 120 nL/sec was employed and space between two bands was 14.0 mm. The mobile phase methanol : ethyl acetate : triethylamine (1.5 : 3.5 : 0.4 v/v) was used. Linear ascending development was carried out in twin trough glass chamber of dimension 10 cm × 10 cm, saturated with mobile phase for 20 min at room temperature (25 ± 2°C). The length of chromatogram run was 70 mm. Subsequent to the development, TLC plates were dried in current of air. Densitometric scanning was performed using Camag TLC scanner 3 in the absorbance mode at wavelength 250 nm (Figure 2).

3.4. Spectrophotometric Conditions. For the selection of analytical wavelength, standard solution of PFH was scanned in the spectrum mode from 400 nm to 200 nm. From the spectrum, λ max of PFH, 247.4 nm (Figure 3) was selected for method 2. Then it was transformed to its first derivative spectrum. It was found that PFH showed maximum amplitude at 219.2 nm (Figure 4) which was selected for method 3.

3.5. Preparation of Stock Standard and Working Solution. For both chromatographic and spectrophotometric methods, methanol was used as a solvent. Stock solution was prepared by dissolving 10 mg of PFH in 10 mL volumetric flask and dissolved in methanol to obtain 1000 μg/mL concentration. Working solution (50 μg/mL) was used for initial spectral scan in spectrophotometric method and further dilutions for linearity were prepared from stock solution.

3.6. Linearity Studies

Method 1. From the stock standard solution (1000 μg/mL) 0.4, 0.8, 1.2, 1.6, 2.0, and 2.4 μL was applied on HPTLC plate to obtain the concentration of 400 to 2400 ng per band and 0.2 μL of internal standard (paracetamol 100 μg/mL) was over spotted on each band. The plate was developed and scanned as per the optimized chromatographic conditions. R_f values

TABLE 1: Estimation of PFH in standard mixture.

Method	Amount taken ($n = 3$)	Amount found (%)	% RSD
1	1500 ng	101.3	0.93
2	15 μg/mL	100.5	0.67
3	15 μg/mL	101.4	1.04

TABLE 2: Estimation of PFH in tablet.

Method	Amount taken ($n = 3$)	Amount found (%)	% RSD
1	1500 ng	98.83	0.84
2	15 μg/mL	99.05	1.24
3	15 μg/mL	99.49	0.97

and peak area were recorded for each concentration of drug. The calibration curve was plotted as drug concentration versus relative peak area. Regression equation, correlation coefficient, slope, and y-intercept are reported.

Method 2. Aliquot portions of about 0.1, 0.2, 0.3, 0.4, and 0.5 mL of standard stock solution (1000 μg/mL) were transferred to five different 10 mL volumetric flask and volume was made up to the mark with methanol (10–50 μg/mL). Absorbances were recorded for each concentration at 247.4 nm. Calibration graph was constructed by plotting absorbances versus concentrations in μg/mL. The regression equation and correlation coefficient are computed.

Method 3. A working standard of 100 μg/mL was prepared from standard stock solution. From this, aliquot portions of about 0.5, 1.0, 1.5, 2.0, 2.5, and 3.0 mL were transferred to six different 10 mL volumetric flask and volume was made to the mark with methanol (5–30 μg/mL). These solutions were scanned in the spectrum mode from 400 nm to 200 nm. Then it was transformed to its first derivative spectrum and amplitudes were noted at 219.2 nm. Calibration graph was constructed by plotting amplitudes versus concentrations in μg/mL. The regression equation and correlation coefficient were calculated.

3.7. Estimation of PFH in Standard Mixture and Tablet Dosage Form. PFH was estimated by preparing standard mixture of label claim 150 mg and it was diluted to make 15 μg/mL for methods 2 and 3. For method 1, the concentration of 1500 ng was applied on plate and then percentage amount was calculated (Table 1).

For the estimation in tablet dosage form, the internal formulation (tablet) containing PFH equivalent to 150 mg was used. Twenty tablets were weighed and average weight was noted. Then the tablets were triturated and powder equivalent to average weight was weighed and transferred to 10 mL volumetric flask and diluted with 6 mL methanol. It was shaken for five minutes and this solution was filtered immediately through Whatman filter paper No. 41. Volume was adjusted up to the mark with methanol to obtain concentration 15000 μg/mL. From this solution, dilutions were made according to respective method and percentage amount was calculated (Table 2).

3.8. Determination of Variable Parameters for Method 1. Variable parameters for method 1 were designed as Ishikawa diagram (Figure 5). For all the variable parameters plates were developed according to optimized method except changing particular variable parameter. Plates were scanned for each

variable parameter and relative R_f values and peak area were noted. Saturation time for the plate development was varied at various time intervals such as 0 min, 10 min, 15 min, 20 min, 25 min, 30 min, 45 min, and 60 min. Band lengths were varied as 2 mm, 4 mm, 6 mm, 8 mm, 10 mm, and 12 mm. Also, development of chromatographic plate was performed by varying solvent front position of mobile phase. Various solvent fronts selected include 50 mm, 60 mm, 70 mm, 80 mm, and 90 mm. Mobile phase volume was varied as 5 mL, 10 mL, and 15 mL. Development of plates was performed in three different types of twin trough chamber that are 10 cm × 10 cm, 20 cm × 10 cm, and 20 cm × 20 cm. Duration of scanning after the development of plate was varied. Scanning time was altered in portions of time intervals that are 5 min, 10 min, 20 min, 30 min, 45 min, 60 min, 2 hour, 5 hour, 1 day, 2 days, and 3 days. After the development of plate, scanning wavelength was also varied. The plate was scanned at 247 nm, 250 nm, and 254 nm. The composition of mobile phase was modified slightly and the variation in relative R_f values and peak area was observed. Four mobile phase compositions were used as listed in the following.

Mobile phase 1: methanol : ethyl acetate : triethylamine (1.5 : 3.5 : 0.1 v/v).

Mobile phase 2: methanol : ethyl acetate : triethylamine (1.5 : 3.5 : 0.2 v/v).

Mobile phase 3: methanol : ethyl acetate : triethylamine (2 : 3 : 0.4 v/v).

Mobile phase 4: methanol : ethyl acetate : triethylamine (2.5 : 2.5 : 0.4 v/v).

Different detection mode at the stage of detection was selected. Detection was done in absorbance mode, reflectance mode, and fluorescent mode. Plate used for the development was not prewashed with methanol and plate development was carried out as such. Plate for the development was not activated in oven. Such plate was developed and results were recorded as relative R_f values and peak area. Glass plate was used for development by maintaining all the standard conditions and results were observed.

3.9. Determination of Variable Parameters for Methods 2 and 3. According to QbD approach, the first step is to determine the variable parameters for the respective method. Thus, the variable parameters for both the spectrophotometric methods were designed as Ishikawa diagram (Figures 6 and 7).

For all the variable parameters as stated in Ishikawa diagram, the absorbances were recorded over the concentration range according to respective method. Working solution (50 μg/mL) was scanned from 400 to 200 nm and three

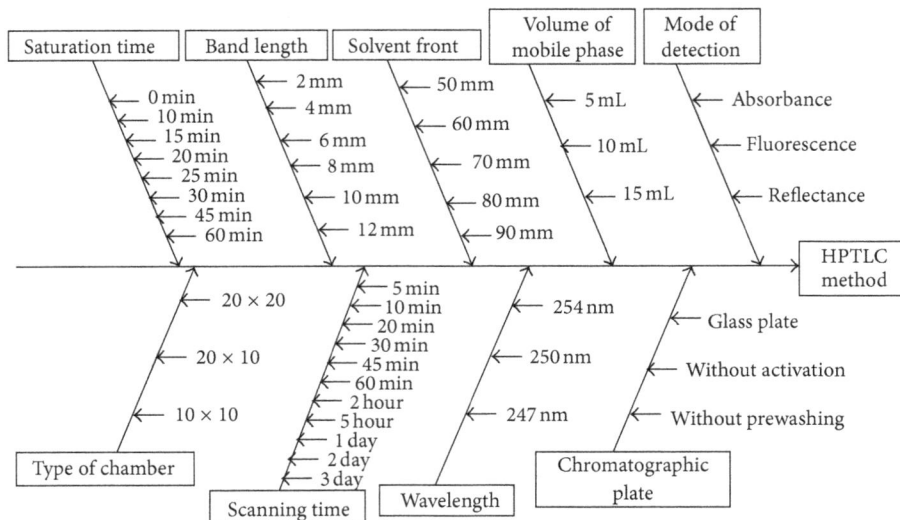

FIGURE 5: Ishikawa diagram for method 1.

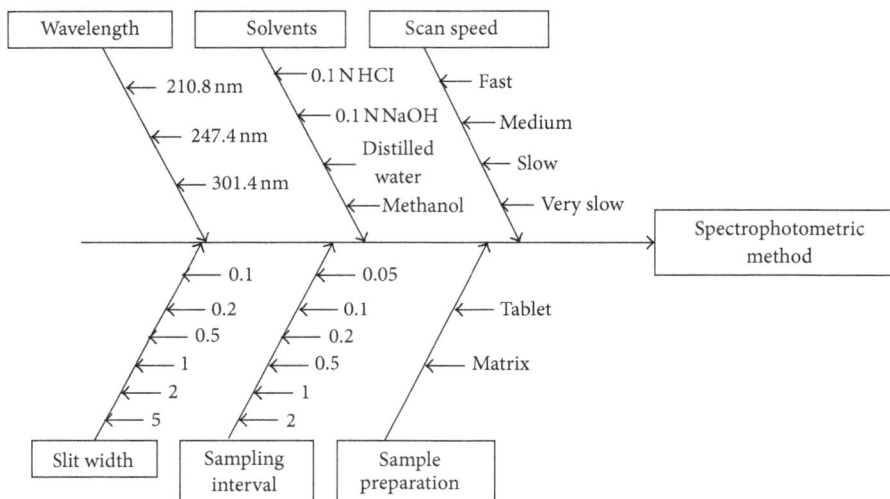

FIGURE 6: Ishikawa diagram for method 2.

peaks were observed at wavelengths 210.8 nm, 247.4 nm, and 301.4 nm. These three wavelengths were used as variable parameters. Also, the solubility was studied in various solvents including distilled water, 0.1 N NaOH, 0.1 N HCl, and methanol. The sharpness of spectra was compared for selection of critical parameter. Scan speed was varied as fast, medium, slow, and very slow over the range 400–200 nm, while slit width and sampling interval were varied in particular ranges of 0.1, 0.2, 0.5, 1.0, 2.0, and 5.0 nm and 0.05, 0.1, 0.2, 0.5, 1.0, and 2.0 nm, respectively.

For the estimation of PFH, two types of sample preparations were selected and evaluated. Tablets were formulated as per the master formula and were used in method development. Average weight of tablets was noted and tablets were triturated. Tablet powder equivalent to average weight was taken for study. Matrix containing excipients like lactose monohydrate, microcrystalline cellulose, and corn starch was

mixed with PFH and evaluated for the method development. Recovery study was carried out at three levels 80%, 100%, and 120%. For the optimization of first order derivative method, scaling factor and delta lambda were varied as 2, 3, 4, and 5 and 2, 4, 8, and 16, respectively. Amplitudes of linear concentrations were noted and evaluated.

3.10. Extraction of Critical Parameters. From the evaluated variable parameters, critical parameters were extracted by two ways, observation and principal component analysis using SPSS software. By comparing the spectral shape, sharpness, and absorbances of linearity and range, few parameters were selected as critical parameters. In principal component analysis, all these parameters were entered in variable entry window of SPSS software. Simultaneously, all the values of variable parameters were arranged in a datasheet. This datasheet was then substituted in the data entry window of

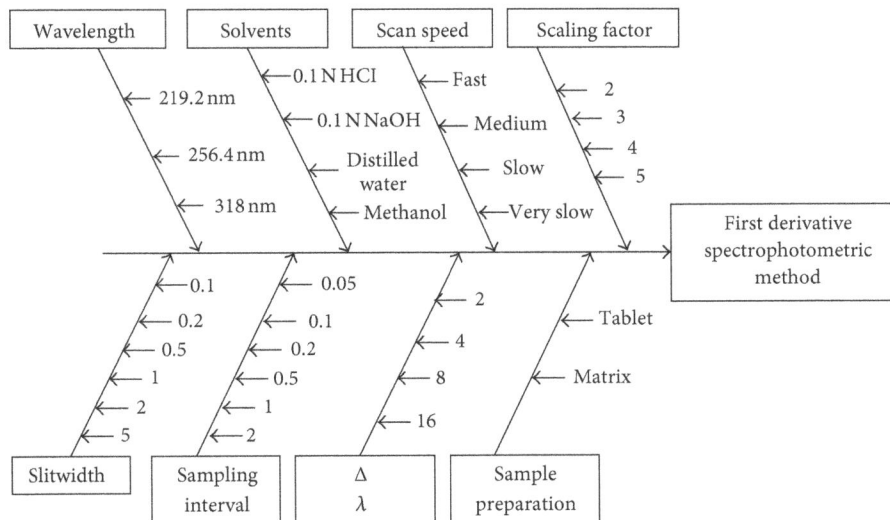

FIGURE 7: Ishikawa diagram for method 3.

SPSS software. Then the program was run to get principal components (critical parameters).

3.11. Validation. Validation was performed according to ICH Q2 (R1) guidelines. The parameters studied were linearity, precision, accuracy, specificity, LOD, LOQ, and robustness.

3.11.1. Precision. The reproducibility of all these three methods was determined by repeating the above methods at different time intervals (morning, afternoon, and evening) on the same day (Intraday precision) and on three consecutive days (interday precision). The intraday and interday variation for the estimation of PFH was carried out at three different concentration levels of 800, 1200, and 1600 ng/band for method 1, then 20, 30, and 40 μg/mL for method 2, and 10, 15, and 20 μg/mL for method 3 (Table 3).

3.11.2. Accuracy. The accuracy of the method was performed by calculating % recovery of PFH by the standard addition method. Known amounts of standard solutions of PFH were added at 80%, 100% and 120% levels to prequantified sample solutions of PFH (15 μg/mL). At each level of the amount three determinations were performed. The amount of PFH was estimated by applying obtained values to regression equation. In method 1, the known amount of standard stock solution of PFH was added to preanalyzed sample (750 ng/band). From the sample solution of tablet (150 μg/mL), 5 μL was applied and from the working standard solution (100 μg/mL), 6 μL, 7.5 μL, and 9 μL was over spotted for 80%, 100% and 120% levels, respectively. The plate was developed and analyzed by the proposed method (Table 4).

3.11.3. Specificity. The interference of other excipients was evaluated by adding 20 μg/mL of microcrystalline cellulose (MCC), corn starch, and lactose monohydrate to the 15 μg/mL of sample solution separately and absorbance was

TABLE 3: Precision data.

Method	Concentration	Intraday (% RSD) ($n = 3$)	Interday (% RSD) ($n = 3$)
1	800 ng	0.11	0.60
	1200 ng	0.10	0.38
	1600 ng	0.10	0.32
	Mean	0.10	0.41
2	20 μg/mL	0.15	0.10
	30 μg/mL	0.10	0.07
	40 μg/mL	0.08	0.06
	Mean	0.11	0.08
3	10 μg/mL	0.55	1.15
	15 μg/mL	0.59	1.19
	20 μg/mL	0.48	1.08
	Mean	0.54	1.14

TABLE 4: Accuracy data.

Method	Level	Concentration Tablet	Concentration Standard	% Recovery	Mean
1	80%	750	600	99.9	
	100%	750	750	100.2	100.2
	120%	750	900	100.6	
2	80%	15	12	100.1	
	100%	15	15	99.9	100.1
	120%	15	18	100.3	
3	80%	15	12	99.9	
	100%	15	15	100.8	100.4
	120%	15	18	100.6	

recorded. In method 1, from the working standard solution (1000 μg/mL), 1 μL was applied (1000 ng) and 1 μL (1000 ng)

TABLE 5: Critical parameters extracted for method 1.

| | | Critical parameters extracted | |
| By observation | | By principal component analysis | |
Parameter	Extracted result	Parameter	Extracted result
Solvent	Methanol	Wavelength	250 nm
		Saturation time	20 min
		Band length	8 mm
Mode of detection	Absorbance	Solvent front	70 mm
		Volume of mobile phase	5 mL
		Type of chamber	10 cm × 10 cm
Plate	Precoated plate	Scanning time	10 min
		Mobile phase composition	Methanol : ethyl acetate : triethylamine (1.5 : 3.5 : 0.4 v/v)

of each excipients was over spotted. The percentage of concentration was calculated.

3.11.4. Sensitivity.
The limit of detection (LOD) and limit of quantification (LOQ) were calculated by using (1):

$$
\begin{aligned}
\text{LOD} &= 3.3 \times \frac{\sigma}{S}, \\
\text{LOQ} &= 10 \times \frac{\sigma}{S},
\end{aligned}
\tag{1}
$$

where σ is the standard deviation of intercept and S is the slope of the calibration curve.

3.11.5. Robustness.
Robustness of the method was assessed by making variations in solvent (ethanol) and wavelength (248 nm) for spectrophotometric methods. Linear concentration range was subjected to analysis and results were evaluated. In method 1, mobile phases having different compositions like methanol : ethyl acetate : triethylamine (2 : 3 : 0.4 v/v) and methanol : ethyl acetate : triethylamine (2.5 : 2.5 : 0.4 v/v) were tried and chromatograms were run. The duration of chamber saturation time was also varied as 20 min, 25 min, and 30 min.

4. Results and Discussion

Implementation of QbD approach was carried out by studying variable parameters in the analytical method development. Critical parameters were extracted by observation of results as well as performing principal component analysis. Also, each method was validated according to ICH Q2 (R1) guidelines.

4.1. Method 1.
Method 1 has been developed and validated for the determination of PFH in pharmaceutical formulation. QbD approach was carried out by varying 46 parameters and critical parameters were extracted by using principal component analysis and by observation. The extracted critical parameters are summarized in Table 5.

PFH obeyed linearity in concentration range of 400–2400 ng/band. The R_f for PFH was found to be 0.60 ± 0.02

TABLE 6: Critical parameters extracted for method 2.

| | | Critical parameters extracted | |
| By observation | | By principal component analysis | |
Parameter	Extracted result	Parameter	Extracted result
Solvent	Methanol	Wavelength	247.4 nm
		Scan speed	Medium
Sample preparation	Tablet	Slit width	1.0
		Sampling interval	0.2

TABLE 7: Critical parameters extracted for method 3.

| | | Critical parameters extracted | |
| By observation | | By principal component analysis | |
Parameter	Extracted result	Parameter	Extracted result
Solvent	Methanol	Wavelength	219.2 nm
		Scan speed	Medium
		Scaling factor	5
Sample preparation	Tablet	Slit width	1.0
		Sampling interval	0.2
		Delta lambda	4

and paracetamol was used as an internal standard with R_f value 0.80 ± 0.02. The proposed method was applied for pharmaceutical formulation and percentage label claim for PFH was found to be 98.83%. The amount of drug estimated by proposed method was in good agreement with the label claim. The method was validated for accuracy, precision, specificity, LOD, LOQ, and robustness. Accuracy of the method was checked by recovery study at three different levels, that is, 80%, 100%, and 120%. The % recovery of PFH was found to be 100.2%. The method was found to be precise as indicated by the interday and intraday analysis showing % RSD less than 2. There was no any interference of excipients showing that method was specific. Limit of detection and limit of quantitation were 35.4 and 107.4 ng/band, respectively. The result did not show any statistical difference

Implementation of QbD Approach to the Analytical Method Development and Validation for the Estimation of Propafenone Hydrochloride in Tablet Dosage Form

155

TABLE 8: Statistical data of Validation.

Parameters	Method 1	Method 2	Method 3
λ_{max}	250	247.4 nm	219.2 nm
Linearity range	400–2400 ng	10–50 μg/mL	5–30 μg/mL
Regression equation	$Y = 0.339\,X + 0.131$	$Y = 0.019\,X + 0.0544$	$Y = 0.029\,X + 0.035$
Correlation coefficient (R^2)	0.99	0.99	0.99
% Recovery ($n = 3$)	100.2	100.07	100.4
LOD	35.4 ng	0.09 μg/mL	0.48 μg/mL
LOQ	107.4 ng	0.27 μg/mL	1.44 μg/mL
Molar absorptivity (lit/mole/cm)	—	6537.84	13793.72
Sandell's sensitivity (μg/sqcm/0.001)	—	0.058	0.027
Precision (% R.S.D.)			
Intra-day ($n = 3$)	0.10	0.11	0.54
Inter-day ($n = 3$)	0.41	0.08	1.14
Standard error	0.168×10^{-3}	2.65×10^{-4}	2.38×10^{-3}
Specificity (% R.S.D.)	0.55	0.57	0.28
Robustness (% R.S.D.)	0.19	0.84	0.57

between different solvents and different wavelengths suggesting that the method developed was robust. The statistical data of validation is summarized in Table 8.

4.2. Method 2. A zero order spectrophotometric method has been developed and validated for the determination of PFH in pharmaceutical formulation. QbD approach was carried out by varying 19 parameters and critical parameters were extracted by using principal component analysis and by observation. The extracted critical parameters are summarized in Table 6.

PFH followed linearity in the concentration range of 10–50 μg/mL. The proposed method was applied for pharmaceutical formulation and percentage label claim was found to be 99.05%. The amount of drug estimated by proposed method was in good agreement with the label claim. The % recovery for PFH was found to be 100.1%. The method was found to be precise as indicated by the inter-day and intra-day analysis showing % RSD less than 2. There was no any interference of excipients showing that the method was specific. Limit of detection and limit of quantitation were 0.09 and 0.27 μg/mL, respectively. The result did not show any statistical difference between different solvents and different wavelengths suggesting that the method developed was robust. The statistical data of validation is summarized in Table 8.

4.3. Method 3. A first order derivative spectrophotometric method has been developed and validated for the determination of PFH in pharmaceutical formulation. QbD approach was carried out by varying 27 parameters and extracted critical parameters by using principal component analysis and by observation. The extracted critical parameters are summarized in Table 7.

PFH followed linearity in the concentration range of 5–30 μg/mL. The proposed method was applied for pharmaceutical formulation and percentage label claim was found to be 99.49%. The amount of drug estimated by proposed method was in good agreement with the label claim. The % recovery for PFH was found to be 100.4%. The method was found to be precise as indicated by the inter-day and intra-day analysis showing % RSD less than 2. There was no any interference of excipients and no statistical difference between different conditions showing that the method was specific and robust. Limit of detection and limit of quantitation were 0.48 and 1.44 μg/mL, respectively. The statistical data of validation is summarized in Table 8.

5. Conclusion

By comparing all three methods, it can be concluded that spectrophotometric method was found to be simple and rapid. First order spectrophotometric method was found to be more accurate as compared to zero order spectrophotometric method. Chromatographic method was found to be more accurate, precise, robust, and more sensitive as compared to both spectrophotometric methods. Statistical analyses prove that all the developed methods can be used for routine analysis of PFH in the tablet dosage form. Implementation of QbD approach resulted in more robust methods which can produce consistent, reliable, and quality data throughout the process and also save time and money.

Acknowledgments

The authors are thankful to the management, Principal Dr. R. S. Bhambar, and the staff of M.G.V's Pharmacy College, Panchavati, Nashik,for their kind help and support.

References

[1] ICH Topic Q8 (R2), "ICH harmonised tripartite guideline," in *Proceedings of the International Conference on Harmonisation of Technical Requirements for Registration of Pharmaceuticals for Human Use (ICH '09)*, Pharmaceutical Development, 2009.

[2] A. B. Godfrey and R. S. Kenett, "Joseph M. Juran, a perspective on past contributions and future impact," *Quality and Reliability Engineering International*, vol. 23, no. 6, pp. 653–663, 2007.

[3] P. Nethercote, P. Borman, T. Bennett et al., *QbD for Better Method Validation & Transfer*, Pharmaceutical Manufacturing, 2010.

[4] ICH Topic Q2 (R1), "ICH harmonised tripartite guideline," in *Proceedings of the International Conference on Harmonisation of Technical Requirements for Registration of Pharmaceuticals for Human Use (ICH '94)*, Validation of Analytical Procedures, 1994.

[5] ICH Topic Q9, "ICH harmonised tripartite guideline," in *Proceedings of the International Conference on Harmonisation of Technical Requirements for Registration of Pharmaceuticals for Human Use (ICH '94)*, Quality Risk Management, 2005.

[6] P. Borman, P. Nethercote, M. Chatfield et al., *The Application of Quality by Design to Analytical Methods*, PharmTech, 2007.

[7] L. X. Yu, "Pharmaceutical quality by design: product and process development, understanding, and control," *Pharmaceutical Research*, vol. 25, no. 4, pp. 781–791, 2008.

[8] M. Pohl, M. Schweitzer, G. Hansen et al., "Implications and opportunities of applying the principles of QbD to analytical measurements," *Pharmaceutical Technology Europe*, vol. 22, no. 2, pp. 29–36, 2010.

[9] *The United States Pharmacopoeia 27, the National Formulary 22*, Asian Edition, United States Pharmacopoeial Convection, 2004.

[10] C. M. de Gaitani, V. L. Lanchote, and P. S. Bonato, "Enantioselective analysis of propafenone in plasma using a polysaccharide-based chiral stationary phase under reversed-phase conditions," *Journal of Chromatography B*, vol. 708, no. 1-2, pp. 177–183, 1998.

[11] Y. Wu, M. Ma, and S. Zeng, "Enantioselective assay of S(+)- and R(-)-propafenone in human urine by using RP-HPLC with pre-column chiral derivatization," *Journals of Zhejiang University-Science A*, vol. 5, no. 2, pp. 226–229, 2004.

[12] D. Zhong and X. Chen, "Enantioselective determination of propafenone and its metabolites in human plasma by liquid chromatography-mass spectrometry," *Journal of Chromatography B*, vol. 721, no. 1, pp. 67–75, 1999.

[13] L. R. P. de Abreu, V. L. Lanchote, C. Bertucci, E. J. Cesarino, and P. S. Bonato, "Simultaneous determination of propafenone and 5-hydroxypropafenone enantiomers in plasma by chromatography on an amylose derived chiral stationary phase," *Journal of Pharmaceutical and Biomedical Analysis*, vol. 20, no. 1-2, pp. 209–216, 1999.

[14] U. Hofmann, M. Pecia, G. Heinkele, K. Dilger, H. K. Kroemer, and M. Eichelbaum, "Determination of propafenone and its phase I and phase II metabolites in plasma and urine by high-performance liquid chromatography-electrospray ionization mass spectrometry," *Journal of Chromatography B*, vol. 748, no. 1, pp. 113–123, 2000.

[15] M. Schweitzer, M. Pohl, M. Hanna-Brown et al., "Implications and opportunities of applying QbD principles to analytical measurements," *Pharmaceutical Technology*, vol. 34, no. 2, pp. 52–59, 2010.

[16] R. M. Bianchini, P. M. Castellano, and T. S. Kaufman, "Development and validation of an HPLC method for the determination of process-related impurities in pridinol mesylate, employing experimental designs," *Analytica Chimica Acta*, vol. 654, no. 2, pp. 141–147, 2009.

[17] K. Monks, I. Molnar, H. J. Rieger, B. Bogati, and E. Szabo, "Quality by design: multidimensional exploration of the design space in high performance liquid chromatography method development for better robustness before validation," *Journal of Chromatography A*, vol. 1232, pp. 218–230, 2012.

[18] G. Srinubabu, C. A. I. Raju, N. Sarath, P. K. Kumar, and J. V. L. N. S. Rao, "Development and validation of a HPLC method for the determination of voriconazole in pharmaceutical formulation using an experimental design," *Talanta*, vol. 71, no. 3, pp. 1424–1429, 2007.

[19] S. M. Khamanga and R. B. Walker, "The use of experimental design in the development of an HPLC-ECD method for the analysis of captopril," *Talanta*, vol. 83, no. 3, pp. 1037–1049, 2011.

[20] E. M. Sheldon and J. B. Downar, "Development and validation of a single robust HPLC method for the characterization of a pharmaceutical starting material and impurities from three suppliers using three separate synthetic routes," *Journal of Pharmaceutical and Biomedical Analysis*, vol. 23, no. 2-3, pp. 561–572, 2000.

[21] P. F. Gavin and B. A. Olsen, "A quality by design approach to impurity method development for atomoxetine hydrochloride (LY139603)," *Journal of Pharmaceutical and Biomedical Analysis*, vol. 46, no. 3, pp. 431–441, 2008.

[22] B. Dejaegher and Y. vander Heyden, "Experimental designs and their recent advances in set-up, data interpretation, and analytical applications," *Journal of Pharmaceutical and Biomedical Analysis*, vol. 56, no. 2, pp. 141–158, 2011.

[23] D. Awotwe-Otooa, C. Agarabia, P. J. Faustinoa et al., "Application of quality by design elements for the development and optimization of an analytical method for protamine sulfate," *Journal of Pharmaceutical and Biomedical Analysis*, vol. 62, pp. 61–67, 2012.

[24] B. Vasselle, G. Gousset, and J. P. Bounine, "Development and validation of a high-performance liquid chromatographic stability-indicating method for the analysis of Synercid in quality control, stability and compatibility studies," *Journal of Pharmaceutical and Biomedical Analysis*, vol. 19, no. 5, pp. 641–657, 1999.

23

CZE/PAD and HPLC-UV/PAD Profile of Flavonoids from *Maytenus aquifolium* and *Maytenus ilicifolia* "espinheira santa" Leaves Extracts

Cristina A. Diagone, Renata Colombo, Fernando M. Lanças, and Janete H. Yariwake

Instituto de Química de São Carlos, Universidade de São Paulo, Caixa Postal 780, 13560-970 São Carlos, SP, Brazil

Correspondence should be addressed to Janete H. Yariwake, janete@iqsc.usp.br

Academic Editor: Irena Vovk

This paper describes the application of HPLC and CZE to analyze flavonoids in the leaves of *Maytenus ilicifolia* and *Maytenus aquifolium*, which are species widely used in Brazilian folk medicine. The two species showed different flavonoid profiles, but acidic hydrolysis of the *Maytenus* extracts confirmed that all these compounds are quercetin or kaempferol derivatives. A comparison of the CZE and HPLC profiles of *Maytenus* extracts showed numerous flavonoid peaks using HPLC. However, the advantages of CZE such as analysis without requiring clean-up and less generation of chemical waste than with HPLC point to the potential of the CZE technique for the quality control (routine analysis) of "espinheira santa" phytopharmaceuticals.

1. Introduction

Flavonoids are a heterogeneous group of polyphenols (about 4000 substances) present in all plants and responsible for their color, growth, development, and immunity [1, 2] and can occur in free form (aglycones) or linked to sugars (glycosides) [3]. Many flavonoids found in plants have biological and pharmacological activities, such as antimicrobial, antiinflammatory, and antiallergic action [4–7]. The antioxidant property of these substances has also been established and correlated to their protective effects on cardiovascular disease and some forms of cancer [8–10].

Maytenus ilicifolia and *M. aquifolium* (Celastraceae) are Brazilian medicinal plants known as "espinheira santa", which are used in Brazil as phytopharmaceuticals due to their antiulcer activity [11, 12]. Several studies focus on the bioactivity of *Maytenus* extracts, whose main compounds include flavonoid derivatives of quercetin and kaempferol [13, 14] and tannins [15]. These polyphenolic compounds can be correlated with the diverse pharmacological activities of these extracts [16, 17]. Due to the structural characteristics of polyphenolic compounds, most of the procedures described in the literature for the analysis of *M. aquifolium* and

M. ilicifolia extract are based on RP-HPLC (reverse-phase high-performance liquid chromatography). Recently, however, a two-dimensional LC (size-exclusion—reverse-phase) procedure was employed for the LC-MS analysis of flavonol glycosides from *M. ilicifolia* leaves [18].

Due to its robustness, sensitivity, and versatility, HPLC-UV/PAD (high performance liquid chromatography-ultraviolet detection using a photodiode array detector) is the technique of choice for the analysis of flavonoids and other phenolic compounds in natural products [19, 20]. However, more recently, CE (capillary electrophoresis) techniques, including CZE (capillary zone electrophoresis), have been introduced as an analytical tool in studies of many secondary plant metabolites, mainly due to the method's faster development, lower operating cost and solvent consumption, and higher separation efficiencies [19, 21].

This work compares the HPLC and CZE techniques applied in the analysis of flavonoids contained in these two *Maytenus* species. Analytical methods for these two species that are suitable for application in agronomic studies or the quality control of phytopharmaceuticals, for example, require numerous analyses. In the development of these analytical methods, one must also keep in mind that the two aforementioned *Maytenus* species are known by the same popular name, "espinheira santa", but only *M. ilicifolia* is registered in the 4th Edition of the Brazilian Pharmacopoeia (2003) [22].

2. Materials and Methods

2.1. Plant Material. Leaves of *Maytenus aquifolium* Mart. and *Maytenus ilicifolia* (Schrad.) Planch. (Celastraceae) were supplied by Dr. Ana Maria Soares Pereira (UNAERP—Universidade de Ribeirão Preto, Ribeirão Preto, SP, Brazil). These leaves were picked from specimens cultivated on the farm of the UNAERP campus; voucher specimens were deposited at the UNAERP herbarium and identified as HPMU-0755 (*M. aquifolium*) and HPMU-0266 (*M. ilicifolia*). Immediately after the leaves were picked, they were dried at 40°C to constant weight, ground in domestic blender, and pulverized. Only particles of 0.5–1.0 mm were used for the extractions and were stored in glass flasks protected from light and humidity until required for analysis.

2.2. Reagents and Materials. Rutin, quercetin, and kaempferol standards were obtained from Sigma (St. Louis, MO, USA). HPLC-grade acetonitrile (ACN) and trifluoroacetic acid (TFA) were purchased from Mallinckrodt (Paris, Kentucky, USA). Analytical grade methanol (MeOH) and ethyl acetate (EtOAc) were purchased from Mallinckrodt (Xalostoc, State of Mexico, Mexico). Analytical grade chloroform (CHCl$_3$) was purchased from Merck (Rio de Janeiro, Brazil). TLC plates of silica gel 60, without fluorescent indicator, were purchased from Merck (Darmstadt, Germany). Analytical grade monobasic potassium phosphate (KH$_2$PO$_4$) and sodium tetraborate decahydrate (NaB$_4$O$_7 \cdot 10$ H$_2$O) were purchased from Reagen (Rio de Janeiro, Brazil). Analytical grade formic acid (HCOOH), phosphoric acid (H$_3$PO$_4$), hydrochloric acid (HCl), sodium hydroxide (NaOH), and polyethylene-glycol (PEG 400) were obtained from Synth (São Paulo, Brazil). Diphenylboric acid 2-aminoethylester (C$_{14}$H$_{16}$BNO) was purchased from Sigma (St. Louis, MO, USA). Water was purified in a Millipore Milli-Q Water Purification System (Eschborn, Germany). Hydrophobic Fluoropore (HF-PTFE) membranes (0.5 μm) and HA membranes (0.45 μm) in cellulose ester media were purchased from Millipore (São Paulo, Brazil).

2.3. Preparation of Samples. 1.0 g of the *Maytenus* leaves was extracted by maceration agitation with 10 mL of MeOH/H$_2$O (1 : 1 v/v) for 30 min at 50°C. The hydromethanolic extracts were filtered, and their final volume was adjusted to 10 mL with MeOH/H$_2$O (1 : 1 v/v). No clean-up was necessary for the CZE analysis: the hydromethanolic extracts were simply filtered through 0.5 μm HF-PTFE membranes (Millipore) and analyzed. For the HPLC analysis, the extracts were subjected to liquid-liquid extraction using 5 mL of CHCl$_3$; the organic layer was discarded, and the hydromethanolic layer was filtered through 0.5 μm HF-PTFE membranes (Millipore) before the HPLC analysis.

2.4. Preparation of Standards. 0.01 g of each flavonol standard (rutin, quercetin, or kaempferol) was dissolved separately in 10 mL of MeOH. An aliquot of 0.1 mL of each stock solution was diluted to 10 mL with MeOH to obtain a stock solution containing the three flavonols; this stock solution was utilized in the HPLC and CZE analyses.

2.5. Thin Layer Chromatography. Analyses were carried out on silica gel 60 aluminum sheets precoated with EtOAc/HCOOH/H$_2$O (6 : 1 : 1 v/v). After developing the plates, the solvent was dried and the flavonoids were visualized with diphenylboric acid 2-aminoethylester-PEG 400 under UV at $\lambda = 360$ nm [23].

2.6. Acid Hydrolysis. *Maytenus* extract was evaporated to 8.3 mL and mixed with 1.7 mL of 2.0 mol/L HCl. The solution was refluxed for 10 min at 95°C. The resulting extracts were filtered through 0.5 μm HF-PTFE membranes (Millipore) and analyzed by HPLC.

2.7. CZE Analysis. The CZE analysis was performed in an HP^{3D} Capillary Electrophoresis System (Hewlett Packard, Waldbronn, Germany) equipped with a photodiode array (Hewlett Packard) and an HP Chem Station data processing system. Separations were performed using an uncoated fused silica capillary tube (Polymicro Technologies, Phoenix, AZ, USA) with a total length of 64.5 cm, effective length of 56.0 cm, and i.d. of 50.0 μm. Samples were injected in hydrostatic mode at 500 mbar for 7 s. The analysis was performed at 25°C and an applied voltage of 20 kV, and the samples were introduced into the system in hydrostatic mode at 500 mbar pressure for 7 s. Capillary conditioning was carried out by first washing with H$_2$O for 10 min, followed by 1.0 mol/L NaOH for 5 min, 0.1 mol/L NaOH for 5 min, and finally with the running buffer for 10 min. Between consecutive runs, the capillary tube was flushed with 0.1 mol/L NaOH for 5 min and running buffer for 5 min. Buffer solutions of sodium tetraborate and potassium phosphate in water were prepared, and the pH was adjusted using phosphoric acid or NaOH solutions. Optimal separation conditions were determined after testing different buffer conditions: concentration of tetraborate-phosphate (resp., 50 : 5; 30 : 5; 30 : 25; 30 : 50; 10 : 5 mmol/L) and pH values (8.0; 8.5; 9.0; 9.3; 9.5, 10.0), as well as the percentage of methanol (2.0; 5.0; 8.0, 12.0%) used as organic solvent.

2.8. HPLC-UV-PAD Analysis. This analysis was performed in a modular LC System (Shimadzu, Kyoto, Japan) consisting of two LC-10 AD pumps; a CTO-10A column oven; an SPD-M10A variable wavelength diode array detector; the LC-10 Workstation Class data processing system. Supelcosil columns (Supelco, Bellefonte, PA, USA) with stationary phase C-18 and C-8 columns (250 mm × 4.6 mm, 5 μm) protected by guard columns filled with the same stationary phase (20 mm × 4.6 mm, 5 μm) were utilized. The column oven was thermostat controlled at 35°C, and the flow rate was 1.0 mL/min. The injection volume was 10 μL (Rheodyne loop). Detection was monitored at 254 and 350 nm. The mobile phases tested were: (A) 2.0, 2.5 and 3.0% formic acid in water and 0.3% trifluoroacetic acid in water; (B) ACN or MeOH.

CZE/PAD and HPLC-UV/PAD Profile of Flavonoids from Maytenus aquifolium and Maytenus ilicifolia "espinheira santa" Leaves Extracts

159

3. Results and Discussion

Prior to the HPLC analysis, the *Maytenus* extracts were subjected to TLC analysis. *M. aquifolium* extracts showed two spots with lower Rf values (= more polar compounds) than quercetin and kaempferol standards. The fluorescence of these spots indicated the presence of quercetin derivatives (orange fluorescent spots) and kaempferol derivatives (green fluorescent spots) [23]. *M. ilicifolia* extracts exhibited six glycoside flavonols derivatives of quercetin (one of them with Rf identical to that of rutin) and two glycoside flavonol derivatives of kaempferol. These compounds have higher Rf values and are therefore less polar than the two glycoside flavonols reported in *M. aquifolium* extracts [24, 25].

3.1. HPLC-UV-PAD Analysis. Optimization of the chromatographic conditions showed that the C-18 and C-8 columns were highly efficient in the separation of flavonoids from *Maytenus*. However, for *M. aquifolium* extracts, the C-18 column provided better resolution in the separation of flavonoids. The amount of formic acid (2.0% in water, solvent A) was chosen because the increase in the percentage of formic acid (2.5 and 3.0%) and its replacement with trifluoroacetic acid did not improve the resolution and led to similar separation efficiencies. Acetonitrile showed better results than methanol and was therefore selected as the organic solvent in the optimized HPLC conditions for the extracts of the two *Maytenus* species.

The HPLC-UV/PAD analysis led to the detection of two flavonoids in *M. aquifolium* leaves (Figure 1).

The flavonoid peaks can be identified by their characteristic UV/PAD spectral pattern with two bands, Band I, λ_{max} around 300–380 nm and Band II, λ_{max} around 240–280 nm. Moreover, quercetin derivatives ($\lambda_{max} = 354$ nm) can be distinguished from kaempferol derivatives ($\lambda_{max} = 344$ nm) also considering the data obtained by TLC and the acid hydrolysis of *Maytenus* extracts [26]. Therefore, the comparison of the material obtained by acid hydrolysis (Figure 2(a)) with authentic standards (Figure 2(b): retention time of the aglycones and UV-PAD spectra) confirmed quercetin and kaempferol as the aglycones of *M. aquifolium* flavonoids.

In the chromatogram of *Maytenus ilicifolia* leaf extracts (Figure 3), twelve peaks show UV/PAD spectra characteristic of flavonoids. Peaks 1 to 4, 7, and 9 to 12 are quercetin derivatives ($\lambda_{max} \sim 354$ nm) while peaks 5 and 6 are kaempferol derivatives ($\lambda_{max} \sim 344$ nm).

Peak 8 was identified as rutin by direct comparison (retention time and UV-DAD spectra) with an authentic commercial standard (Figure 4). The acid hydrolysis of extract also confirmed quercetin and kaempferol as aglycones of *M. ilicifolia* flavonoids, which are identified in Figure 5.

3.2. CZE Analysis. Figures 6 and 7 illustrate the optimized conditions for CZE analysis of *M. aquifolium* and *M. ilicifolia*, respectively. The CZE/DAD-UV electropherogram of *M. aquifolium* showed the presence of two major compounds, peaks 1 and 2, respectively, identified as kaempferol and quercetin derivatives (Figure 6), plus other minor

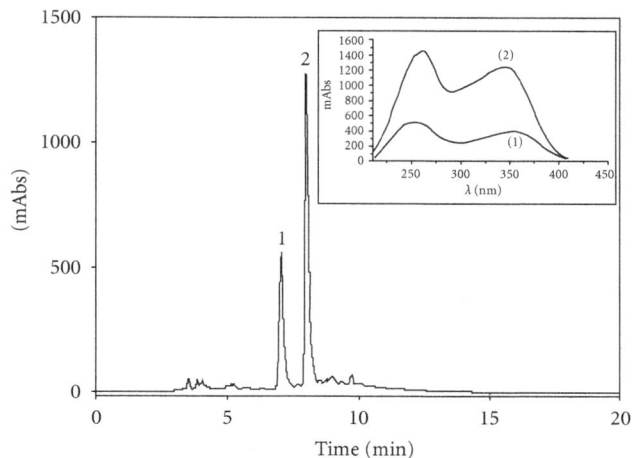

FIGURE 1: HPLC/DAD-UV ($\lambda = 270$ nm) chromatogram of flavonoids from *M. aquifolium* leaves. (1) Quercetin derivative and (2) kaempferol derivative. Mobile phase: 0–20 min 15–80% acetonitrile (solvent B); for other chromatographic conditions, see experimental part.

flavonoids not detected in the HPLC-UV/DAD chromatogram. The electropherogram of *M. ilicifolia* in Figure 7, which was obtained at $\lambda = 380$ nm due to the interference of other compounds at $\lambda = 270$ nm (possibly phenolic compounds), indicates the presence of ten flavonoids, including rutin. The presence of rutin was suggested by TLC analysis and confirmed by spiking *M. ilicifolia* extract. Moreover, the longer migration time of this compound compared to the two major flavonoids (peaks 1 and 2, Figure 7) indicates that these major peaks are more polar compounds, possibly the triglycosylated flavonoids reported in *M. aquifolium* extracts [24, 25].

The CZE separation was optimized based on the parameters of pH, buffer concentration, and the effect of modifier. An important parameter is pH, which changes the electroosmotic flow (EOF) and affects the degree of ionization of the solutes. The electrophoretic mobility (μ_{ef}) and migration times (t_M) of three flavonol standards—rutin, quercetin, and kaempferol—were calculated to verify the electrophoretic behavior of *Maytenus* extracts (Table 1). The results indicate that the increase in pH values augmented both the μ_{ef} and migration times of all flavonoids, while lower values pH showed a decrease in μ_{ef}, resulting in a decrease in the negative charges of the compounds.

Figure 8 illustrates the effect of pH on the μ_{ef} of flavonol standards. The differences in their μ_{ef} were attributed to differences in molecular size and in the number and acidity (pKa) of the free phenolic groups attached to the flavonoid skeleton, which contribute to different levels of charge in flavonol molecules due to differences in acidity. A pH of 8.5 was chosen for the CZE analysis of both *Maytenus* extracts due to the higher efficiency and resolution and faster analysis. An analysis was made of the influence of tetraborate and phosphate concentrations on the CZE analysis (Table 2).

The results showed that the decrease in tetraborate concentration diminished the resolution in the separation of

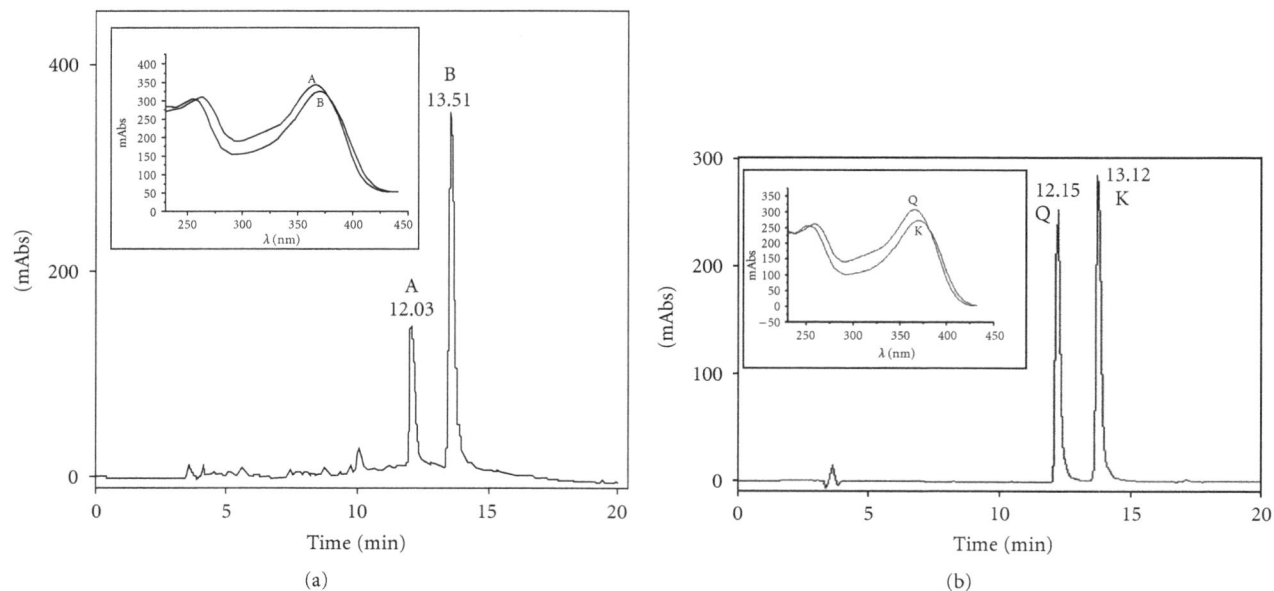

FIGURE 2: HPLC/DAD-UV ($\lambda = 270$ nm) chromatogram and UV-PAD spectra of (a) *M. aquifolium* leaves extract after acid hydrolysis and (b) standards quercetin (Q) and kaempferol (K). Mobile phase: 0–20 min 15–80% acetonitrile; for other chromatographic conditions, see experimental part.

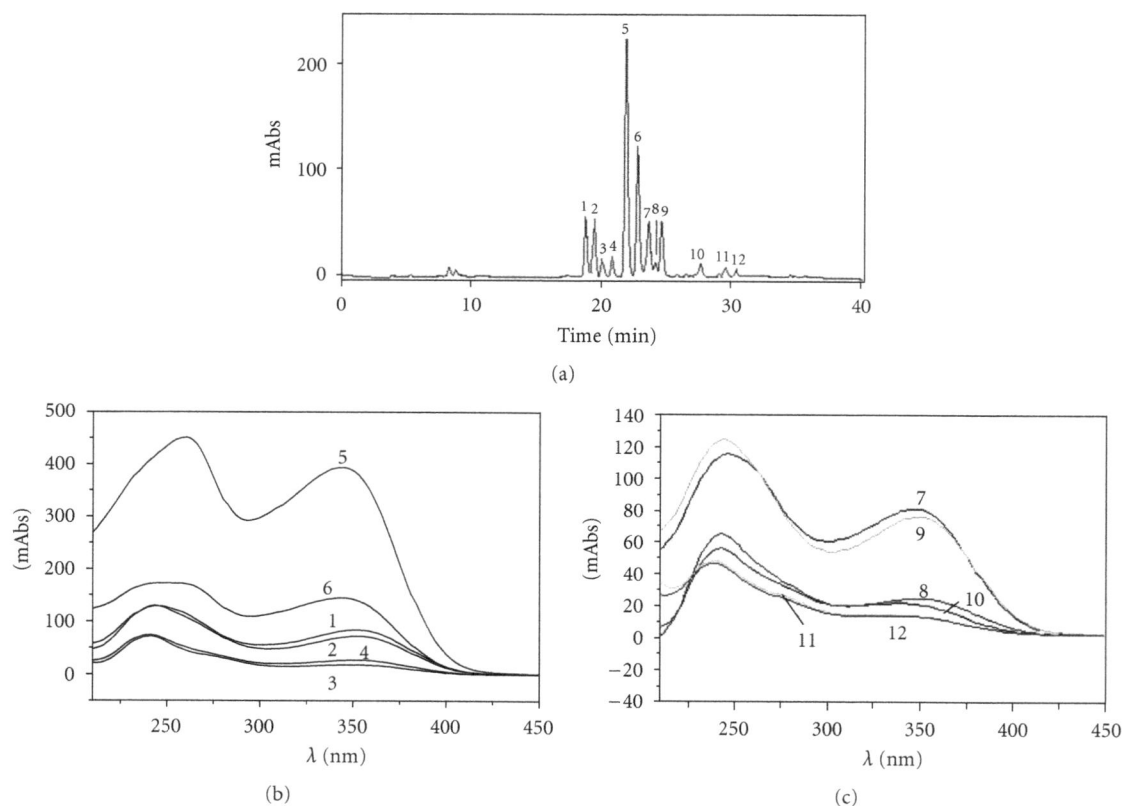

FIGURE 3: HPLC/DAD-UV ($\lambda = 380$ nm) chromatogram and UV-PAD spectra of flavonoids from *M. ilicifolia* leaves. Peaks 1–4, 7, and 9–12: quercetin derivatives, peaks 5 and 6: kaempferol derivatives. Peaks 8: rutin. Mobile phase: 0–2 min 10% acetonitrile (solvent B), 2–15 min 10–15% B, 11–22 min 15–18% B, 22–37 min 18–30% B, 37–42 min 30–40% B; for other chromatographic conditions, see experimental part.

TABLE 1: Effect of pH variation on the values of migration time (t_M) and electrophoretic mobility (μ_{ef}) of flavonoid standards: (1) rutin, (2) kaempferol, and (3) quercetin. Buffer: tetraborate 30 mmol/L/phosphate 5 mmol/L. For detailed CZE conditions, see experimental section.

pH	$t_{M\,(1)}$ (min)	$t_{M\,(2)}$ (min)	$t_{M\,(3)}$ (min)	$\mu_{ef\,(1)} \times 10^{-4}$ (cm^2 V · s)	$\mu_{ef\,(2)} \times 10^{-4}$ (cm^2 V · s)	$\mu_{ef\,(3)} \times 110^{-4}$ (cm^2 V · s)
8.0	9.625	8.983	12.399	1.671	1.433	2.372
8.5	9.512	9.597	12.613	1.651	1.679	2.430
9.0	9.347	10.587	12.869	1.850	2.228	2.733
9.3	10.040	12.537	14.576	1.837	2.435	2.772
9.5	10.248	13.742	15.375	1.880	2.628	2.861
10.0	11.475	18.693	19.741	1.940	2.954	3.040

$\mu_{ef} = (L_t \times L_{ef})/(t_m \times V) - (L_t \times L_{ef})/(t_{nm} \times V)$, where L_t is the total lenght of the capillary, L_{ef} is the effective lenght of the capillary, t_m is the migration time of the analyte, t_{nm} is the migration time of the neutral marker (methanol), and V is the applied voltage.

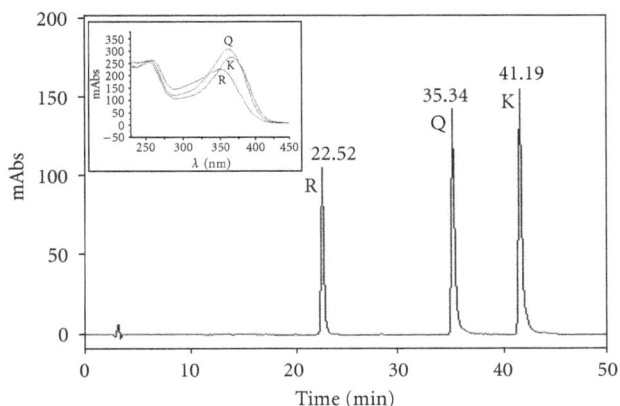

FIGURE 4: HPLC/DAD-UV ($\lambda = 380$ nm) chromatogram and UV-PAD spectra of standards rutin, quercetin, and kaempferol. Mobile phase: 0–2 min 10% acetonitrile (solvent B), 2–15 min 10–15% B, 11–22 min 15–18% B, 22–37 min 18–30% B, 37–42 min 30–40% B; for other chromatographic conditions, see experimental part.

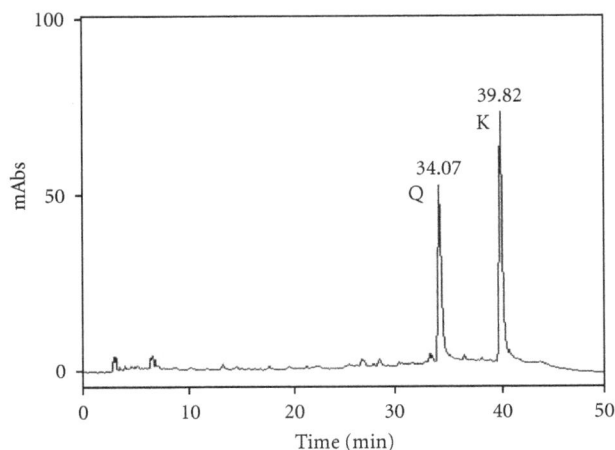

FIGURE 5: HPLC/DAD-UV ($\lambda = 380$ nm) chromatogram of *M. ilicifolia* leaves extract after acid hydrolysis. Mobile phase: 0–2 min 10% acetonitrile (solvent B), 2–15 min 10–15% B, 11–22 min 15–18% B, 22–37 min 18–30% B, 37–42 min 30–40% B; for other chromatographic conditions, see experimental part.

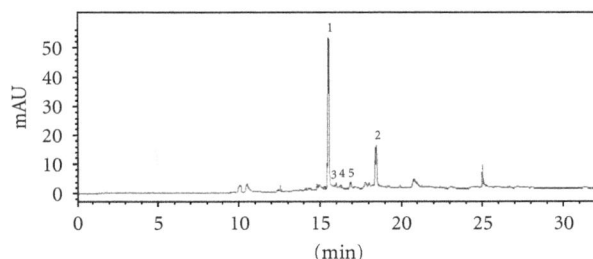

FIGURE 6: CZE/DAD-UV electropherogram of *Maytenus aquifolium* leaves extract ($\lambda = 270$ nm). Peak 1: kaempferol derivative; peak 2: quercetin derivative; peaks 3–5: other minor flavonoids. Conditions: buffer 30 mmol/L tetraborate, 50 mmol/L phosphate, pH = 8.5, 20 kV, and 12% MeOH; for other electrophoretic conditions, see experimental part.

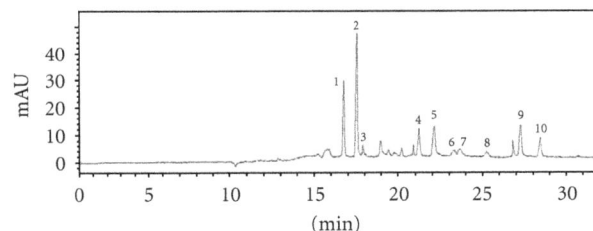

FIGURE 7: CZE/DAD-UV electropherogram of *Maytenus ilicifolium* leaves extract ($\lambda = 380$ nm). Conditions: buffer 30 mmol/L tetraborate, 50 mmol/L phosphate, pH = 8.5, 20 kV, and 12% MeOH, for other electrophoretic conditions, see experimental part.

the flavonol glycosides due to the minor presence of tetraborate complexes at this concentration. On the other hand, increasing the tetraborate and phosphate concentrations led to a decrease in EOF and an increase in migration time due to the higher viscosity of the buffer. The resolution was calculated using the peaks of kaempferol and quercetin derivatives (major flavonoids), and the best results were achieved with 50/50 mmol/L tetraborate/phosphate. However, 30/50 mmol/L tetraborate/phosphate showed better separation if one also considers the minor flavonoids, so the latter proportion was chosen as the optimum condition for both *Maytenus* extracts.

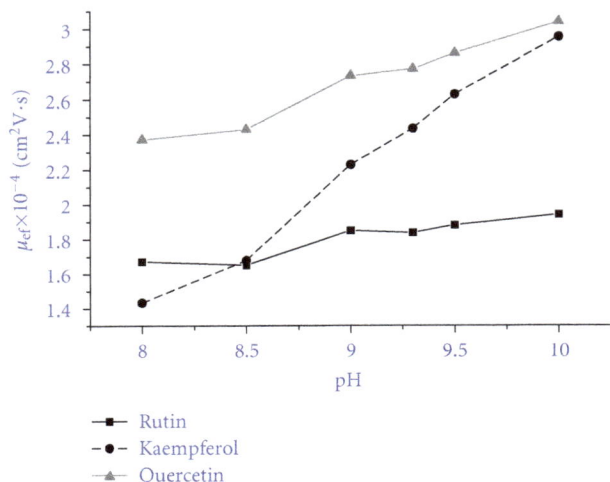

FIGURE 8: Effect of pH on electrophoretic mobility of flavonols standards (rutin, quercetin, kaempferol). Conditions: buffer 30 mmol/L tetraborate, 50 mmol/L phosphate; for other electrophoretic conditions, see experimental part.

TABLE 2: Resolution (Rs) between the peaks corresponding to the quercetin and kaempferol derivatives found in *Maytenus aquifolium* extracts (resp., peaks 1 and 2 at Figure 1), at pH 8.5 and with variation of buffer tetraborate/phosphate concentration.

Concentration of tetraborate/phosphate (mmol/L)	$Rs_{1,2}$
10/5	Coelution
30/5	5.506
50/5	6.320
30/25	5.941
30/50	6.311

$Rs = (1/4)N^{1/2}(\Delta\mu_{ef}/(\overline{\mu_{ef}} + \mu_{eof}))$, where $\Delta\mu_{ef}$ is the difference on the electrophoretic mobility of the two analytes; μ_{ef} is the mean of mobility of compounds corresponding to peaks 1 and 2; μ_{eof} is the mobility of the eletroosmotic flow (neutral marker: methanol).

Figures 9 and 10 illustrate the effect of different percentages of methanol as organic modifier: the use of 12% methanol increased the migration times of the analytes. Moreover, methanol increased the resolution for some flavonoids that coeluted in the absence of organic modifier (peaks 3 to 5, Figure 6, possible flavonols) in *M. aquifolium*. Similar results were observed in *M. ilicifolia* extracts, with the separation of peaks 6 (rutin) and 7; hence, the optimized conditions for both extracts (Figures 6 and 7) include 12% methanol.

4. Conclusions

The HPLC and CZE techniques can both be used in the analysis of flavonoids in *Maytenus aquifolium* and *Maytenus ilicifolia* extracts. The comparison of the results obtained by these techniques showed that CZE offers some advantages, for example, higher efficiency and resolution, shorter separation time, and the fact that CZE does not require cleanup of the extracts. Furthermore, the CZE method is an

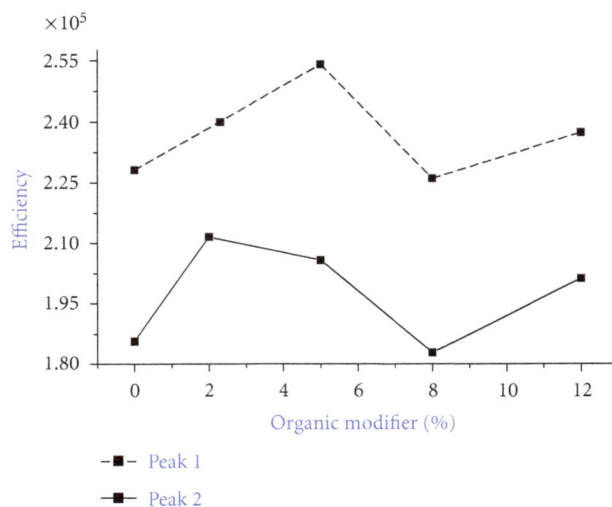

FIGURE 9: Effect of percentage organic modifier on efficiency (N) in the CZE analysis of *Maytenus aquifolium* leaves extract (see Figure 6 for electropherogram and identification of the peaks).

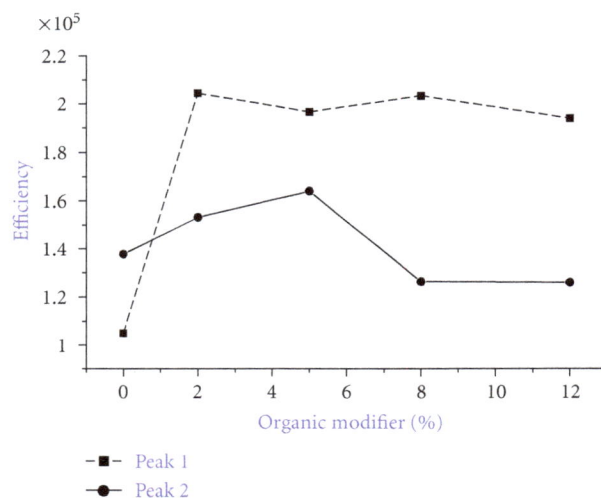

FIGURE 10: Effect of percentage organic modifier on efficiency (N) in the CZE analysis of *Maytenus ilicifolium* leaves extract (see Figure 7 for electropherogram and identification of the peaks).

"ecofriendly", "green" analytical method, which was confirmed by the fact that the optimized conditions allowed for the elimination of acetonitrile from the mobile phase, a significant benefit considering its toxicity. These advantages suggest that CZE should be more widely exploited as an analytical method, for example, in the quality control of "espinheira santa" phytopharmaceuticals, particularly considering the huge amounts of chemical waste produced by the pharmaceutical industry in routine analyses. On the other hand, HPLC showed greater efficacy in the detection of flavonols, since twelve flavonols were detected using this technique while only ten flavonols were detected in the optimized CZE conditions.

CZE/PAD and HPLC-UV/PAD Profile of Flavonoids from Maytenus aquifolium and Maytenus ilicifolia "espinheira santa"
Leaves Extracts

163

Acknowledgments

The authors thank Dr. Ana Maria Soares Pereira for kindly supplying plant material, Professor Dr. Emanuel Carrilho for the discussions about CZE, and FAPESP (98/04334-9, 00/11645-2, 02/00493-2, and 06/59457-6), CNPq, and CAPES for granting fellowships and financial support.

References

[1] J.B. Harborne, *The Flavonoids*, Chapman and Hall, London, UK, 1994.

[2] M. T. L. Ielpo, A. Basile, R. Miranda et al., "Immunopharmacological properties of flavonoids," *Fitoterapia*, vol. 71, no. 1, pp. 101–109, 2000.

[3] K. R. Markham, *Techniques of Flavonoid Identification*, Academic Press, New York, NY, USA, 1982.

[4] C. Salvatore, "The role of quercetin, flavonols and flavones in modulating inflammatory cell function," *Inflammation & Allergy*, vol. 9, pp. 263–285, 2010.

[5] J. Gonzalez-Gallego, M. V. Garcia-Mediavilla, S. Sánchez-Campos, and M. J. Tuñó, "Fruit polyphenols, immunity and inflammation," *Journal of Nutrition*, vol. 104, supplement 3, pp. S15–S27, 2010.

[6] S. Kesarkar, A. Bhandage, S. Deshmukh, K. Shevkar, and M. Abhyankar, "Flavonoids: an overview," *Journal of Pharmacy Research*, vol. 2, pp. 1148–1154, 2009.

[7] M. A. S. Coutinho, M. F. Muzitano, and S.S. Costa, "Flavonoids: potential therapeutic agents for the inflammatory process," *Revista Virtual de Quimica*, vol. 1, pp. 241–256, 2009.

[8] E. P. Scholz, E. Zitron, H. A. Katus, and C. A. Karle, "Cardiovascular ion channels as a molecular target of flavonoids," *Cardiovascular Therapeutics*, vol. 28, no. 4, pp. e46–e52, 2010.

[9] T. R. Cibin, D. Gayathri Devi, and A. Abraham, "Chemoprevention of skin cancer by the flavonoid fraction of Saraca asoka," *Phytotherapy Research*, vol. 24, no. 5, pp. 666–672, 2010.

[10] R. K. Parabathina, G. V. Raja, M. N. Rao, G. S. Rao, and K. S. Rao, "Cardioprotective effects of vitamin E, morin, rutin and quercetin against doxorubicin induced oxidative stress of rabbits: a biochemical study," *Journal of Chemical and Pharmaceutical Research*, vol. 2, pp. 754–765, 2010.

[11] R. M. Jorge, J. P. V. Leite, A. B. Oliveira, and C. A. Tagliati, "Evaluation of antinociceptive, anti-inflammatory and antiulcerogenic activities of *Maytenus ilicifolia*," *Journal of Ethnopharmacology*, vol. 94, no. 1, pp. 93–100, 2004.

[12] M. L. O. Souza-Formigoni, M. G. M. Oliveira, M. G. Monteiro, N. G. da Silveira-Filho, S. Braz, and E. A. Carlini, "Antiulcerogenic effects of two *Maytenus* species in laboratory animals," *Journal of Ethnopharmacology*, vol. 34, no. 1, pp. 21–27, 1991.

[13] L. A. Tiberti, J. H. Yariwake, K. Ndjoko, and K. Hostettmann, "Identification of flavonols in leaves of *Maytenus ilicifolia* and *M. aquifolium* (Celastraceae) by LC/UV/MS analysis," *Journal of Chromatography B*, vol. 846, no. 1-2, pp. 378–384, 2007.

[14] J. H. Y. Vilegas, F. M. Lanças, J. N. Wauters, and L. Angenot, "Characterization of adulteration of 'Espinheira santa' (*Maytenus ilicifolia* and *Maytenus aquifolium*, Celastraceae) hydroalcoholic extracts with Sorocea bomplandii (Moraceae) by high-performance thin layer chromatography," *Phytochemical Analysis*, vol. 9, no. 6, pp. 263–266, 1998.

[15] L. M. Souza, T. R. Cipriani, M. Iacomini, P. A. J. Gorin, and G. L. Sassaki, "HPLC/ESI-MS and NMR analysis of flavonoids and tannins in bioactive extract from leaves of *Maytenus ilicifolia*," *Journal of Pharmaceutical and Biomedical Analysis*, vol. 47, no. 1, pp. 59–67, 2008.

[16] R. Bruni, D. Rossi, M. Muzzoli et al., "Antimutagenic, antioxidant and antimicrobial properties of *Maytenus krukovii* bark," *Fitoterapia*, vol. 77, no. 7-8, pp. 538–545, 2006.

[17] J. C. R. Vellosa, N. M. Khalil, V. A. F. Formenton et al., "Antioxidant activity of *Maytenus ilicifolia* root bark," *Fitoterapia*, vol. 77, no. 3, pp. 243–244, 2006.

[18] L. M. de Souza, T. R. Cipriani, C. F. Sant'Ana, M. Iacomini, P. A. J. Gorin, and G. L. Sassaki, "Heart-cutting two-dimensional (size exclusion × reversed phase) liquid chromatography-mass spectrometry analysis of flavonol glycosides from leaves of *Maytenus ilicifolia*," *Journal of Chromatography A*, vol. 1216, no. 1, pp. 99–105, 2009.

[19] I. Molnár-Perl and Z. Füzfai, "Chromatographic, capillary electrophoretic and capillary electrochromatographic techniques in the analysis of flavonoids," *Journal of Chromatography A*, vol. 1073, no. 1-2, pp. 201–227, 2005.

[20] F. Xu, Y. Liu, R. Song, H. Dong, and Z. Zhang, "HPLC/DAD comparison of sixteen bioactive components between Da-Cheng-Qi decoction and its parent herbal medicines," *Natural Product Communications*, vol. 5, no. 6, pp. 893–896, 2010.

[21] L. Chi, Z. Li, S. Dong, P. He, Q. Wang, and Y. Fang, "Simultaneous determination of flavonoids and phenolic acids in Chinese herbal tea by beta-cyclodextrin based capillary zone electrophoresis," *Microchimica Acta*, vol. 167, no. 3-4, pp. 179–185, 2009.

[22] *Farmacopéia Brasileira*, fascículo 4, parte II, Atheneu, São Paulo, Brazil, 4th edition, 2003.

[23] P. Poukens-Renwart, M. Tits, J. N. Wauters, and L. Angenot, "Densitometric evaluation of spiraeoside after derivatization in flowers of *Filipendula ulmaria* (L.) Maxim," *Journal of Pharmaceutical and Biomedical Analysis*, vol. 10, no. 10–12, pp. 1085–1088, 1992.

[24] M. Sannomiya, W. Vilegas, L. Rastrelli, and C. Pizza, "A flavonoid glycoside from *Maytenus aquifolium*," *Phytochemistry*, vol. 49, no. 1, pp. 237–239, 1998.

[25] W. Vilegas, M. Sanommiya, L. Rastrelli, and C. Pizza, "Isolation and structure elucidation of two new flavonoid glycosides from the infusion of *Maytenus aquifolium* leaves. Evaluation of the antiulcer activity of the infusion," *Journal of Agricultural and Food Chemistry*, vol. 47, no. 2, pp. 403–406, 1999.

[26] T. J. Mabry, K. R. Markham, and M. B. Thomas, *The Systematic Identification of Flavonoids*, Academic Press, New York, NY, USA, 1970.

24

Analysis of Some Biogenic Amines by Micellar Liquid Chromatography

Irena Malinowska and Katarzyna E. Stępnik

Chair of Physical Chemistry, Department of Planar Chromatography, Faculty of Chemistry, Maria Curie Skłodowska University, M. Curie—Skłodowska Sq. 3, 20-031 Lublin, Poland

Correspondence should be addressed to Irena Malinowska, irena.malinowska@poczta.umcs.lublin.pl

Academic Editor: Burcu Dogan-Topal

Micellar liquid chromatography (MLC) with the use of high performance liquid chromatography (HPLC) was used to determine some physicochemical parameters of six biogenic amines: adrenaline, dopamine, octopamine, histamine, 2-phenylethylamine, and tyramine. In this paper, an influence of surfactant's concentration and pH of the micellar mobile phase on the retention of the tested substances was examined. To determine the influence of surfactant's concentration on the retention of the tested amines, buffered solutions (at pH 7.4) of ionic surfactant—sodium dodecyl sulfate SDS (at different concentrations) with acetonitrile as an organic modifier (0.8/0.2 v/v) were used as the micellar mobile phases. To determine the influence of pH of the micellar mobile phase on the retention, mobile phases contained buffered solutions (at different pH values) of sodium dodecyl sulfate SDS (at 0.1 M) with acetonitrile (0.8/0.2 v/v). The inverse of value of retention factor ($1/k$) versus concentration of micelles (C_M) relationships were examined. Other physicochemical parameters of solutes such as an association constant analyte—micelle (K_{ma})—and partition coefficient of analyte between stationary phase and water (hydrophobicity descriptor) ($P_{sw}\Phi$) were determined by the use of Foley's equation.

1. Introduction

Micellar liquid chromatography (MLC) is a mode of conventional reversed-phase liquid chromatography which uses a surfactant solution (anionic, cationic, or nonionic) above the critical micellar concentration (cmc) as a mobile phase [1, 2]. The retention of a compound in MLC depends on the type of interaction (electrostatic and/or hydrophobic) with the micelles and with a surfactant-modified stationary phase [2–5]. There is an assumption that there exist three different partition equilibria between: micellar mobile phase and water, micellar mobile phase and stationary phase, and stationary phase and aqueous mobile phase [1]. There are a lot of equations which describe interactions between these equilibria.

From the beginning of MLC in 1980, the technique has evolved up to become a real alternative in some instances to classical RPLC with hydro-organic mixtures, owing to its peculiar features and unique advantages [6]. MLC is an effective technique for analysis of the organic compounds with biological activity [7–9]. Amongst many advantages of MLC the following can be specified as the most important [1, 10–12]:

  (i) low price of a mobile phase,

  (ii) little toxicity in comparison with classical mobile phases in RP-HPLC,

  (iii) possibility of separation of ionic and nonionic substances in the same column packing and the same distributing cycle,

  (iv) possibility of control of system's selectivity through the modification of simple parameters of mobile phase such as surfactant's concentration, pH, or ionic strength,

  (v) possibility of fast modification of concentration gradient.

The biogenic amines are the group of compounds which are formed during many important intracellular metabolic processes [13]; these amines are the products of transformation of amino acids or their derivatives. In the living organism, the biogenic amines take physiological roles which are important for the correct course of many biological processes [14]. The biogenic amines are necessary for keeping an existence of cell. They are the part of coenzymes and might be hormones. In cells, among other things, amines have an effect on DNA replication and cell permeability and might control mammal's lactation. Some of these are carcinogenic.

In this paper, some biogenic amines (adrenaline, dopamine, octopamine, 2-phenylethylamine, tyramine, and histamine) were determined.

Biogenic amines are damaging to health because they are substrates in the carcinogenic compound synthesis—nitrosoamines. The presence of histamine in food (i.e., in cheese or fish) can result in strong alimentary intoxication [15, 16].

Adrenaline is a hormone and neurotransmitter which takes a role of stress mechanism that is lightning reactions of living organism to stress. Adrenaline can increase heart rate, constrict blood vessels, dilate air passages, and participate in the fight-or-flight response of the sympathetic nervous system. Adrenaline also increases cerebral blood flow and hence oxygen delivery to the brain [17, 18].

Dopamine is called the hormone of happiness. Dopamine is available as an intravenous medication acting on the sympathetic nervous system, producing effects such as increased heart rate and blood pressure. It is involved in the regulation of a variety of functions, including locomotor activity, emotion and affect, and neuroendocrine secretion [19].

Histamine is a powerful biologically active substance, and therefore it can exert many responses within the body. Histamine exerts its effects by binding to receptors on cellular membranes which are found in the cardiovascular system and in various secretory glands [20].

Octopamine is an endogenous biogenic amine that is closely related to noradrenaline. It has effects on the adrenergic and dopaminergic systems. Octopamine has been reported to influence animal behavior, such as avoidance conditioning or motor activity [21, 22].

Many derivates of 2-phenylethylamine are psychoactive and psychedelic. This amine as a hormone and neurotransmitter takes a fundamental role in nervous system. Moreover, it is present in food, mainly in chocolate, and also in some species of mushrooms [23, 24].

Tyramine acts mainly indirectly by releasing noradrenaline from the sympathetic nervous system which causes an increase of the blood pressure by peripheral vasoconstriction and by increasing the cardiac output [15].

Due to biogenic amines are a large group of naturally occurring biologically active compounds, which are present in food products and can act as hormones and neurotransmitters, there are many different analytical techniques for their quantitative and qualitative determination. Amongst chromatographic techniques the most commonly used are RP-LC [25–28], MEKC [29, 30], and IEC [31–33]. MLC technique for the determination of biogenic amines is rather rarely applied.

Paleologos and coworkers successfully separated nine biogenic amines (cadaverine, tyramine, putrescine, agamatine, spermidine, tryptamine, phenylalanine, spermine, and histamine) in the form of benzoyl derivatives by the use of MLC with gradient elution. Investigated biogenic amines were quantitatively and qualitatively determined in fish products [34].

Bose and coworkers used MLC technique for the determination of dopamine and serotonine and their metabolites (homovalinic acid (HVA) and hydroxyindoleacetic acid (HIAA)) and tyramine in the serum samples [35]. Other biogenic amines (such as tyramine and tryptamine and their precursors: thyrosine and tryptophan) were determined in MLC systems in vine samples [36].

2. Experimental

2.1. Apparatus. Chromatographic data were obtained using Shimadzu Vp liquid chromatographic system equipped with an LC 10AT pump, an SPD 10A UV-VIS detector, an SCL 10A system controller, a CTO-10 AS chromatographic oven, and a Rheodyne injector valve with $20\,\mu L$ loop. This system utilizes a class-Vp computer program to control hardware, acquire and store data, and determine retention times.

A stainless-steel column (125×4 mm, I.D.) packed with $5\,\mu m$ RP-8 endcapped Purospher (Merck) was used in each experiment.

The flow rate was $1.0\,mL\,min^{-1}$. All measurements were carried out at $20°C$. The test compounds were detected at $230\,nm$.

2.2. Reagents and Materials. Stock standards of six biogenic amines (Table 1): adrenaline, dopamine, histamine, octopamine, 2-phenylethylamine, and tyramine were purchased from Sigma Aldrich (St. Louis, USA).

All stock solutions of the analytes were prepared with the concentration of about $0.1\,mg\,mL^{-1}$ in methanol.

For the determination of the influence of surfactant's concentration on the retention, mobile-phases-buffered solutions at pH 7.4 of sodium dodecyl sulfate (SDS) at the concentrations of 0.02 M, 0.04 M, 0.06 M, 0.08 M, and 0.1 M + acetonitrile as an organic modifier (for HPLC) at the concentration of 20% v/v were used. After measuring of SDS concentration, the organic modifier was added, so the real concentrations of SDS were correspondingly 0.016 M, 0.032 M, 0.048 M, 0.064 M, and 0.08 M.

The buffer was prepared with Na_2HPO_4 (0.02 M) and citric acid (0.01 M) and was vacuum-filtered through $0.45\,\mu m$ membrane filter before use. pH value of this buffer was measured before the preparation of the mobile phases. Distilled water was obtained from Direct-Q UV apparatus (Millipore). Acetonitrile (for HPLC) at the concentration of 20% v/v was utilized as an organic modifier. Micellar mobile phases were degassed in the ultrasonic bath for about two minutes before use.

For the determination of the influence of pH of the micellar mobile phase on the retention, mobile-phases-buffered

TABLE 1: Structures of tested substances.

Number	Name	Structures of tested substances
1	Adrenaline	
2	Dopamine	
3	Octopamine	
4	Histamine	
5	2-phenylethylamine	
6	Tyramine	

solutions at pH 3, 4, 5, 6, 7.4, 8, and 9 of sodium dodecyl sulfate (SDS) at the concentration of 0.1 M were used. The buffer in the same way as presented above was prepared. Acetonitrile (for HPLC) at the concentration of 20% v/v was utilized as an organic modifier.

3. Result and Discussion

Besides that MLC is an acknowledged analytical technique, it has high significance in elementary research.

There are many agents in MLC which have an effect on retention such as [5, 37]:

(i) type of organic modifier,

(ii) concentration of organic modifier,

(iii) concentration of surfactant (concentration of micelles),

(iv) pH of mobile phase.

In this paper, an influence of surfactant's concentration and pH of the micellar mobile phase on the retention of tested substances were examined.

3.1. An Influence of Surfactant's Concentration on the Retention of Tested Amines. For this purpose, the retention of

tested substances was defined in the following mobile phase: buffer (pH 7.4) + acetonitrile + SDS (of variable concentration).

The buffer solution at pH 7.4 was used for the reason that in biology and medicine, this value is often referred to as physiological (the pH of blood is usually slightly basic with a value of pH 7.365) [38] and tested substances are very closely connected with many processes in the living organism.

The research of variation of retention factor in relation to the concentration of surfactant can give much important information applying to interactions of tested substances with components of the micellar mobile phase.

The analysis of the relationship: inverse of value of retention factor $(1/k)$ versus concentration of micelles (C_M) can give information about the type of interaction between tested solutes and micelles [39] and can determine such physicochemical parameters of solutes as an association constant analyte—micelle (K_{ma})—and can compare partition coefficient between micellar mobile phase and water $(P_{sw}\Phi)$ [12, 40].

If substance does not work with micelle, the change of retention was not observed with the change of concentration of surfactant.

If substance acts with micelle, decrease of retention was observed with increase of surfactant's concentration (the so-called binding analyte) [39].

In case of the increase of retention, it was noticed that substance from micelle was eliminated with the increase of surfactant's concentration (the so-called antibinding analyte) [39].

Linear, ascending relationship between $1/k$ value and concentration of micelles with high coefficient of determination ($R^2 > 0.96$) was obtained for all investigated solutes (Figure 1). The character of relationships involves the fact that all tested biogenic amines act with micelles of the mobile phase (are so-called binding analytes).

3.2. Calculations of Chosen Physicochemical Parameters. Linear relationships $1/k = f(C_M)$ can be presented in the following way [12, 40] (Foley's equation):

$$\frac{1}{k} = \frac{K_{ma}C_M}{P_{sw}\Phi} + \frac{1}{P_{sw}\Phi}, \tag{1}$$

where C_M: concentration of micelles, K_{ma}: analyte-micelle association constant, P_{sw}: partition coefficient of analyte between stationary phase and water; hydrophobicity descriptor, and Φ—volume ratio of stationary phase to volume of mobile phase.

The slope of a straight line from (1) is formulated as $K_{ma}/(P_{sw}\Phi)$ and free term in an expression as $1/P_{sw}\Phi$.

The slope of a straight line informs about substance's sensitivity to the change of concentration of surfactant, therefore shows the possibility of modification retention through the change of surfactant's concentration. Moreover, the slope of a straight line $1/k = f(C_M)$ depends on the interaction of the substance with micelle (K_{ma}) and on partition coefficient between stationary phase and water (P_{sw}) (Table 2).

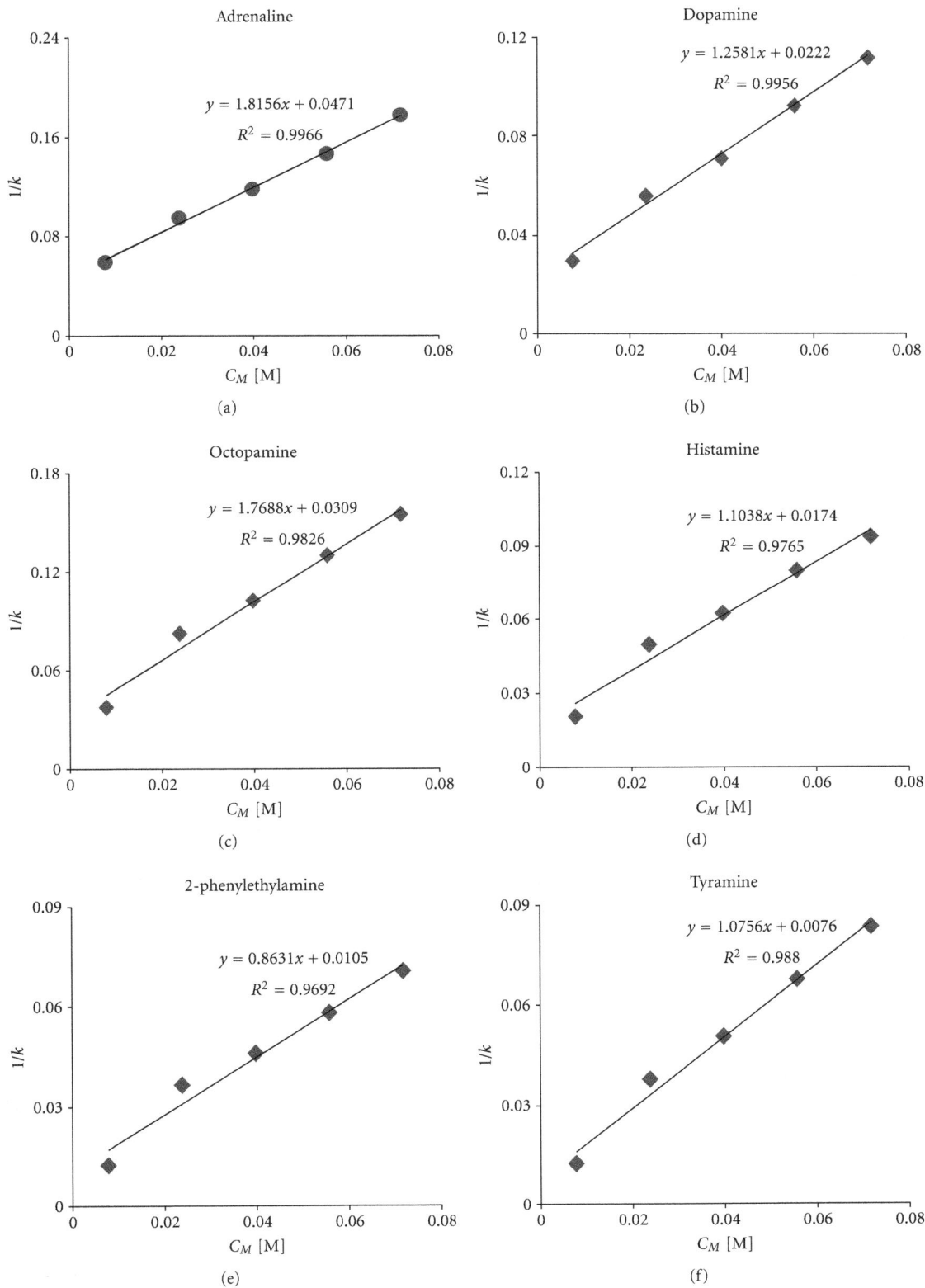

FIGURE 1: $1/k$ versus C_M relationships obtained for all tested biogenic amines.

TABLE 2: $K_{ma}/P_{sw}\Phi$, K_{ma} and $P_{sw}\Phi$ values of tested biogenic amines according to (1).

Solute number	Name	$K_{ma}/P_{sw}\Phi$	K_{ma}	$P_{sw}\Phi$
1	adrenaline	1.82	38.55	21.23
2	dopamine	1.26	56.67	45.05
3	octopamine	1.77	57.24	32.36
4	histamine	1.10	63.44	37.47
5	2-phenylethylamine	0.86	82.20	95.24
6	tyramine	1.08	141.53	131.58

Obtained values of K_{ma} (Table 2) demonstrate that tyramine acts the most strongly and adrenaline the most weakly with SDS micelle. Almost all of calculated values are larger than 1, which means that analyte-micelle interaction has higher influence than analyte-modified stationary phase on the retention of biogenic amines. Only 2-phenylethylamine demonstrates other type of interaction which suggests higher impact of analyte-modified stationary phase interaction on the retention of this amine.

Moreover, $1/b = P_{sw}\Phi$ value can inform also about hydrophobicity relations of tested substances.

Because of large difficulty to determine Φ value, it is given as $P_{sw}\Phi$ value, because at the given chromatographic system Φ value is constant. $P_{sw}\Phi$ value informs about relative hydrophobicity of tested substances. When analyte acts more strongly with micelle, it acts more weakly with water.

On the basis of the $P_{sw}\Phi$ values, it was demonstrated that tyramine is the most hydrophobic solute and adrenaline the least. The obtained K_{ma} values show that the increase of solutes hydrophobicity increases K_{ma} values.

In the micellar systems, the retention of the solutes depends on their hydrophobic properties.

Therefore, $\log k$ versus $\log P_{sw}\Phi$ relationships were determined for all chromatographic systems (Figure 2).

Very good $\log k$ versus $\log P_{sw}\Phi$ linear relationships ($R^2 > 0.98$) suggest that $P_{sw}\Phi$ parameter should be considered as good hydrophobicity descriptor for tested solutes despite using different concentrations of surfactant in the micellar mobile phases.

3.3. An Influence of pH of the Micellar Mobile Phase on the Retention of Tested Substances.

Due to possible existence of many different forms of biogenic amines [41], the pH of the micellar mobile phase is a very important agent which can influence the retention of tested substances [42].

For this purpose, the retention of tested substances was defined in the following mobile phases: acetonitrile (4 : 1 v/v) + SDS at the concentration of 0.1 M + citric buffer at different pH (3, 4, 5, 6, 7.4, 8, and 9). Due to stability of chromatographic column, the range of pH was as present above.

In Figure 3 there is a comparison of obtained relationships (k versus pH) with the literature data [41] related to the range of pH in which particular forms of tested substances exist for chosen biogenic amines.

As we can observe, depending on the pH value, the forms of tested substances are replaced. These changes influence the retention value.

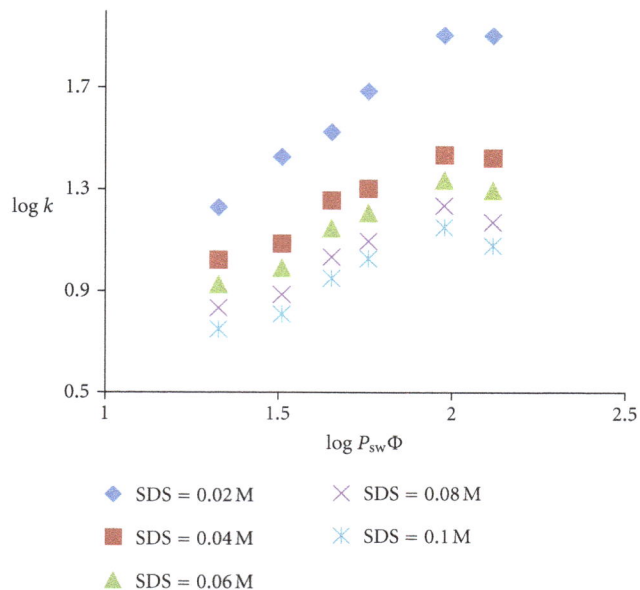

FIGURE 2: $\log k$ versus $\log P_{sw}\Phi$ relationships for all types of micellar mobile phases.

TABLE 3: pKa values of tested biogenic amines in the examined pH range.

Substance name	pKa value
adrenaline	8.91
dopamine	9.27
histamine	9.58
octopamine	8.98
2-phenylethylamine	9.79
tyramine	9.66

MLC technique can be used as an instrument to verify or even determine such physicochemical parameters as pKa in examined pH range [41] (Table 3). In this case, the limitation of pH range, caused by the stability of chromatographic column, makes the calculation of pKa impossible.

3.4. The Analytical Aspect of MLC.

Except for defined physicochemical parameters through the data obtained from the retention, the MLC performs a role as the analytical method. In this paper, it was also examined what was a separation possibility of analyzed biogenic amines in the tested systems. Because surfactant's concentration (concentration of micelles) effects on retention of tested substances, in the mobile phases at the different concentration of SDS (but at the same amount of other components), the longest time of retention was observed at the lowest concentration of surfactant.

For the micellar mobile phase: 0.02 M SDS + acetonitrile (4 : 1 v/v) at pH 7.4, the best separation was achieved for five tested amines out of six that were tested (except 2-phenylethylamine which retention time is almost the same as the retention time of tyramine). However, time of analysis

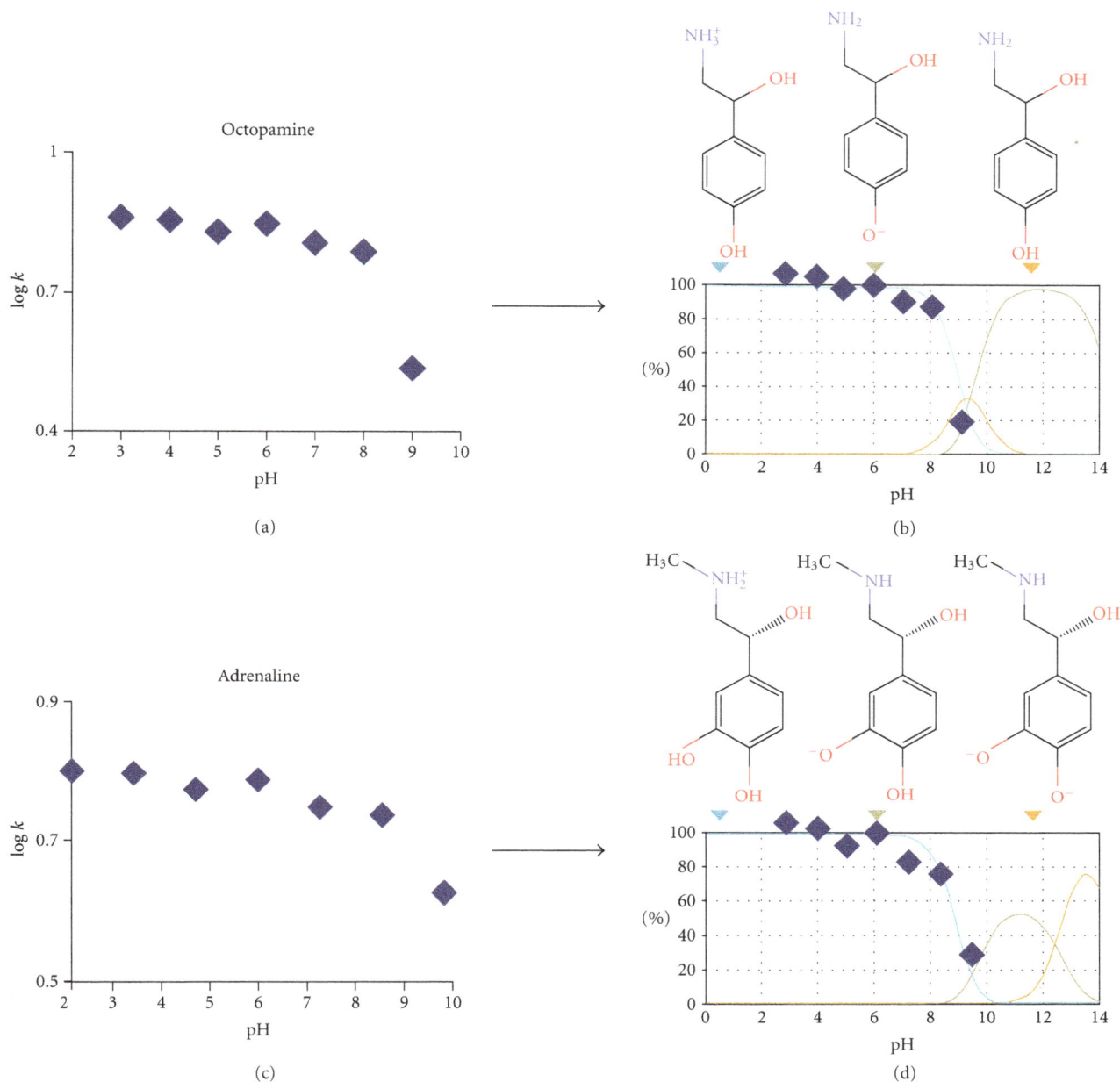

FIGURE 3: The comparison of obtained relationships $\log k$ versus pH of the micellar mobile phase with the pKa of tested amines.

using this mobile phase is about 50 minutes and peaks of the substances at the longest retention time (histamine and tyramine), are diffused. It is comprehensible, because it is a result of a diffusion. On the other hand, the separation of adrenaline (peak 1), octopamine (peak 2), and dopamine (peak 3) is satisfactory ($R_{s12} = 5.04$, $R_{s23} = 3,30$). Time of analysis for these three compounds is about 25 minutes. Therefore, for the determination of these amines mobile phase with 0.02 M SDS will be the most appropriate (Figure 4).

The retention time decreases with the surfactant's concentration. The increase of surfactant's concentration to 0.04 M induces the worse separation of adrenaline and

octopamine ($R_{s12} = 0.42$), but separation of octopamine and dopamine is acceptable ($R_{s23} = 1.29$) (to the zero line). The improvement of the shape of peaks 4 and 5 (histamine and tyramine) is noticeable in comparison with the mobile phase presented above, peaks 3 and 4 overlap.

The successive increase of SDS concentration to 0.06 M does not have an effect on the systems' resolution, but the time of analysis is observably shorter.

The next increase of SDS concentration results in no separation of adrenaline and octopamine, but dopamine, histamine, and tyramine to the zero line are separated and for such mixture, mobile phase with 0.08 M SDS would be better than mobile phase with 0.06 M SDS. In this mobile phase,

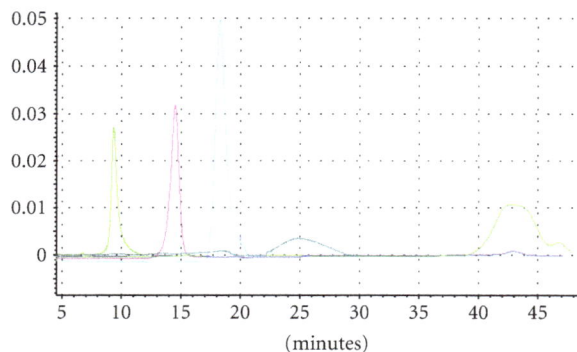

FIGURE 4: The chromatogram of the separation of tested amines: 0.02 M SDS (pH 7.4) + acetonitrile (20% v/v).

the separation and the identification of 4 out of 5 tested amines is practically possible.

At the mobile phase with the highest concentration of SDS (0.1 M) (retention times are the shortest), the overlapping of all peaks is observed, therefore this phase is not suitable for the analytical purpose.

4. Conclusions

For all tested biogenic amines, the decrease of retention (increase of $1/k$) with the increase of surfactant's concentration in the micellar mobile phase was observed. Very good $1/k$ value versus C_M linear relationships with high coefficient of determination confirm the above dependence. An impact of the surfactant's concentration in the micellar mobile phase on retention of tested amines was proved. The slope of the change of retention line $1/k = f(C_M)$ depends mainly on hydrophobicity of analytes. Linear, ascending relationship between $1/k$ value and concentration of micelles with high coefficient of determination was obtained for all investigated solutes (Figure 1). The character of relationships involves the fact that all tested biogenic amines act with micelles of the mobile phase (are binding analytes).

To determine this attribute P_{sw} and K_{ma} values were calculated. On the basis of the $P_{sw}\Phi$ values it was demonstrated that tyramine is the most hydrophobic solute and adrenaline the least. The obtained K_{ma} values show that the increase of solutes hydrophobicity, involves the increase of K_{ma} values. Moreover, between K_{ma} values significant differences for different tested amines were visible. The rise of $P_{sw}\Phi$ value responds to the increase of analytes retention. The above results prove that the concentration of surfactant in the micellar mobile phase is a very important agent that influences the retention of organic compounds. Therefore, the optimization of chromatographic systems in MLC through the choice of concentration of surfactant in the micellar mobile phase is necessary.

The investigated micellar chromatographic systems can be used for the separation and the determination of tested biogenic amines.

References

[1] D. W. Armstrong and F. Nome, "Partitioning behavior of solutes eluted with micellar mobile phases in liquid chromatography," *Analytical Chemistry*, vol. 53, no. 11, pp. 1662–1666, 1981.

[2] A. Berthod and M. C. García-Álvarez, *Coque, Micellar Liquid Chromatography*, vol. 83, Marcel Dekker, New York, NY, USA, 2000, Edited by J. Cazes.

[3] J. M. Sanchis Mallols, R. M. Villanueva Camañas, S. Sagrado, and M. J. Medina-Hernández, "Quantitative retention—structure and Retention—activity relationship studies of ionic and non-ionic catecholamines by micellar liquid chromatography," *Chromatographia*, vol. 46, no. 11-12, pp. 605–612, 1997.

[4] M. J. Medina-Hernández and M. C. García-Álvarez-Coque, "Solute-mobile phase and solute-stationary phase interactions in micellar liquid chromatography. A review," *Analyst*, vol. 117, pp. 831–837, 1992.

[5] A. Berthod, I. Girard, and C. Gonnet, "Micellar liquid chromatography. Retention study of solutes of various polarities," *Analytical Chemistry*, vol. 58, no. 7, pp. 1359–1362, 1986.

[6] M. J. Ruiz-Ángel, S. Carda-Broch, J. R. Torres-Lapasió, and M. C. García-Álvarez-Coque, "Retention mechanisms in micellar liquid chromatography," *Journal of Chromatography A*, vol. 1216, no. 10, pp. 1798–1814, 2009.

[7] G. Maiti, A. P. Walvekar, and T. S. Krishnamoorthy, "Separation and determination of inorganic anions by high-performance liquid chromatography using a micellar mobile phase," *Analyst*, vol. 114, no. 6, pp. 731–733, 1989.

[8] F. P. Tomasella, P. Zuting, and L. J. Chin Love, "Determination of sun-screen agents in cosmetic products by micellar liquid chromatography," *Journal of Chromatography*, vol. 587, no. 2, pp. 325–328, 1991.

[9] M. G. Khaleda and E. D. Breyer, "Quantitation of hydrophobicity with micellar liquid chromatography," *Analytical Chemistry*, vol. 61, no. 9, pp. 1040–1047, 1989.

[10] P. Yarmchuck, R. Weinberger, R. F. Hirsch, and L. J. Cline-Love, "Selectivity in liquid chromatography with micellar mobile phases," *Analytical Chemistry*, vol. 54, pp. 2233–2238, 1982.

[11] R. A. Barford and B. Sliwinski, "Micellar chromatography of proteins," *Analytical Chemistry*, vol. 56, no. 9, pp. 1554–1556, 1984.

[12] J. P. Foley, "Critical compilation of solute-micelle binding constants and related parameters from micellar liquid chromatographic measurements," *Analytica Chimica Acta*, vol. 23, pp. 237–247, 1990.

[13] S. Oguri, "Electromigration methods for amino acids, biogenic amines and aromatic amines," *Journal of Chromatography B*, vol. 747, no. 1-2, pp. 1–19, 2000.

[14] G. Suzzi and F. Gardini, "Biogenic amines in dry fermented sausages: a review," *International Journal of Food Microbiology*, vol. 88, no. 1, pp. 41–54, 2003.

[15] H. Zhai, X. Yang, L. Li et al., "Biogenic amines in commercial fish and fish products sold in southern China," *Food Control*, vol. 25, no. 1, pp. 303–308, 2012.

[16] H. M. L. J. Joosten, "The biogenic amine contents of Dutch cheese and their toxicological significance," *Netherlands Milk and Dairy Journal*, vol. 42, no. 1, pp. 25–42, 1988.

[17] R. E. Jackson, K. Joyce, S. F. Danosi, B. C. White, D. Vigor, and T. J. Hoehner, "Blood flow in the cerebral cortex during cardiac resuscitation in dogs," *Annals of Emergency Medicine*, vol. 13, no. 9, pp. 657–659, 1984.

[18] J. Johansson, R. Gedeborg, S. Basu, and S. Rubertsson, "Increased cortical cerebral blood flow by continuous infusion of adrenaline (epinephrine) during experimental cardiopulmonary resuscitation," *Resuscitation*, vol. 57, no. 3, pp. 299–307, 2003.

[19] M. Jaber, S. W. Robinson, C. Missale, and M. G. Caron, "Dopamine receptors and brain function," *Neuropharmacology*, vol. 35, no. 11, pp. 1503–1519, 1996.

[20] A. R. Shalaby, "Significance of biogenic amines to food safety and human health," *Food Research International*, vol. 29, no. 7, pp. 675–690, 1996.

[21] J. Delacour, J. F. Coulon, J. C. David, and C. Guenaire, "Brain octopamine and strain differences in avoidance behavior," *Brain Research*, vol. 288, no. 1-2, pp. 169–176, 1983.

[22] T. Bungo, T. Higaki, H. Ueda, and M. Furuse, "Intracerebroventricular administration of octopamine stimulates food intake of chicks through α2-adorenoceptor," *Physiology and Behavior*, vol. 76, no. 4-5, pp. 575–578, 2002.

[23] B. Pfundstein, A. R. Tricker, E. Theobald, B. Spiegelhalder, and R. Preussmann, "Mean daily intake of primary and secondary amines from foods and beverages in West Germany in 1989-1990," *Food and Chemical Toxicology*, vol. 29, no. 11, pp. 733–739, 1991.

[24] M. H. Silla Santos, "Biogenic amines: their importance in foods," *International Journal of Food Microbiology*, vol. 29, no. 2-3, pp. 213–231, 1996.

[25] S. Jia, Y. P. Kang, J. H. Park, J. Lee, and S. W. Kwon, "Simultaneous determination of 23 amino acids and 7 biogenic amines in fermented food samples by liquid chromatography/quadrupole time-of-flight mass spectrometry," *Journal of Chromatography A*, vol. 1218, no. 51, pp. 9174–9182, 2011.

[26] H. M. Mao, B. G. Chen, X. M. Qian, and Z. M. Liu, "Simultaneous determination of twelve biogenic amines in serum by high performance liquid chromatography," *Microchemical Journal*, vol. 91, no. 2, pp. 176–180, 2009.

[27] G. Sagratini, S. Vittor, M. Fernandez-Franzon, G. Font, and F. De Berardinis, "Simultaneous determination of eight underivatised biogenic amines in fish by solid phase extraction and liquid chromatography-tandem mass spectrometry," *Food Chemistry*, vol. 132, no. 1, pp. 537–543, 2012.

[28] T. Tang, T. Shi, K. Qian, P. Li, J. Li, and Y. Cao, "Determination of biogenic amines in beer with pre-column derivatization by high performance liquid chromatography," *Journal of Chromatography B*, vol. 877, no. 5-6, pp. 507–512, 2009.

[29] I. Rodriguez, H. K. Lee, and S. F. Y. Li, "Separation of biogenic amines by micellar electrokinetic chromatography," *Journal of Chromatography A*, vol. 745, no. 1-2, pp. 255–262, 1996.

[30] G. Nouadje, N. Siméon, F. Dedieu, M. Nertz, P. Puig, and F. Couderc, "Determination of twenty eight biogenic amines and amino acids during wine aging by micellar electrokinetic chromatography and laser-induced fluorescence detection," *Journal of Chromatography A*, vol. 765, no. 2, pp. 337–343, 1997.

[31] M. Triki, F. Jimenez-Colmenero, A. M. Herrero, and C. Ruiz-Capillas, "Optimisation of a chromatographic procedure for determining biogenic amine concentrations in meat and meat products employing a cation-exchange column with a post-column system," *Food Chemistry*, vol. 130, no. 4, pp. 1066–1073, 2012.

[32] B. M. De Borba and J. S. Rohrer, "Determination of biogenic amines in alcoholic beverages by ion chromatography with suppressed conductivity detection and integrated pulsed amperometric detection," *Journal of Chromatography A*, vol. 1155, no. 1, pp. 22–30, 2007.

[33] G. Saccani, E. Tanzi, P. Pastore, S. Cavalli, and M. Rey, "Determination of biogenic amines in fresh and processed meat by suppressed ion chromatography-mass spectrometry using a cation-exchange column," *Journal of Chromatography A*, vol. 1082, no. 1, pp. 43–50, 2005.

[34] E. K. Paleologos, S. D. Chytiri, I. N. Savvaidis, and M. G. Kontominas, "Determination of biogenic amines as their benzoyl derivatives after cloud point extraction with micellar liquid chromatographic separation," *Journal of Chromatography A*, vol. 1010, no. 2, pp. 217–224, 2003.

[35] D. Bose, A. Durgbanshi, M. E. Capella-Peiró, M. Gil-Agustí, J. Esteve-Romero, and S. Carda-Broch, "Micellar liquid chromatography determination of some biogenic amines with electrochemical detection," *Journal of Pharmaceutical and Biomedical Analysis*, vol. 36, no. 2, pp. 357–363, 2004.

[36] M. Gil-Agustí, S. Carda-Broch, L. Monferrer-Pons, and J. Esteve-Romero, "Simultaneous determination of tyramine and tryptamine and their precursor amino acids by micellar liquid chromatography and pulsed amperometric detection in wines," *Journal of Chromatography A*, vol. 1156, no. 1-2, pp. 288–295, 2007.

[37] A. Berthod, M. F. Borgerding, and W. L. Hinze, "Investigation of the causes of reduced efficiency in micellar liquid chromatography," *Journal of Chromatography*, vol. 556, no. 1-2, pp. 263–275, 1991.

[38] W. F. Boron and E. L. Boulpaep, *Medical Physiology: A Cellular and Molecular Approach*, Saunders Elsevier, Philadelphia, Pa, USA, 2nd edition, 2009.

[39] D. W. Armstrong and G. Y. Stine, "Selectivity in pseudophase liquid chromatography," *Analytical Chemistry*, vol. 55, no. 14, pp. 2317–2320, 1983.

[40] A. Szymański, Badania retencyjne i oznaczanie wybranych sufloamidów w środkach spożywczych metodą micelarnej chromatografii cieczowej, Wydawnictwo Naukowe Uniwersytetu Adama Mickiewicza w Poznaniu, Poznań, 2006.

[41] http://www.chemicalize.org/.

[42] A. H. Rodgers, J. K. Strasters, and M. G. Khaledi, "Simultaneous optimization of pH and micelle concentration in micellar liquid chromatography," *Journal of Chromatography*, vol. 636, no. 2, pp. 203–212, 1993.

25

A New Validated Stability Indicating RP-HPLC Method for Simultaneous Estimation of Pyridoxine Hydrochloride and Meclizine Hydrochloride in Pharmaceutical Solid Dosage Forms

Md. Saddam Nawaz

Quality Assurance Department, ACI Ltd., Narayanganj 1400, Bangladesh

Correspondence should be addressed to Md. Saddam Nawaz; nawazdu@gmail.com

Academic Editor: Irene Panderi

A simple, specific, accurate, precise stability indicating reversed-phase high-performance liquid chromatographic (RP-HPLC) method was developed and validated for the simultaneous determination of pyridoxine hydrochloride (PYH) and meclizine hydrochloride (MEH). An isocratic separation of PYH and MEH were achieved on C 18, 250 × 4.6 mm ID, 5 μm particle size columns at column oven temperature 37°C with a flow rate of 0.5 mL min^{-1} and using a diode array detector to monitor the detection at 254 nm. The mobile phase consisted of buffer : acetonitrile : trifluoroacetic acid at a ratio of 30 : 70 : 0.1 (v/v). The retention times of PYH and MEH was found to be 5.25 and 10.14 min, respectively. Suitability, specificity, linearity, accuracy, precision, stability, and sensitivity of this method for the quantitative determination of the drugs were proved by validation in accordance with the requirements laid down by International Conference on Harmonization (ICH) Q2 (R1) guidelines. The proposed method is reliable and robust and can be used as quality control tool for the estimation of these drugs in combined pharmaceutical solid dosage forms.

1. Introduction

Pyridoxine hydrochloride (PYH) is chemically 3, 4-pyridine-diacetonitrile, 5-hydroxy-6-methyl, hydrochloride (Figure 1). It is a water-soluble vitamin and involved principally in amino acid, carbohydrate, and fat metabolism [1]. It is also required for the formation of hemoglobin [2]. Meclizine hydrochloride (MEH) (Figure 2) is often used as "meclozine" which is chemically 1-[(4-Chlorophenyl)(phenyl)methyl]-4-(3-methylbenzyl)piperazine, dihydrochloride monohydrate, a first-generation antihistamine of the piperazine class. Meclizine is effective in inhibiting the symptoms of motion sickness, such as nausea, vomiting, and dizziness. PYH is official in IP [3], BP [4, 5], and USP [6], and MEH is also official in BP [4, 5] and USP [6]. The pharmacopeias describe potentiometric, spectrophotometrics and HPLC method for the determination of PYH and MEH individually from the bulk and tablet dosage form. No reversed-phase high-performance liquid chromatography (RP-HPLC) method is reported in pharmacopeias for the simultaneous estimation of PYH and MEH from their combined formulation, though there are various methods for the determination of PYH and MEH by spectrophotometric [7–12], voltammetric [13], HPLC [14–16], electrophoresis [17], GLC [18], HPTLC [19], and TLC [20] methods in different pharmaceutical dosages forms. The present work describes the development and validation of stability indicating RP-HPLC method, which can quantify PYH and MEH simultaneously in pharmaceutical solid dosage form. The confirmation of the applicability of this developed method was validated according to the International Conference on Harmonization (ICH) Q2 (R1) [21].

2. Experimental

2.1. Reagents and Chemicals. Pure standards of PYH and MEH were obtained from Jianxi Sentai Pharmaceutical Co. Ltd., China, and M/S FDC Ltd., India, having purities of 99.23% and 99.10%, respectively. The formulated film-coated tablets (pyrimac tablet) were prepared from Advanced Chemical Industries (ACI) Limited, Narayanganj,

A New Validated Stability Indicating RP-HPLC Method for Simultaneous Estimation of Pyridoxine Hydrochloride and
Meclizine Hydrochloride in Pharmaceutical Solid Dosage Forms

173

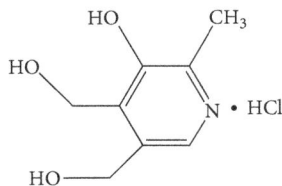

FIGURE 1: Chemical structure of pyridoxine hydrochloride.

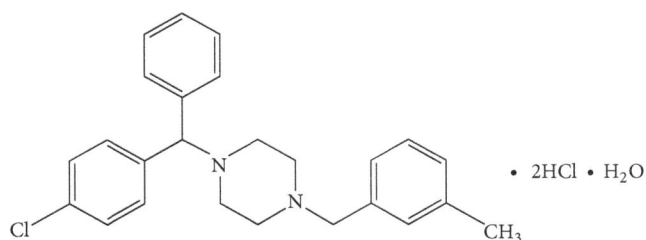

FIGURE 2: Chemical structure of meclizine hydrochloride.

Bangladesh. Each tablet contains combined 50 mg pyridoxine hydrochloride and 25 mg meclizine hydrochloride (based on 100% potency) as active ingredients, as well as microcrystalline cellulose (avicel PH 102), sodium starch glycolate (SSG), Colloidal silicon dioxide (aerosil-200), croscarmellose sodium, Allura red lake, and magnesium stearate as excipients in addition to Opadry II 85G54426 pink as coating material.

Acetonitrile HPLC grade (Scharlau), anhydrous sodium dihydrogen phosphate (Scharlau), phosphoric acid AR grade, (Merck) and trifluoroacetic acid (Fisher Scientific) were used for analytical purposes. Ultrapure water was used to prepare the mobile phase. Ultrapure water was prepared by using Labconco WaterPro PS purification system.

2.2. Instrument and Chromatographic Condition.
Chromatographic separation was achieved by using Shimzadu prominence LC-20AD high-performance liquid chromatography, equipped with degasser PGU-20A 5, variable wavelength programmable diode array detector SPD-M20A, autosampler SIL-20 AC HT, and column oven CTO-10 A5 VP. ProntoSIL C 18, 250×4.6 mm ID, 5μm particle size was used as the stationary phase. The column temperature was kept at 37°C, and the mobile phase flow rate was maintained at 0.5 mL min^{-1}. The detection was monitored at 254 nm. The injection volume was 20μL, and the run time was 13 min for each injection. Other instruments such as pH meter (Jenway 3510), electronic weighing balance (Mettler-Toledo), and ultrasonic bath (Clifton) were also used.

Dionex ultimate 3000 series HPLC and Hichrom C18, 250×4.6 mm ID, 5μm particle size column were used for ruggedness study.

2.2.1. Mobile Phase.
A mixture of buffer, acetonitrile, and trifluoroacetic acid (TFA) at a ratio of $30 : 70 : 0.1$ (v/v) was

prepared. The resulting solution was sonicated for 5 min using ultrasonic bath, and finally the mixture was filtered using 0.2μm membrane filter.

2.2.2. Preparation of Buffer for Mobile Phase.
12 g of anhydrous sodium dihydrogen phosphate was dissolved in 900 mL of ultrapure water. Then the pH was adjusted to 3.0 with orthophosphoric acid and volumed up to 1000 mL with ultrapure water and sonicated for 5 min using ultrasonic bath then filtered through 0.2μm membrane filter.

2.2.3. Diluting Solution.
A mixture of buffer, acetonitrile, and trifluoroacetic acid (TFA) at a ratio of $35 : 65 : 0.1$ (v/v) was used as the diluents.

2.2.4. Standard Preparation (at Nominal Concentration).
PYH and MEH working standards were accurately weighed and were transferred into a clean and dry 100 mL standard volumetric flask and dissolved to prepare 0.50 mg mL^{-1} and 0.25 mg mL^{-1} concentrations of PYH and MEH stock solution, respectively, with the diluting solution and finally volumed up to the mark. The solution was sonicated for 5 min using ultrasonic bath and then filtered through 0.2μm disk filter.

2.2.5. Sample Preparation.
Twenty tablets (pyrimac tablet) were crushed and then powdered finely. To prepare assay sample solution, powdered sample equivalent to 50 mg pyridoxine hydrochloride and 25 mg meclizine hydrochloride was weighed accurately and taken into 100 mL volumetric flask. About 40 mL of diluting solution was added and shaken thoroughly to extract the drug from the excipients and then sonicated for 5 min to complete dissolution of drug. The solution was allowed to cool at room temperature and then volume up to the mark. The solution was filtered through Whatman filter paper (no. 42) and then finally filtered through 0.2μm disk filter.

2.3. Method Validation Parameters

2.3.1. System Suitability.
To assess system suitability of the method, the repeatability, theoretical plates, tailing factor, and retention time of six replicate injections of standard PYH and MEH of concentrations 0.50 mg mL^{-1} and 0.25 mg mL^{-1}, respectively, were used, and the percent relative standard deviation (%RSD) values were calculated in each case.

2.3.2. Linearity.
The linearity of the method was determined at five different concentration levels (80%, 90%, 100%, 110%, and 120%) ranging from 0.395 to 0.592 mg mL^{-1} of PYH and 0.203–0.304 mg mL^{-1} of MEH, respectively. The linearity was evaluated by peak area versus concentration, which was calculated by the least-square regression analysis, and the respective regression equation was computed.

2.3.3. Specificity.
The specificity of the developed RP-HPLC method for the determination of PYH and MEH in bulk

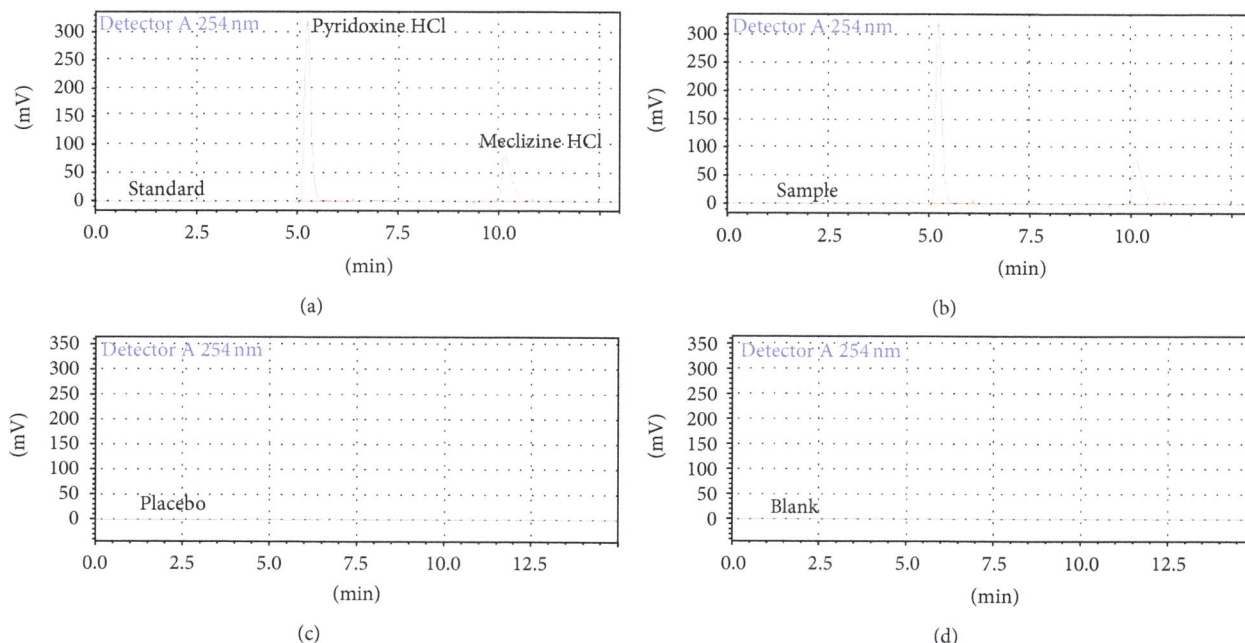

FIGURE 3: RP-HPLC chromatogram of blank, placebo, tablet sample, and standard of pyridoxine HCl and meclizine HCl.

drug and pharmaceutical preparation (pyrimac tablet) was investigated by chromatographic analysis of the following.

Noninterference of Placebo. To check the noninterference of placebo, placebo solution was prepared in the same way of the sample solution in the presence of all inactive ingredients of the pyrimac tablet formulation but without PYH and MEH.

Degradation Studies. Degradation studies were carried out under acid hydrolysis, alkali hydrolysis, oxidation, and reduction. One formulated full tablet was taken as a sample for degradation study. For acid hydrolysis 5 mL of 0.1 N HCl was added to sample, kept for 8 hours (at $25 \pm 2°C$), then neutralized with 0.1 N NaOH and volumed up to mark with diluent. For alkali hydrolysis 5 mL of 0.1 N NaOH was added to sample, kept for 8 hours (at $25 \pm 2°C$), then neutralized with 0.1 N HCl and made to volume with diluent. A 10% H_2O_2 solution and 10% sodium bisulphate solution were used for oxidative and reductive study, consecutively. The final solution was injected for assay analysis, and the presence of interfering peak(s) eluted at or near the retention time of PYH and MEH was also checked. All determinations were conducted in triplicate. The Peak purity tool was used to check the purity of the test solution.

2.3.4. Accuracy. Accuracy was carried out for drug-matrix solutions. Accuracy parameter was determined by the recovery test, which consisted of adding known amounts of PYH and MEH in to the placebo sample solutions. This test was conducted by three different concentrations (80, 100, and 120%) of test sample in three replicate sample preparations, and the percent recoveries (mean ± %RSD of three replicates) of PYH and MEH in drug-matrix form were calculated.

TABLE 1: Chromatographic characteristics of system suitability study.

Parameters	Value (Mean ± %RSD)[*]	
	PYH	MEH
Peak area	3485193 ± 0.08	1347970 ± 0.08
Tailing factors	1.29 ± 1.06	1.53 ± 0.45
Retention time	5.25 ± 0.02	10.14 ± 0.04
Theoretical plates	5198 ± 0.10	8127 ± 0.08

[*] Mean and %RSD of six replicate.

TABLE 2: Parameters of regression analysis.

Parameters	PYH	MEH
Linearity range (mg mL^{-1})	0.395–0.592	0.203–0.304
Correlation coefficient	0.999	1.000
% Y intercept	1.61	0.17

The accuracy was also evaluated by linear regression analysis and computed.

2.3.5. Precision. Precision of the method was studied by analysis of three replicates of standard solution in three different concentrations (80, 100, and 120%). It was demonstrated by repeatability (intraday precision) and intermediate precision (interday precision) of standard solutions. The results were expressed as %RSD of the measurements.

2.3.6. Stability of Solution. The solution stability is tested by allowing the prepared drug matrix tested sample to stand exposed to room light and ambient room temperature for

A New Validated Stability Indicating RP-HPLC Method for Simultaneous Estimation of Pyridoxine Hydrochloride and Meclizine Hydrochloride in Pharmaceutical Solid Dosage Forms

175

TABLE 3: Results showing different degradative outcomes.

Parameters	Pyrimac tablet (mg tab^{-1})		Assay at forced condition (mg tab^{-1})*		Percent of degradation (%)	
	PYH	MEH	PYH	MEH	PYH	MEH
Normal	50.00	25.00				
Acidic			49.39 ± 0.10	23.84 ± 0.09	1.22	4.64
Alkaline			49.40 ± 0.08	23.75 ± 0.11	1.20	5.00
Oxidation			48.11 ± 0.06	23.85 ± 0.07	3.78	4.60
Reduction			49.68 ± 0.12	23.90 ± 0.15	0.64	4.40

*Mean of three replicate, tab: Tablet.

TABLE 4: Accuracy studies of PYH and MEH in drug-matrix solutions.

	Amount added (mg mL^{-1})	Peak area	Amount recovered (mg mL^{-1})	% Recovery	% Recovery (mean ± %RSD)	Over all (mean ± %RSD)
PYH	0.395	2804505	0.392	99.24		
	0.395	2805301	0.402	101.65	100.85 ± 1.38	
	0.395	2804670	0.402	101.65		
	0.494	3481689	0.497	100.71		
	0.494	3487433	0.495	100.20	100.57 ± 0.33	100.32 ± 1.06
	0.494	3485039	0.498	100.81		
	0.592	4181905	0.595	100.51		
	0.592	4181507	0.590	99.66	99.55 ± 1.02	
	0.592	4183455	0.583	98.48		
MEH	0.203	1084674	0.207	101.88		
	0.203	1084350	0.206	101.64	101.76 ± 0.17	
	0.203	1085947	0.207	101.88		
	0.252	1345809	0.253	100.38		
	0.252	1342589	0.252	99.81	100.00 ± 0.33	100.81 ± 1.01
	0.252	1348349	0.252	99.81		
	0.304	1624589	0.308	101.41		
	0.304	1623470	0.308	101.25	100.63 ± 1.21	
	0.304	1625347	0.302	99.22		

three consecutive days. The sample is to be assayed daily and compared to freshly prepared standard solutions.

2.3.7. Sensitivity. For sensitivity study the limit of detection (LOD) and limit of quantitation (LOQ) were estimated by determination of signal-noise ratio. The LOD ($\alpha = 3.3$) and LOQ ($\alpha = 10$) of the proposed method were calculated using the following equation:

$$A = \alpha \times C \times \frac{N}{S},\qquad(1)$$

where A is LOD or LOQ, C is the concentration in ppm, and N/S is the signal-noise ratio.

2.3.8. Ruggedness. Ruggedness of the current method was determined by analyzing six assay sample solutions of pyrimac tablet by different instrument, column, and two analysts in the same laboratory to check the reproducibility of the test result.

2.3.9. Robustness. To determine the robustness of the current method, the effect of flow rate was studied at 0.48 and 0.52 mL min^{-1} instead of 0.5 mL min^{-1}. The effect of column temperature was studied at 35 and 39°C instead of 37°C. The effect of mobile phase composition was assessed at (buffer : ACN = 31.3 : 68.7, v/v) and (buffer : ACN = 28.7 : 71.3, v/v) instead of (buffer : ACN = 30 : 70, v/v). The effect of wavelength change was studied at 252 nm and 256 nm instead of at 254 nm.

3. Results and Discussion

3.1. Method Validation

3.1.1. System Suitability. The results (Mean ± %RSD of six replicates) of the chromatographic parameters (Table 1) indicate the good performance of the system.

3.1.2. Linearity. The peak area was dynamic-linear in the concentration ranges of 0.395–0.592 mg mL^{-1} for PYH and

TABLE 5: Intra-day and inter-day precision of the method.

	Sample Conc. (%)	Peak area (day 1)	Peak Area (Mean ± %RSD)	Peak area (day 2)	Peak Area (Mean ± %RSD)	Overall (Mean ± %RSD)
PYH	80	2802395 2803486 2804553	2803478 ± 0.04	2804067 2804588 2804530	2804395 ± 0.01	2803937 ± 0.03
	100	3484270 3487551 3485761	3485861 ± 0.05	3485121 3487734 3485570	3486142 ± 0.04	3486001 ± 0.04
	120	4181755 4181438 4184432	4182542 ± 0.04	4180232 4181345 4183670	4181749 ± 0.04	4182145 ± 0.04
MEH	80	1084378 1084566 1084854	1084599 ± 0.02	1084421 1084666 1084502	1084530 ± 0.01	1084565 ± 0.02
	100	1346372 1345756 1347547	1346558 ± 0.06	1345207 1345391 1346445	1345681 ± 0.04	1346120 ± 0.06
	120	1625634 1623340 1625378	1624784 ± 0.08	1623245 1623167 1624345	1623586 ± 0.04	1624185 ± 0.07

TABLE 6: Stability of analytical sample solution.

Day	Room Temperature (°C)	% Recovery* PYH	MEH
1	25 ± 2	100.94	100.46
2	25 ± 2	100.17	99.65
3	25 ± 2	96.67	99.00

*Mean of three replicates.

0.203–0.304 mg mL^{-1} for MEH, respectively. Highly significant correlation coefficient (R^2) demonstrated the linearity of the method (Table 2).

3.1.3. Specificity. The chromatograms of blank, placebo, test sample, and standard were used to justify the specificity of target analyte. The method was specific since excipients in the formulation did not interfere in the estimation of PYH and MEH (Figure 3).

The samples submitted to acidic and alkaline condition showed significant alteration in the peak area, and also there was no detectable degradation peak(s). Similarly, during oxidative and reductive hydrolysis study, degradation peak(s) was not found. In every case the peak purity was 99.99%. The results acquired from peak purity tool confirmed that the active components' peak response was pure proving no other substances in the same retention time. Percent of degradation is mainly 3.78% for PYH during oxidation study as well as 4.64% for MEH in acidic condition (Table 3).

3.1.4. Accuracy. The results were expressed as percent recoveries of the particular components in the samples. The overall results of percent recoveries (mean ± %RSD) of drug-matrix solutions are indicating good accuracy of the proposed RP-HPLC method (Table 4). Correlation coefficient R^2 = 0.999 and 0.999 established excellent accuracy for the active ingredients PYH and MEH, respectively.

3.1.5. Precision. The values of %RSD for intraday and interday variation were found very well and within 2% limit, indicating that the current method is repeatable (Table 5).

3.1.6. Stability of Solution. In the stability study, the retention time remained unchanged till third day, but peak area of PYH and MEH deviated at third day by more than 2.0% from initial. This indicates that both solutions were stable for at least 48 hours, which was sufficient to complete the analytical procedure (Table 6).

3.1.7. Sensitivity. The LOD and LOQ by the proposed method were found for PYH 1.90 ppm and 5.74 ppm as well as for MYH 3.75 ppm and 11.35 ppm, respectively.

3.1.8. Ruggedness. The results (% of Recovery ± RSD) of six assay samples are indicating the ruggedness of the current method (Table 7).

3.1.9. Robustness. The effects of robustness study under different altered conditions of this proposed method are

A New Validated Stability Indicating RP-HPLC Method for Simultaneous Estimation of Pyridoxine Hydrochloride and
Meclizine Hydrochloride in Pharmaceutical Solid Dosage Forms

177

TABLE 7: Ruggedness of the method.

Amount of PYH (mg tab^{-1})	Amount of MEH (mg tab^{-1})	Analyst 1, instrument 1, column 1		Analyst 2, instrument 2, column 2	
		Amount found PYH (mg tab^{-1}) (Mean ± %RSD)[*]	Amount found MEH (mg tab^{-1}) (Mean ± %RSD)[*]	Amount found PYH (mg tab^{-1}) (Mean ± %RSD)[*]	Amount found MEH (mg tab^{-1}) (Mean ± %RSD)[*]
50	25	50.53 ± 0.3	25.63 ± 0.3	50.97 ± 0.04	25.60 ± 0.50

[*]Mean of six replicates.

TABLE 8: Robustness of the method.

Parameters	Actual Variance	Amount added (mg mL^{-1})		% Recovery (Mean ± %RSD)[*]	
		PYH	MEH	PYH	MEH
Flow Rate	0.48 mL min^{-1}	0.5	0.25	100.55 ± 0.33	100.04 ± 0.04
	0.52 mL min^{-1}	0.5	0.25	100.12 ± 0.02	100.18 ± 0.07
Organic (%) in mobile phase	68.7	0.5	0.25	99.95 ± 0.22	100.15 ± 0.04
	71.3	0.5	0.25	99.75 ± 0.51	100.18 ± 0.07
Detector wavelength	252 nm	0.5	0.25	100.00 ± 0.15	99.81 ± 0.19
	256 nm	0.5	0.25	100.00 ± 0.11	99.53 ± 0.37
Column Temperature	39°C	0.5	0.25	100.17 ± 0.20	100.08 ± 0.11
	35°C	0.5	0.25	100.41 ± 0.46	100.07 ± 0.05

[*]Mean of three replicates.

satisfactory (Table 8). The mean recovery and %RSD of analyzed sample indicate that the current method is robust.

4. Conclusion

The developed RP-HPLC method for the simultaneous determination of pyridoxine hydrochloride and meclizine hydrochloride is simple, precise, accurate, reproducible and highly sensitive. The developed method was validated based on ICH guidelines [21]. Hence, this method can be routinely used for the simultaneous determination of pyridoxine hydrochloride and meclizine hydrochloride in pure and pharmaceutical formulations.

Conflict of Interests

The author wishes to confirm that there is no known conflict of interests associated with this paper. The author confirms that he/she has given due consideration to the protection of intellectual property associated with this work and that there is no impediment to publication, including the trademarks mentioned in my paper.

Acknowledgment

The author is grateful to ACI Limited, Narayanganj, Bangladesh, for providing all kinds of financial and technical support to conduct this research work.

References

[1] A. Pathak and S. J. Rajput, "Simultaneous derivative spectrophotometric analysis of doxylamine succinate, pyridoxine hydrochloride and folic acid in combined dosage forms," The Indian Journal of Pharmaceutical Sciences, vol. 70, no. 4, pp. 513–517, 2008.

[2] J. E. Reynolds, The Extra Pharmacopoeia, Martindale, The Pharmaceutical Press, London, UK, 1996.

[3] Indian Pharmacopoeia, vol. 2, Government of India, The Controller of Publications, New Delhi, India, 1996.

[4] British Pharmacopoeia, vol. 3, British Pharmacopoeia Commission, The Stationery Office, London, UK, 2012.

[5] British Pharmacopoeia, vol. 2, British Pharmacopoeia Commission, The Stationery Office, London, UK, 2012.

[6] The United States Pharmacopoeia, 35, NF 30, vol. 3, United States Pharmacopeial Convention, Rockville, Md, USA, 2011.

[7] M. S. Arayne, N. Sultana, F. A. Siddiqui, M. H. Zuberi, and A. Z. Mirza, "Spectrophotometric methods for the simultaneous analysis of meclezine hydrochloride and pyridoxine hydrochloride in bulk drug and pharmaceutical formulations," Pakistan journal of pharmaceutical sciences, vol. 20, no. 2, pp. 149–156, 2007.

[8] H. H. Abdine, A. A. Gazy, and M. H. Abdel-Hay, "Simultaneous determination of melatonin-pyridoxine combination in tablets by zero-crossing derivative spectrophotometry and spectrofluorimetry," Journal of Pharmaceutical and Biomedical Analysis, vol. 17, no. 3, pp. 379–386, 1998.

[9] S. B. Bari and S. G. Kashedikar, "Simultaneous estimation of meclizine hydrochloride and nicotinic acid in combined dosage form by spectrophotometry," Indian Drugs, vol. 33, no. 8, pp. 411–414, 1996.

[10] J. G. Portela, A. C. S. Costa, and L. S. G. Teixeira, "Determination of Vitamin B$_6$ in pharmaceutical formulations by flow injection-solid phase spectrophotometry," Journal of Pharmaceutical and Biomedical Analysis, vol. 34, no. 3, pp. 543–549, 2004.

[11] C. S. Suresh, C. S. Satish, R. C. Saxena, and K. T. Santosh, "Simultaneous spectrophotometric analysis of a ternary mixture

of pharmaceuticals-assay for meclozine hydrochloride, pyridoxine hydrochloride and caffeine," *Journal of Pharmaceutical and Biomedical Analysis*, vol. 7, no. 3, pp. 321–327, 1989.

[12] S. B. Bari and S. G. Kaskhedikar, "Simultaneous estimation of meclozine hydrochloride and caffeine in solid dosage form by employing multicomponent and derivative spectrophotometry," *Indian Drugs*, vol. 34, no. 2, pp. 85–88, 1997.

[13] B. Habibi, H. Phezhhan, and M. H. Pournaghi-Azar, "Voltammetric determination of vitamin B_6 (pyridoxine) using multi wall carbon nanotube modified carbon-ceramic electrode," *Journal of the Iranian Chemical Society*, vol. 7, supplement, pp. S103–S112, 2010.

[14] A. P. Argekar and J. G. Sawant, "Simultaneous determination of pyridoxine hydrochloride and doxylamine succinate from tablets by ion pair reversed-phase high-performance liquid chromatography (RP-HPLC)," *Drug Development and Industrial Pharmacy*, vol. 25, no. 8, pp. 945–950, 1999.

[15] E. S. Tee and S. C. Khor, "Simultaneous determination of B-vitamins and ascorbic acid in multi-vitamin preparations by reversed-phase HPLC," *Malaysian Journal of Nutrition*, vol. 2, no. 2, pp. 176–194, 1996.

[16] M. S. Arayne, N. Sultana, and F. A. Siddiqui, "Simultaneous determination of pyridoxine, meclizine and buclizine in dosage formulations and human serum by RP-LC," *Chromatographia*, vol. 67, no. 11-12, pp. 941–945, 2008.

[17] M. E. Capella-Peiró, A. Bossi, and J. Esteve-Romero, "Optimization by factorial design of a capillary zone electrophoresis method for the simultaneous separation of antihistamines," *Analytical Biochemistry*, vol. 352, no. 1, pp. 41–49, 2006.

[18] C. K. Wong, J. R. Urbigkit, and N. Conca, "GLC determination of meclizine hydrochloride in tablet formulations," *Journal of Pharmaceutical Sciences*, vol. 62, no. 8, pp. 1340–1342, 1973.

[19] A. P. Argekar and J. G. Sawant, "Simultaneous determination of pyridoxine hydrochloride and doxylamine succinate in tablets by HPTLC," *Journal of Liquid Chromatography and Related Technologies*, vol. 22, no. 13, pp. 2051–2060, 1999.

[20] D. Widiretnani, S. I. Wahyuni, F. Kartinasari, and G. Indrayanto, "Simultaneous determination of pyrathiazine theoclate and pyridoxine HCl by TLC-densitometry in commercial tablets: validation of the method," *Journal of Liquid Chromatography and Related Technologies*, vol. 32, no. 1, pp. 154–165, 2009.

[21] ICH, *Validation of Analytical Procedures: Text and Methodology Q2 (R1)*, Geneva, Switzerland, 1996.

26

New Developments in Liquid Chromatography Mass Spectrometry for the Determination of Micropollutants

Zoraida Sosa-Ferrera, Cristina Mahugo-Santana, and José Juan Santana-Rodríguez

Departamento de Química, Universidad de Las Palmas de Gran Canaria, 35017 Las Palmas de Gran Canaria, Spain

Correspondence should be addressed to José Juan Santana-Rodríguez, jsantana@dqui.ulpgc.es

Academic Editor: Toyohide Takeuchi

The combination of liquid chromatography (LC) with mass spectrometry (MS) in the environmental field has appeared as a valuable tool for the determination of micropollutants. Several groups of compounds have been considered as particularly relevant (e.g., pharmaceuticals, hormones and other endocrine-disrupting, personal care products and their metabolites, flame retardants, surfactants, and plasticizers, among others) since the same ones are continuously being released in the environment mainly as a result of the manufacturing processes, the disposal of unused or expired products, and the excreta. Because these micropollutants are not completely removed in the environment, very specific and sensitive analytical procedures are needed for their identification and quantification. High performance liquid chromatography coupled to tandem mass spectrometry (LC-MS/MS) (or LC-MS^2) and especially time-of-flight mass spectrometry (TOF/MS), has allowed that many environmental contaminants that are highly polar or nonvolatile or have a high molecular weight to be analyzed or identified. In this work we present an overview focused on the developments of liquid chromatography mass spectrometry applied to the analysis of the main classes of micropollutants in aqueous and solid environmental samples. Various aspects of methodologies based on these techniques, including sample preparation (extraction/preconcentration) and matrix effects, are discussed.

1. Introduction

Over the last two decades the use of LC techniques coupling with a high resolution MS to identify unknown contaminants has advanced spectacularly. This progress is mostly due to the development of new instrumentation. LC techniques have replaced gas chromatography (GC) as they present obvious advantages such as reduced sample pretreatment and their capacity to determine polar or thermally stabile compounds.

The combination of LC and MS offers the possibility to take advantages of both LC as a powerful and versatile separation technique and MS as a powerful and sensitive detection and identification technique. The intrinsic properties of these two techniques result in an extremely analytical tool useful with many application areas. There are many different LC-MS systems on the market, that present advantages and limitations according to the type of samples that must be analyzed.

Interface designs have changed considerably and have become much more sophisticated and efficient. Since the introduction of atmospheric pressure ionization techniques (API), LC-MS has played an increasingly important role in environmental analysis allowing to analyze a broad range of compounds, including nonvolatile, thermally labile, and polar species. Today, the interfaces most widely used for the LC-MS analysis are electrospray (ESI) and atmospheric pressure chemical ionisation (APCI), both using atmospheric pressure ionization (API). They produce protonated $[M+H]^+$ or deprotonated $[M-H]^-$ molecules. ESI is particularly well suited for the analysis of polar compounds whereas APCI is highly effective in the analysis of medium- and low-polarity substances. When ESI is operating in the negative ion mode of ionisation (NI) the sensitivity achieved in the analysis of some relevant pollutant compounds is considerably better than that of the ESI interface operating in the positive ion mode of ionisation (PI) and the APCI interface operating in the NI mode [1, 2]. However, some recent studies [3–5] indicate that the APCI interface operating in the PI mode can furnish sensitivities comparable in many cases to that of the negative ion ESI.

Combined ion sources can be considered as an option merging the advantages and application ranges of atmospheric pressure ionization techniques, but on the other hand their sensitivity may be a compromise between both modes. The advantage of combined ESI/APCI [6] ion sources is the possible detection of both polar and nonpolar analytes in one run, which can increase the number of the identified components for highly complex matrices.

These API technologies have been interfaced with a variety of mass analysers, including single-(Q) and triple-quadrupole (QqQ), orthogonal-acceleration time-of-flight (oaTOF), linear ion trap (LIT), and sector-field MS instruments.

For complex samples, containing many compounds, LC-MS is not enough for the unequivocal confirmation of analytes for the final identification. LC-MS^2 takes out this problem and results in a much higher degree of certainty in the identification of the unknowns. Triple-quadrupole (QqQ) mass analyzers have become the most widely used analytical tool in the environmental analysis. Their application has allowed the determination of a great number of compounds, especially polar ones that were previously difficult or even impossible to analyze. More recent approaches in LC-MS^2 are linear ion traps (LITs), new-generation QqQs, and hybrid instruments, such as quadrupole time-of-flight (Qq-TOF) and Q-linear ion traps (Qq-LITs). When the first quadrupole of a QqQ is replaced by a double-focusing mass spectrometer, the instrument is termed a hybrid. Hybrid Qq-TOF-MS technique is the most common application in the structural characterization in the environmental analysis and allows an unequivocal confirmation of the contaminants detected. Moreover, TOF-based mass analyzers allow to find additional nontarget organic contaminants. The elimination of false positives is possible by generating full-scan production spectra with an exact mass. Qq-LIT is considered as a very powerful tool for a rapid identification and confirmation of metabolites in different matrices because of its capability of producing additional spectral information useful for structure clarification [7–9].

Some papers compared a real performance of different types of modern tandem mass analyzers for particular applications, which provides valuable complementary information. LC-MS with QqQ and LIT has been compared for the determination of 6 pesticides in fruits [10]. QqQ provides better linear dynamic range, higher precision, less matrix interferences, and better robustness, while LIT provides an excellent sensitivity for product ion measurements. Four LC-MS systems equipped with Q, QqQ, IT (ion trap) and Q-TOF have been compared in the quantitative analysis (sensitivity, precision, and accuracy) of carbosulfan and its main transformation products [11]. QqQ provides at least 20-fold higher sensitivity compared to other mass analyzers and better linear dynamic range. The repeatability (within a day) is slightly better for Q (5–10%) and QqQ (5–9%) compared to LIT (12–16%) and Q-TOF (9–16%). Although the QqQ is more sensitive and precise, mean values obtained by all instruments are comparable.

The miniaturization is an important issue considered in all fields of analytical instrumentation including both parts of LC-MS coupling. The most widespread and well-established approach is UHPLC [12], which is based on the use of small particle size (sub-2 μm particles) in the stationary phase and short columns, at ultrahigh pressures (up to 1300 bars) yielding fast analyses and narrow chromatographic peaks. Moreover, UHPLC dramatically shortens analysis times, often to 10 min or less [13, 14]. On the other hand, it requires a higher acquisition speed of mass spectrometer to obtain enough sampling points for the reliable peak integration. Typical peak widths in routine LC-MS are 3–10 s [14–16], while peak widths in the fast/ultrafast UHPLC-MS are generally in the range 1–3 s, but they can be narrower than 1 s under well-optimized conditions [17, 18]. Modern TOF-based mass analyzers and also some ion traps are capable of reaching higher acquisition speed points for peak, what is useful to generate more sampling points per peak for a better quantification. Examples of potential applications of these methods have been published. Ibáñez et al. published an overview of the applications of UHPLC with TOF-MS for the rapid screening of multiclass organic pollutants in water [19].

The composition of the mobile phase is an important factor for improving separation in LC. An acidic condition with acetonitrile-water and methanol-water mixtures with a gradient elution is among the most common approach for improved peak shape in chromatography. Modification of the mobile phase, when performed in an attempt to improve the sensitivity of MS detection, has been accomplished with acetic acid, formic acid, or ammonium acetate. Nonvolatile additives, such as oxalic acid, should be avoided when ESI is used and trifluoroacetic (TFA) acid can suppress the ionization in the electrospray source.

The ion suppression/enhancement effects play an important role in LC–MS quantification and extend of these effects needs to be quantitatively assessed. The ion suppression and matrix effects can cause severe problems with the quantification in the trace analysis.

In MS quantification, to eliminate any possible variations during the ionization process and the mass analysis, such as the ion suppression/enhancement, the contamination of the ion source or the mobile phase, extraction losses, or any other unpredictable reasons, an internal standard must be used.

Another important issue is the sample preparation prior to LC-MS analysis [20]. Obviously, the internal standard must be added before any sample preconcentration step. Another alternative approach for the relative quantification is the use of response factors determined from the calibration curves of pure standards and then applied for real samples [21, 22]. The internal standard addition and response factors approach can be combined in one platform together with the well-optimized chromatographic separation.

An overview of the applications of LC techniques coupled to mass spectrometry in the determination of the main classes of micropollutants in aqueous and solid environmental samples is presented. These compounds are present to very low concentrations and due to the high complexity of some environmental samples; very specific and sensitive analytical procedures are needed for their determination. Although these compounds are not currently covered by the existing regulations, the possibility of adverse effects on humans and animals and their extensive environmental distribution has recently attracted an increasing interest. In particular, these compounds include pharmaceuticals, personal care products, flame retardants, surfactants, and plasticizers, among others. Figure 1 shows a scheme summary of the micropollutants considered in this work.

Micropollutants contaminants are released into environment mainly as a result of the manufacturing processes, the disposal of unused or expired products, and the excreta, mostly through urban wastewater and many of them can further spread through the water cycle, even reaching drinking water, due to their hydrophilic character and low removal at wastewater treatment plants (WWTPs) and drinking water treatment plants (DWTPs) [23–25]. They can also enter into the environment due to surface-water runoffs and soil leaching after the agricultural applications of manure. Once released into the environment, micropollutants are subject to different processes, such as biodegradation and chemical and photochemical degradation, which contribute to their elimination. When these transformations take place, degradation products can differ in the environmental behaviour and toxicity. However, they are often more persistent than their corresponding parent compounds [26].

To obtain high recoveries and minimise interference, the determination of these pollutants requires extraction and clean-up steps prior to detection. Solid phase extraction (SPE) is frequently used to extract these compounds from aqueous samples [27]. However, the demand to reduce the solvent volumes and avoid the use of toxic organic solvents has led to substantial efforts to adapt existing sample preparation methods to the development of new approaches. Miniaturisation has been a key factor in the search of these objectives. Microextraction techniques allow high enrichment factors and minimise solvent consumption which avoid environmental pollution. Among these techniques are solid phase microextraction (SPME), stir-bar sorptive extraction (SBSE), and liquid-phase microextraction (LPME) approaches. Although SPME has been the technique most widely used, in recent years LPME approaches, such as single-drop microextraction (SDME), hollow-fiber liquid-phase microextraction (HF-LPME), and dispersive liquid-liquid microextraction (DLLME), have been growing more interest. The extraction of emerging pollutants from solid matrices is carried out by accelerated solvent extraction (ASE) pressurised liquid extraction (PLE), ultrasound assisted extraction (UAE), and microwave-assisted extraction (MAE). These methods have been replaced to Soxhlet extraction, the classical procedure for solid matrices [28–30].

The US Environmental Protection Agency (USEPA) published the final Contaminant Candidate List (CCL-3) in September 2009, which is a drinking water priority contaminant list for regulatory decision making and information collection. The listed contaminants are either known or anticipated to exist in drinking water systems and will be considered for a potential regulation. This final CCL-3 contains 104 chemicals and 12 microbial contaminants, and it includes three pharmaceuticals, eight hormones, and several disinfectant by-products and industrial additives [31]. In the European Union (EU), the Water Framework Directive (WFD) sets the strategy against the pollution of water by dangerous substances. The WFD provisions will be required from Member States and Associated States to establish programs to monitor the quality of water, which implies a review of human activity on the pollutants and an economic analysis of water use. In this context, there is an urgent need for a list of emerging contaminants as possible candidates for introduction into the WFD list of priority substances. This can be amended every four years with revisions and additions of new contaminants [32].

2. Application to the Determination of Micropollutants in Environmental Samples

2.1. Pharmaceuticals Compounds. Among new contaminants, pharmaceuticals belong to a group of an increasing interest due to their pharmacological activity and rising consumption deriving from their use in human and veterinary medicine [33, 34]. Moreover, due to their ubiquitous presence in the environment arising from continual input into the aquatic compartment, they are considered as "pseudo" persistent pollutants [35]. The discharge of therapeutic agents in effluents from production facilities, hospitals and private households, improper disposal of unused drugs, and the direct discharge of veterinary medicines leads to the contamination of environmental waters, and wastewater-treatment plants are considered to be a major source [33, 34, 36–38]. Biological treatment in WWTPs affects only the partial removal of a wide range of microcontaminants, especially polar ones, which are discharged into the final effluent. Thus, it has become evident that the application of more enhanced technologies may be crucial to full the requirements to recycle municipal and industrial wastewaters as drinking water. However, the removal of polar contaminants during drinking water treatment is incomplete. This fact was demonstrated by Ternes et al. when they investigated the elimination of selected pharmaceuticals, such as clofibric acid, bezafibrate, or carbamazepine, during drinking water treatment at the pilot-plant scale and in real waterworks in Germany [39]. The concentration of pharmaceuticals in water can vary between a few nanograms per liter to the micrograms per liter levels. These levels have to be removed in order to achieve the drinking water quality and to protect the water resources. Therefore, the concentrations and identities of these contaminants in water have to be monitored during the entire water purification and transportation process.

Antibiotics, followed by steroid compounds, analgesics, and nonsteroidal anti-inflammatory drugs (NSAIDs), are the most widely studied pharmaceuticals. Table 1 shows the diverse determinations of these compounds in environmental samples.

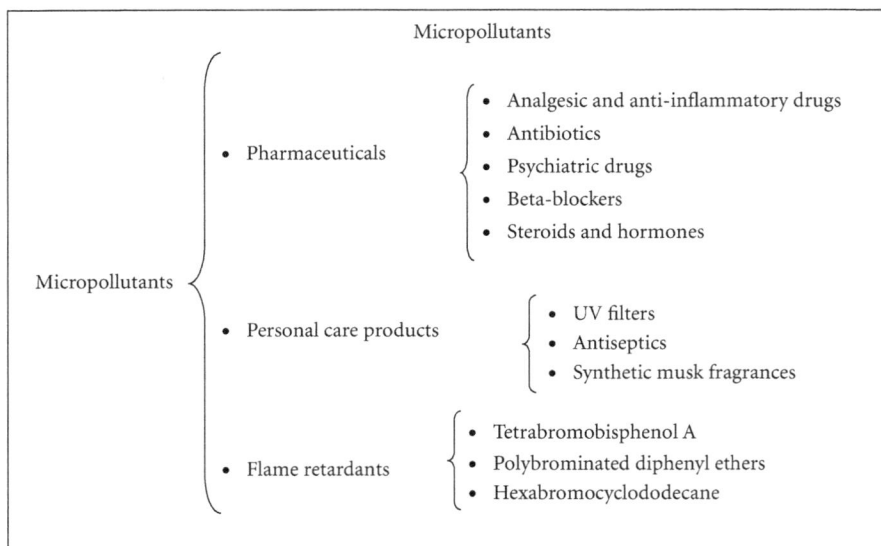

FIGURE 1: Scheme summary of the micropollutants considered in this paper.

2.1.1. Liquid Samples. Methods of sample preparation and extraction for pharmaceuticals have evolved significantly for aqueous phases since they were first described as early as the late 1980s. The traditional sample preparation method, liquid-liquid extraction (LLE), has largely been replaced by solid-phase extraction (SPE) for the aqueous matrices.

Regarding separation and detection techniques, LC, combined with MS or MS^2, is the most suitable technique to separate and detect pharmaceutical residues or metabolites in environmental samples. Most of the pharmaceutical compounds are not very volatile and some are highly polar, containing ionizable functional groups. It requires additional derivatization steps that may involve more labor and time, and cause unwanted contamination in the sample. Many antibiotics are nonvolatile with a high molecular weight, which respond well to ESI interface operating in the positive ion mode of ionization; therefore LC-MS or LC-MS^2 is often selected for their separation and analysis. When using a single MS step, selected-ion monitoring, SIM, is preferred for increased analytical sensitivity and selectivity in complex matrices such as wastewaters. Choi et al. [40] applied this methodology for the simultaneous analysis of seven tetracycline antibiotics and sulfonamide antibiotics from agricultural wastewater samples and sewage effluent samples. They combined the pretreatment technique, SPE, with LC-MS through online connection. This type of connection suppressed the target loss by keeping the cartridge from drying, which resulted in improvement on the recovery and saving of the analytical time. The average LOQ achieved was between 0.09 and 0.11 $\mu g \cdot L^{-1}$ for tetracycline antibiotics and sulfonamide antibiotics, respectively. Chen et al. [41], using IT-MS, reached LODs in the range of 3.2–6.2 $ng \cdot L^{-1}$ when they determined fluoroquinolones in water samples. The target compounds were extracted from samples by molecularly imprinted polymer (MIP) as sorbent. ESI was performed in positive mode and the data acquisition was performed in multiple reaction monitoring (MRM).

Some works can be found in the bibliography which use hybrid quadrupole instruments coupled to conventional liquid chromatography systems for the analysis of pharmaceuticals in waters. For instance, Martínez Bueno et al. [42] developed a method for the determination of 38 pharmaceuticals and 10 metabolites by LC-MS^2 using hybrid triple-quadrupole linear ion trap mass spectrometer in combination with time-of-flight mass spectrometry operating under SRM mode in both positive and negative ESI. This methodology was successfully applied to a monitoring study intended to characterize wastewater effluents of six sewage treatment plants in Spain. A different MS combination, QqLIT, was proposed by Gros et al. [43] for the determination of 73 pharmaceuticals by using both SRM and information-dependent analysis (IDA) acquisition modes with a total analysis time of 87 min. The method developed was applied to the analysis of various influent and effluent wastewaters.

The combination of UHPLC with an MS detector appears to be a suitable approach that fulfills key requirements in terms of sensitivity, selectivity, and peak-assignment certainty for the rapid determination of analytes at low concentrations in complex matrices. Modern QqQ instruments operating in the SRM mode are preferred for targeted analysis, while TOF-MS analyzers are particularly useful for nontargeted analysis. Batt et al. [44] developed a UHPLC method coupled to a QqQ mass spectrometer for the determination of 48 drugs and 6 metabolites in wastewater and surface waters. However, total analysis time was 48 min since four chromatographic conditions were used to determine all the compounds. Langford and Thomas [45] used UHPLC-QqQ with an ESI source; however, chromatographic separation of the 40 pharmaceuticals from hospital effluents took more than 50 min. A different MS analyzer was selected by Petrovic et al. [46] who developed an UHPLC method coupled to a Q-TOF mass spectrometer for the determination of 29 pharmaceuticals in wastewater in 14 min.

Conley et al. [47] described the determination of 13 pharmaceuticals and 1 metabolite in less than 4 min using UHPLC interfaced to a QqQ mass spectrometer with an ESI source. The mass analyzer operated in positive ionization mode for all analytes. The method was applied to samples of surface water collected from the Upper Tennessee River Basin. The same technique was applied by Kasprzyk-Hordern et al. [48] for the determination of 28 basic/neutral pharmaceuticals in river water samples from UK and Poland and fifteen compounds were determined at levels ranging from nanograms to micrograms per liter. In this case they achieved the separation of the target compounds in 16 min.

Considerable attention had been focused on the occurrence of steroid hormones in the environment since recent studies have documented that the exposure of fish to municipal wastewater effluents affects the reproductive physiology and behavior in many fish species at $ng \cdot L^{-1}$ or even $pg \cdot L^{-1}$ levels [49, 50]. Chang et al. [51] developed a method for the simultaneous determination of eighteen androgens and progestogens in environmental waters by using UHPLC-MS^2. Mass spectrometry was performed using a triple-quadrupole detector which was operated with ESI in the positive ion mode. After SPE procedure, a silica cartridge was used to purify the extract and reduce the signal suppression due to coeluting interferences. The developed method was applied to the analysis of these compounds in wastewater and surface-water samples and LODs for the eighteen analytes in the influent, effluent, and surface-water samples were in the ranges 0.20–50, 0.04–20, and 0.01–12 $ng \cdot L^{-1}$, respectively.

Hybrid mass spectrometers have also been used in combination with UHPLC to determine pharmaceutical compounds. Recently, Huerta-Fontela et al. [52] developed a method for the determination of 49 pharmaceuticals and 6 metabolites in six wastewater treatment plants using the dual acquisition modes of a hybrid triple-quadrupole linear ion trap system. The proposed method enabled all the 55 compounds to be separated chromatographically in less than 9 min (6.3 min positive mode and 2.7 min negative mode) with a total analysis time of 18 min.

2.1.2. Solid Samples. Despite the rather low lipophilicity of pharmaceuticals, interaction of the polar functional groups of them with organic matter and/or minerals may result in adsorption to solids. Furthermore, the application of sewage sludge as a fertilizer to the agricultural land and the reuse of manure containing veterinary medicines may also introduce pharmaceuticals into the soil. Animal origin pharmaceuticals, including aquaculture-derived compounds, contribute significantly to the occurrence of pharmaceuticals in solid matrices due to their patterns of application. Sewage sludge is the main solid produced in sewage treatment plants and the European Union (EU) promotes the use of sewage sludge as a fertilizer on agricultural land. Therefore it is important

to know the occurrence of contaminants in sewage-sludge samples as those could reach surface or ground waters.

The presence of pharmaceuticals in sediment, soil, and sewage sludge has been studied extensively. Analytical methods for the determination of specific groups of pharmaceuticals, including NSAIDs, antidepressants, antibiotics, and β-blockers [53, 54], and multiclass methods have been reported in recent years. Soxhlet extraction method for soil or sediment has been replaced by PLE, MAE, and UAE because of the time-consuming nature and high usage of hazardous organic solvents. Sample extracts obtained from solid matrices are most of the time with interfering co-extracts, which dictate an additional cleanup before LC analysis. Soil or sediment sample preparation needs to combine additional cleanup or purification steps, mainly SPE after the extraction step in the solvent due to the complexity of environmental samples [55, 56].

Haller et al. [57] used LC-MS and ESI with SIM mode to measure seven veterinary antibiotics in manure and reported $100 \mu g \cdot kg^{-1}$ as LOQ. This technique was also applied by Sagristà et al. [58] for the direct determination of four NSAIDs in dried sludge from a sewage treatment plant (Figure 2). Extraction experiments were carried out using a magnetic stirrer at 660 rpm for several hours and then a three-phase hollow-fiber liquid-phase microextraction (HF-LPME) was applied. This microextraction procedure allowed for the enrichment factors about 3000 times for all analytes. Data acquisition was performed in negative ion mode with SIM. LODs and LOQs were about 10 and 33 $\mu g \cdot L^{-1}$, respectively.

Determination by Qq-IT, with an ESI source operated in positive mode, has been used to characterize the persistence of tetracyclines in soil fertilized with liquid manure. The analytes were extracted from soil samples by vortex with citrate buffer and ethyl acetate and the LOQ was 5 $\mu g \cdot kg^{-1}$ for all compounds [59]. More recently, microwave-assisted micellar extraction combined with a QqQ mass spectrometer and ESI source, in a positive mode, was applied for the analysis of fluoroquinolone antibiotics in coastal marine sediments and in sewage-sludge samples. This extraction technique, which uses a micellar solution as extractant and LC-MS^2, allowed LODs and LOQs between 0.15–0.55 $\mu g \cdot kg^{-1}$ and 0.49–1.85 $\mu g \cdot kg^{-1}$, respectively [60]. In addition, Jacobsen et al. [61] applied two different LC-MS^2 methods to quantify eight antibiotics from different classes in soil and reported LOQs in the range 1.1–12.8 $\mu g \cdot kg^{-1}$. Löffler and Ternes [62] used two different APCI-MS^2 methods for ten acidic pharmaceuticals in negative mode and ESI-MS^2 for seven antibiotics in positive mode to determine residues in river sediment. This study illustrated that different ionization methods can be adapted to the characteristics of the compounds being examined.

Estrogenic compounds are medium polar to relatively nonpolar substances, with log Kow values in the range 2.5–5.3. Consequently, we can expect sorption of estrogens to the suspended matter and a tendency of them to accumulate in soil and sediments. Several estradiol-mimicking compounds, including 17β-estradiol, estriol, and 17α-ethinylestradiol, were determined from sewage-sludge samples by using MAE

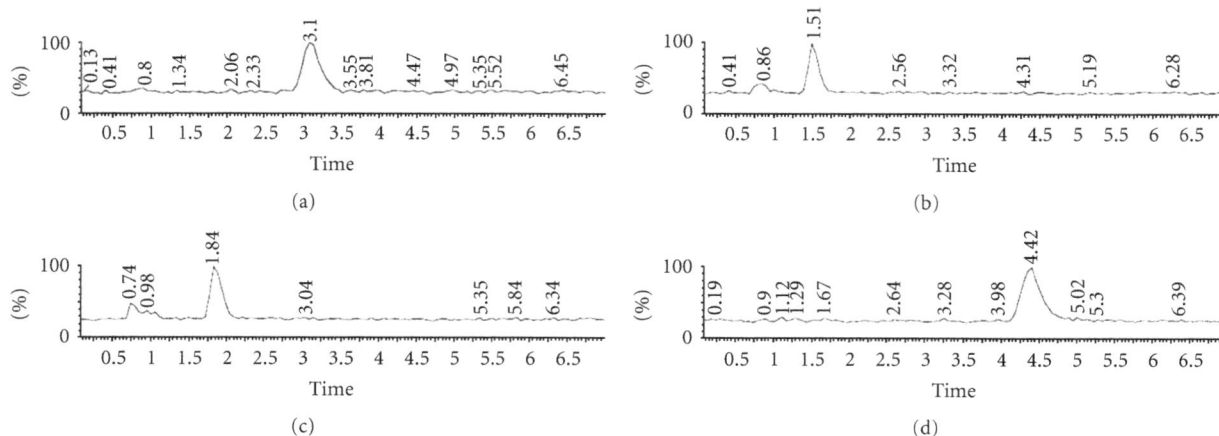

FIGURE 2: Single ion monitoring chromatograms obtained by LC-MS from reagent water spiked at 0.4 mg·L^{-1}. (a) m/z = 294 (diclofenac), (b) m/z = 253 (ketoprofen), (c) m/z = 229 (naproxen), and (d) m/z = 205 (ibuprofen). From reference [58].

followed by LC-MS^2 with ESI in a positive mode. The method provided LODs ranging from 0.6 to 3.5 μg·kg^{-1} [63]. A greater group of steroids, including natural and synthetic estrogens, androgens, progestogens, and glucocorticoids, were determined in the same type of sample by using UAE followed, with analysis; by rapid resolution LC-MS^2. In this case, a triple quadrupole detector was used, which was operated with ESI in both negative and positive modes. LODs for the 28 analytes were 0.08–2.06 μg·kg^{-1} [64]. LC-ESI(PI)-MS^2 and PLE as an extraction technique madeit possible to determine the traces of steroid hormones (including oestrogen, androgens, and progestogens) in soil with LODs in the range 0.08–0.89 μg·kg^{-1}. The results obtained showed ionization suppression for all the analytes in proportions ranging up to nearly 50% [65]. Using the same technique, with ESI (NI), Nieto et al. [66] achieved the determination of a greater number of natural and synthetic estrogens in sewage sludge. The MRM mode enabled LODs lower than 26 μg·kg^{-1} of the dry weight of sewage sludge for most of target analytes. Using an IT-MS equipped with ESI (NI) source, after extraction by MAE, Matjıcek et al. [67] carried out the simultaneous separation and determination of five hormones and their sulfate, glucuronide, and acetate conjugates in river sediments reaching LODs lower than 1 μg·kg^{-1}.

2.2. Personal Care Products. Personal care products (PCPs) constitute a group of emerging contaminants which have received a considerable attention in recent years. PCPs are regarded as being potentially hazardous compounds as many of them are ubiquitous and persistent and due to their continuous introduction might cause unwanted effects in the environment.

The principal pathway by which PCPs enter the environment is disposal in urban receiving waters from individual households, after showering and bathing. A variety of PCPs have been detected everywhere at the ng·L^{-1} concentration level in the effluents of WWTPs, since conventional water-treatment processes do not seem to be sufficient to remove

PCPs from sewage water (30–90% efficiency) [68–74]. The occurrence of PCPs in municipal sewage effluent and other environmental samples could negatively impact the health of the ecosystem and the health of humans, due to the persistent and long-term chronic exposure of aquatic organisms to the concentrations of PCPs [68]. Moreover, there is some evidence of potential interactive effects of PCPs, so that low doses may lead to cumulative stress and synergic toxicity effects in exposed organisms [68, 75]. PCPs, such as UV screens, insect repellents, and some synthetic musk fragrances, have also been suspected endocrine-disrupting compounds (EDCs) (i.e., compounds that can mimic the natural hormones of animals) [71, 73, 74].

Nowadays, in order to achieve greater protection to solar radiation, UV filters are added not only to cosmetics to be used for sunbathing but also to daily cosmetic products, such as face day creams, after-shave products, makeup formulations, lipsticks, and shampoos, thus resulting in an increase in the use of UV filters. Moreover, they can be found as additives in textiles, plastics, paints, car polishes, and so forth [76]. This excessive use of UV filters has led to their presence in the aquatic environment and increased their potential for endocrine and developmental toxicity [77, 78]. The increasing usage of these compounds, combined with their moderate-to-high water solubility, has led to the appearance of some of them in the aquatic environment. As regards toxicological effects, in vivo and in vitro studies have demonstrated that some hydroxylated benzophenones exert estrogenic and antiandrogenic actions [79].

Triclosan, triclocarban and methyl-triclosan are bactericides widely used in household and personal-care products, for example, shampoos, soaps, creams, mouthwash, and toothpaste [80]. Triclosan is found to be acutely toxic to some aquatic organisms and it has been also shown to photo transform into members of the dioxin family, which is known as the most carcinogenic chemicals in the world [81]. Although a relatively few data exist about the toxicity of triclocarban, it has been found to impair reproduction in laboratory rats and that some of its degradation products are carcinogenic. Methyl-triclosan, a metabolite of triclosan, is more lipophilic and environmentally persistent, suggesting its relatively high bioaccumulation potential in aquatic organisms [82].

Table 2 illustrates some applications of LC-MS to the determination of PCPs in environmental samples.

2.2.1. Liquid Samples. Negreira et al. [83] determined six derivatives of 2-hydroxybenzophenone, which are extensively employed as UV absorbers, in water samples by LC-QqQ using ESI in positive and negative modes, except for one compound (2-hydroxy-4-methoxybenzophenone-5-sulphonic acid) which could be ionized only in a negative mode. Benzophenones were recorded in MRM mode using two transitions per compound. Recoveries from the SPE step remained unaffected by the nature of the matrix; however, the efficiency of ESI was compound and sample dependant. Under optimized conditions, the proposed method provided LOQs from less than 1 to 32 ng·L^{-1}, depending on the compound and the type of water sample.

Triple quadrupole is the most common and most useful tool for determining PCPs in high-sensitivity target analysis. This mass spectrometer fitted with an ESI source operating in a negative mode has been used by Zhao et al. to determine triclosan and triclocarban in wastewater and tap water samples (Figure 3). Enrichment of target analytes before analysis was carried out by using ionic liquid dispersive liquid-phase microextraction. The sensitivity of the proposed method allowed for LODs in the range 0.04–0.58 μg·L^{-1} [84]. Klein et al. determined triclocarban in wastewater effluents by LC-QqQ after stir-bar sorptive extraction (SBSE) obtaining an LOQ of 10 ng·L^{-1} for the target analyte [85]. ESI source was also operated in a negative mode and MRM mode was applied.

Pedrouzo et al. [86] determined eleven PCPs, including hydroxylated benzophenones, triclocarban and triclosan, and parabens, (another type of preservatives used in personal care products) by SPE and UHPLC-MS^2 in surface and wastewaters in 9 minutes of chromatographic separation. A triple-quadrupole mass spectrometer and ESI in both PI and NI modes were applied. LOQ was 5 ng·L^{-1} for all the compounds, except for methylparaben (3 ng·L^{-1}). Most of PCPs determined were found in influent waters being methylparaben and propylparaben found at the highest concentration. Both are the most widely used parabens and they are normally used together due to their synergistic preservative effects [87, 88].

By using MRM to monitor two transitions between precursor and product ions, it is possible to confirm and quantify the presence of PCPs in waters at very low levels [89]. For example, Rodil et al. [90] developed a method to determine a group of 53 multiclass emerging organic pollutants (included the types mentioned above) by LC-MS^2, using ESI in both PI and NI modes, after SPE. The proposed method allowed LODs between 0.3 and 30 ng·L^{-1}. The method was used for the simultaneous determination of target analytes in water samples, including tap, surface, and wastewater. LC-QqIT was the technique chosen for determining the presence of 84 pollutants of different classes in wastewaters, including some PCPs such as sunscreen agents and synthetic musks. Previous to LC-QqIT analysis (ESI in PI and NI modes), wastewater samples were preconcentrated by SPE [91].

2.2.2. Solid Samples. Several PCPs (e.g., triclosan, triclocarban, and most UV-filtering compounds) show affinity to solid matrices. As a consequence, to allow the correct evaluation of the ecological impact of these substances, evaluation of their prevalence in solid matrices is important. Several analytical approaches were therefore reported recently.

LC-MS^2, with ESI operated in negative mode and MRM, was applied by Zhang et al. [92] to analyse benzophenone UV filters in sediment and sludge. The method developed allowed LOQs in the ranges of 0.06–0.33 ng·g^{-1} dry weight (dw) and 0.1–1.65 ng·g^{-1} dw for sediment and sludge samples, respectively. ESI and APCI sources, operated in the positive and in the negative ion mode using MRM, were applied by Wick et al. [93] to determine different classes of compounds such as, biocides, UV filters, and benzothiazoles in sludge samples. ESI exhibited a strong ion suppression for most target analytes, while APCI was generally less susceptible to ion suppression which led to higher signal intensities in the samples and consequently to lower LOQs as long as the background noise was not increasing.

UHPLC-MS^2 was applied by Nieto et al. [94] for the determination of a group of parabens and two UV filters in sewage sludge. In the chromatographic step, after pressurized liquid extraction, the compounds were detected by using tandem mass spectrometry with a triple-quadrupole analyzer with ESI in positive and negative modes. The use of small diameter particles in the chromatographic column allowed the compounds to be eluted in 9 min. LODs and LOQs were lower than 8 μg·kg^{-1} and 12.5 μg·kg^{-1} of dw, respectively.

2.3. Flame Retardants. Flame retardant (FR) compounds are a structurally diverse group of chemicals that are added to or reacted with polymers, and they are used in plastics, textiles, electronic circuitry, and other materials to reduce the risk of fire. One of these groups of compounds comprises brominated FRs (BFRs), some of which are ubiquitous, and many of which have been detected in biota, sediments, air, water, marine mammals, and even human milk [95, 96]. BFRs are mainly represented by tetrabromobisphenol A (TBBPA), polybrominated diphenyl ethers (PBDEs), which are aromatic compounds, and the cycloaliphatic compound hexabromocyclododecane (HBCD). Regarding TBBPA, one of the most important and commonly used flame retardants, although the use of TBBPA as an additive is estimated to account for about 10% of the total amount used, excessive non-polymerized TBBPA can be emitted, contaminating the environment [97]. Mainly due to its low bioaccumulation potential, it presents concentration levels lower than those of PBDEs or HBCD in the environment [98]. However, TBBPA, being a phenolic compound, may have a greater adverse effect on humans and wildlife. Likewise, TBBPA can be considered a potential EDC due to the similarities in its structure with 17β-estradiol and thyroxine.

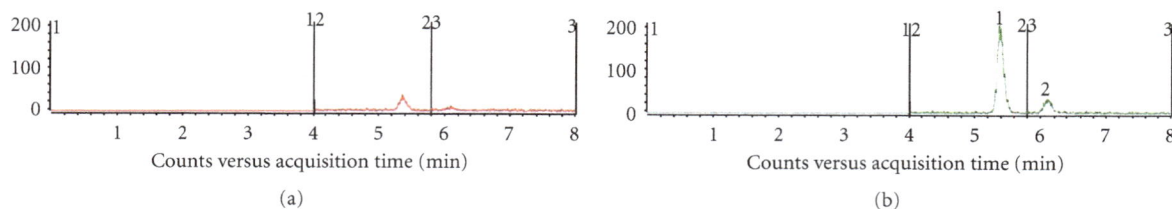

Figure 3: LC-ESI-MS^2 chromatogram obtained from wastewater. (a) Wastewater (b) Wastewater spiked with 0.40 μg$\cdot$L^{-1} triclocarban (1) and 2.0 μg$\cdot$L^{-1} triclosan (2). From reference [84].

LC, coupled to tandem MS, and ESI, APCI or atmospheric pressure photoionization (APPI), represents a valid tool to the determinations of these compounds, since that the determination of some congeners is known to be difficult due to thermal degradation problems and also because a derivatisation step is needed [99–102].

Some applications of the use of LC-MS to the determination of these micropollutants in environmental samples are shown in Table 3.

2.3.1. Liquid Samples.

2.3.1. Liquid Samples. The use of LC-MS in the determination of TBBPA provides several different detection modes and eliminates the need for the derivatization of the phenolic group. Moreover, it facilitates the use of ^{13}C-labelled TBBPA. Tollbäck et al. [103] reported that the most suitable LC-MS interface for TBBPA analysis is ESI operating in a negative ionization mode. ESI gave 30–40 times lower LODs compared to APCI. In addition, it permits the monitoring of the intact TBBPA molecule through the soft ionization of ESI, resulting in improved method selectivity and accuracy. Frederiksen et al. [104] compared LC-MS^2 to GC-MS for the determination of TBBPA and concluded that LC-MS^2 is the method of choice, not only because derivatisation is not needed, but also because of its higher sensitivity and better detection. LC-MS^2 using QqQ and APCI source was performed for the analysis of 38 BFRs in wastewater, finding decabromodiphenylethane (DBDPE), bis(2-ethyl-1-hexyl)tetrabromophthalate (BEHTBP), and TBBPA at ng$\cdot$L^{-1} levels [105]. Bacaloni et al. [106] proposed an LC-MS^2 method for the simultaneous determination of TBBPA and five PBDEs in water samples. LIT mass spectrometer, coupled with an APPI source, was operated in the negative ion mode and each compound was quantified operating in MRM obtaining LOQs of 0.2–3.3 ng$\cdot$L^{-1}, except for one compound. PBDEs were poorly retained by SPE from river water and sewage treatment plant effluent samples; thus LLE by n-hexane was used for these samples.

2.3.2. Solid Samples.

2.3.2. Solid Samples. Due to its low solubility in water (0.72 mg$\cdot$L^{-1}) and high log Kow (4.5), TBBPA is likely to be associated with suspended particulate matter once released in the water column and ultimately buried in sediments [107, 108]. However, due to the lower bioaccumulation potential, TBBPA presents lower concentrations than PBDEs and HBCD in the environment. IT-MS was reported for the determination of TBBPA in sediment and sewage sludge after

LC separation [108]. Although the ion suppression of the TBBPA signal due to matrix components in the ESI process was not high, sewage-sludge extracts suffered greatly from ion suppression and an extensive cleanup was required to minimize this effect.

The distribution of HBCD isomers in suspended sediments from Detroit River was analyzed by QqQ, with an ESI source operated in negative mode and detection by MRM but using ASE as extraction technique [109]. LOD of 10 pg on column was estimated for individual HBCD diastereoisomers. An LC-IT-MS method, employing ESI operated in a negative ionization mode, was developed to determine HBCD diastereoisomers in marine sediment samples, obtaining LOQs ranged from 25 to 40 pg$\cdot$g^{-1} (dw). Target analytes were extracted from sediment samples by MAE. Efficiency of this technique was compared with Soxhlet extraction and PLE and the results obtained showed that MAE provides better extraction efficiencies than either PLE or Soxhlet extraction [110].

LC-ESI-MS^2-based method was developed by Chu et al. [107] for the simultaneous determination of TBBPA, as well as lower brominated BPA analogues, in sediment and sludge samples. LOQs for both kind of samples were in the range 0.02–0.15 ng$\cdot$g^{-1} (dw).

Hybrid mass spectrometer (LC-QqLIT) with an ESI interface was proposed by Guerra et al. to analyze TBBPA and related compounds (bisphenol A (BPA), monobromobisphenol A (MonoBBPA), dibromobisphenol A (DiBBPA), and tribromobisphenol A (TriBBPA)) in sewage-sludge and sediment samples. Sample extraction was based on the use of ultrasonication SPE, allowing for LODs in the range of 0.6–2.7 ng$\cdot$g^{-1} and 1.4–66 ng$\cdot$g^{-1} for sediment and sludge samples, respectively [111].

ESI, APPI, and APCI sources were tested in the determination of HBCDs and TBBPA in sewage-sludge samples [112]. In this study, involving the use of UHPLC-MS^2 and PLE, APCI gave a higher sensitivity than APPI while for TBBPA-bis, APCI and APPI showed a similar performance. ESI was the best option for TCBPA, TBBPA, and HBCDs. Figure 4 shows the total ion current transition of target analytes using APCI as an ionization source.

3. Conclusions and Trends

The applications of advanced LC-MS technologies to an environmental analysis have allowed the determination of

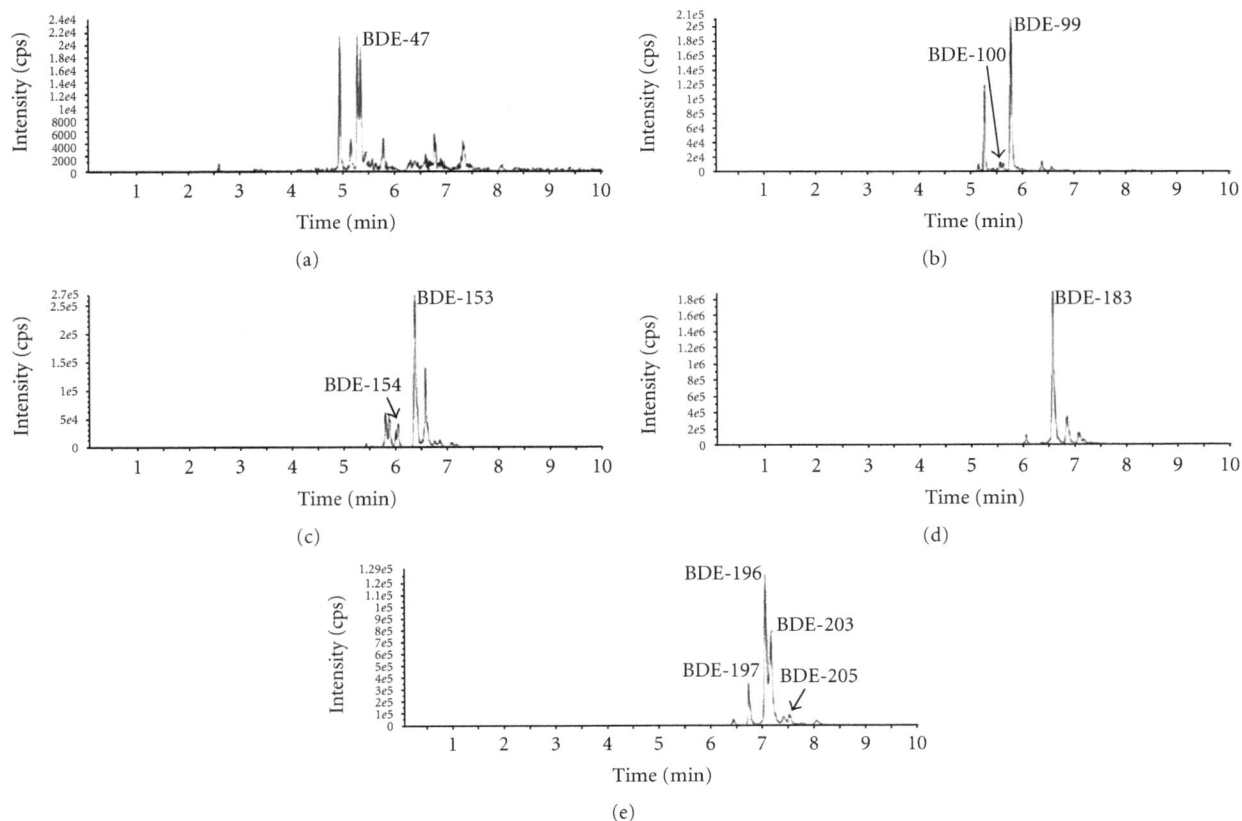

FIGURE 4: TIC of transitions of tetra-BDE (a), penta-BDE (b), esa-BDE (c), epta-BDE (d), and octa-BDE (e) obtained from PLE extraction and UHPLC-APCI/MS/MS analysis of NIST Standard Reference Material New York/New Jersey Waterway Sediment 1944. From reference [112].

a great number of compounds, especially polar compounds, that were previously difficult or even impossible to analyze. In particular, the introduction of API interfaces and triple-quadrupole analyzers has greatly improved the sensitivity and selectivity of detection and today, the analysis of many micropollutants in the environment samples is possible at the $ng \cdot L^{-1}$ and $ng \cdot g^{-1}$ levels, and even at the $pg \cdot L^{-1}$ and $pg \cdot g^{-1}$ levels in the routine bases. Because of the improved sensitivity and selectivity of the detection systems, a sample preparation is becoming easier, and the probe of it is the current trend towards a more extensive application of automated online methodologies with simple sample pretreatment and high sample throughput. However, despite the high selectivity of LC-MS systems, false negative findings can still occur due to the often high complexity of environmental matrices. Therefore, the application of stringent confirmation and identification criteria [113], in terms of retention time, base peak and diagnostic ions, relative abundances, and so forth, is essential.

The recent introduction of tandem mass spectrometry can help eliminate the false identification and quantification of coeluting compounds that can occur with single ion monitoring while it also reduces the amount of background noise present. More recent, the possibility to couple liquid chromatography to ion trap or the new generation of triple-quadrupole and hybrid instruments such as quadrupole time-of-flight and linear traps, more and more applications for the determination of micropollutants have been described for liquid chromatography.

These new approaches are a powerful analytical technique with excellent capabilities due to their high sensitivity in a full-spectrum acquisition mode together with their resolving power and accurate-mass measurements. These features make these techniques very attractive in qualitative analysis, especially for the wide-scope screening of a large number of organic contaminants and residues at trace levels in different fields.

The fastest growing chromatography trend continues to be the use of the ultrahigh performance liquid chromatography. In addition to providing narrow peaks and improved chromatographic separations, it dramatically shortens analysis times.

In the area of sample pretreatment, an important progress has been made also with regards to the preparation techniques. The efforts have gone directed towards to obtain high recoveries and minimise interference, as well as to reduce solvent volumes and avoid the use of toxic organic solvents. Microextraction techniques allow high enrichment

TABLE 1: Determination of pharmaceutical compounds in environmental samples by LC coupled to mass spectrometry.

Compounds	Samples	Extraction	Determination	Analytical parameters	Reference
SMMX, SDMX, SMXZ, STZL, SCPD, SMRZ, SMTZ, OXTC, MCLN, DOXN, MECN, CTCL, DEMN, and TCLN	River water, sewage and agricultural wastewater	Online SPE	LC-MS ESI (PI)	Recoveries: 74.3–116.5% LOQs: 0.09–0.11 μg·L⁻¹	[40]
ENR, LOM, LVFX, FLX, SPFX, AMX, OTC, and SQX	Lake water, river water and sewage effluent	MIP	LC-Qq-IT ESI (PI) MRM	Recoveries: 76.3–94.2% RSD < 9.1% LODs: 3.2–6.2 ng·L⁻¹	[41]
4-Acetamidophenol, carbamazepine, (+)-cis-diltiazem hydrochloride, norfluoxetine hydrochloride, ranitidine hydrochloride, sulfamethoxazole, caffeine, trimethoprim, atorvastatin, lovastatin, levofloxacin, and sertraline	River water	SPE	UHPLC-QqQ ESI (PI) MRM	Mean recovery: 77.9% Mean RSD: 7.3% LODs < 4.1 ng·L⁻¹	[47]
28 basic/neutral pharmaceuticals	River water	SPE	UHPLC-QqQ ESI (PI) MRM	LOQs: 0.3–50 ng·L⁻¹	[48]
18 androgens and progestogens	River water and wastewater	SPE	UHPLC-QqQ ESI (PI) MRM	Recoveries: 77–100% RSD: 1.7–12% LODs: 0.01–50 ng·L⁻¹	[51]
49 pharmaceuticals and 6 metabolites	Wastewater	SPE	UHPLC-QqLIT ESI (PI and NI) MRM and IDA	Recoveries: 55–111% RDS < 5% LODs: 0.02–50 ng·L⁻¹	[52]
Ketoprofen, naproxen, diclofenac, and ibuprofen	Sewage sludge	HF-LPME	LC-MS ESI (NI) SIM	Recoveries: 52–63% RSDs: 0.5–7% LODs < 10 ng·L⁻¹	[58]
Levofloxacin, norfloxacin, ciprofloxacin, and enrofloxacin	Marine sediments and sewage sludge	MAME	LC-QqQ ESI (PI) MRM	Recoveries > 73% RSD < 8% LODs: 0.15–0.55 μg·kg⁻¹	[60]
Ibuprofen, diclofenac, and salicylic acid	Wastewater	HF-LPME	LC-MS/MS	Recoveries: 52–100% RSDs: 1.1–1.6%. LODs: 20–300 ng·L⁻¹	[114]
Ibuprofen, naproxen, clofibric acid, piroxicam, ketorolac, bezafibrate, fenoprofen, diclofenac, and indomethacin	Wastewater	HF-LPME	LC-MS/MS	Recoveries: 80–111% RSDs: 3.4–32% LOQs: 0.5–42 ng·L⁻¹	[115]
Estrone, 17β-estradiol, estriol, ethynyl estradiol, and diethylstilbestrol	Surface water and wastewater	In-tube SPME	LC/MS/MS	Recoveries: 86.1–106.8 LODs: 2.7–11.7 ng·L⁻¹	[116]
36 endocrine-disrupting chemicals, including estrogens and progestogens	Potable and river water	SPE	UHPLC-QqTOF	Recoveries: 46%–134% LOD < 0.72 ng·L⁻¹	[117]
17β-estradiol, estriol, and 17α-ethinylestradiol	Sewage sludge	MAE	LC-QqQ ESI (PI) MRM	Recoveries: 71.7–103.1% RSD < 12% LODs: 0.6–3.5 μg·kg⁻¹	[63]

TABLE 1: Continued.

Compounds	Samples	Extraction	Determination	Analytical parameters	Reference
28 estrogenic compounds	Sewage sludge	UAE	LC-QqQ ESI (PI and NI) MRM	Recoveries: 44–200% RSD: 0.6–11.6% LODs: 0.08–2.06 $\mu g \cdot kg^{-1}$	[64]
Oestrone, testosterone, androstenedione, norethindrone, levonorgestrel, and progesterone	Soil	PLE	LC-QqQ ESI (PI) MRM	Recoveries: 45–100% RSD: 2.2–10.3% LODs: 0.08–2.84 $\mu g \cdot kg^{-1}$	[65]
Estrone, 17α-estradiol, 17β-estradiol, estriol, 17α-ethinylestradiol and diethylstilbestrol, and five conjugates	Sewage sludge	PLE	LC-QqQ ESI (PI) MRM	Recoveries > 81% RSD < 6% LODs < 26 $\mu g \cdot kg^{-1}$	[66]
α-estradiol, β-estradiol, estriol, estrone, ethynylestradiol and their sulfate, glucuronide, and acetate conjugates	River sediments	MAE	LC-Qq-IT ESI (NI) MRM	Recoveries: 83–107% RSD: 4.9–9.6% LODs < 1 $\mu g \cdot kg^{-1}$	[67]

Abbreviations: LC: liquid chromatography; UHPLC: ultrahigh pressure liquid chromatography; MS: mass spectrometry; QqQ: triple quadrupole; IT: ion trap; LIT: lineal ion trap; ESI: electrospray ionization; APCI: atmospheric pressure chemical ionization; APPI: atmospheric pressure photoionization; NI: negative ion mode of ionisation; PI: positive ion mode of ionisation; SIM: selected ion monitoring; MRM: multiple reaction monitoring; SRM: selected reaction monitoring.
SPE: solid phase extraction; HF-LPME: hollow-fiber liquid-phase microextraction; MIP: molecularly imprinted polymer; UAE: ultrasound assisted extraction; PLE: pressurised liquid extraction; MAE: microwave-assisted extraction; MAME: microwave-assisted micellar extraction.
RSD: relative standard deviation; LOQ: limit of quantification; LOD: limit of detection.
SMMX: sulfamonomethoxine; SDMX: sulfadimethoxine; SMXZ: sulfamethoxazole; STZL: sulfathiazole; SCPD: sulfachloropyridazine; SMRZ: sulfamerazine; SMTZ: sulfamethazine; OXTC: oxytetracycline HCl; MCLN: minocycline HCl; DOXN: doxycycline hyclate; MECN: meclocycline sulfosalicylate; CTCL: chlortetracycline HCl; DEMN: democlocycline HCl; TCLN: tetracycline.
ENR: enrofloxacin; LVFX: levofloxacin; FLX: fleroxacin; SPFX: sparfloxacin; AMX: amoxicillin; OTC: oxytetracycline; SQX: sulfaquinoxaline.
LOM: lomefloxacin;

TABLE 2: Determination of personal care products in environmental samples by LC coupled to mass spectrometry.

Compounds	Samples	Extraction	Determination	Analytical parameters	Reference
UV filters: BP-1, BP-2, BP-3, BP-4, BP-6, and BP-8	River water and wastewater	SPE	LC-QqQ ESI (NI and PI) MRM	Recoveries: 83–105% LOQs 1–32 ng·L^{-1}	[83]
Triclosan and triclocarban	Wastewater and tap water	IL-DLLME	LC-QqQ ESI (NI) MRM	Recoveries: 70.0–103.5% RSD 7.0–8.8% LODs: 0.040–0.58 μg·L^{-1}	[84]
Triclocarban	Wastewater effluents	SBSE	LC-QqQ ESI (NI) MRM	Recoveries: 92–96% RSD: 2% LOQ: 10 ng·L^{-1}	[85]
BP-1, BP-3, BP-8, OC, OD-PABA triclocarban, triclosan, methylparaben, ethylparaben, benzylparaben, propylparaben	Surface water and wastewaters	SPE	UHPLC-QqQ ESI (NI and PI) MRM	Recoveries: 20–101% RSD: 1–14% LODs: 20–200 ng·L^{-1}	[86]
53 multiclass emerging pollutants (UV filters and insect repellents, among others)	Tap water, surface water, and wastewater	SPE	LC-qqQ ESI (NI and PI) MRM	Recoveries > 60% RSD < 15% LODs: 0.3–30 ng·L^{-1}	[90]
Benzotriazoles (UVP, UV 329, UV 326, UV 328, UV 327, UV 571, and UV 360)	Coastal marine water and wastewater	SPE online	UHPLC-QqQ ESI (PI) MRM	Recoveries: 65–94% RSD: 6.2–10 % LODs: 0.6–4.1 ng·L^{-1}	[118]
4-Hydroxybenzophenone BP-1, BP-2, BP-3, and BP-8	Sediments and sludge	LLE	LC-QqQ ESI (NI) MRM	Recoveries: 70–116% RSD: 3.3–13.8% LOQs: 0.06–1.65 ng·g^{-1}	[92]
Biocides, UV filters, and benzothiazoles	Sludge sample	PLE	LC-QqQ ESI, APCI (NI and PI) MRM	Recoveries: 74–119% RSD < 25%	[93]
Triclosan, triclocarban, methyl paraben, ethyl paraben, propyl paraben, benzyl paraben, OD-PABA, OC, PMDSA, BP-1, BP-3, and BP-8	Sewage sludge	PLE	UHPLC-QqQ ESI (NI and PI) MRM	Recoveries: 15–100% RSD $\leq$ 9% LODs < 8 ng·g^{-1}	[94]

Abbreviations: LC: liquid chromatography; UHPLC: ultrahigh pressure liquid chromatography; QqQ: triple quadrupole; IT: ion trap; LIT: lineal ion trap; ESI: electrospray ionization; APCI: atmospheric pressure chemical ionization; NI: negative ion mode of ionisation; PI: positive ion mode of ionisation; MRM: multiple reaction monitoring; SRM: selected reaction monitoring.
IL: ionic liquid; LLE: liquid-liquid extraction; SDME: single-drop microextraction; DLLME: dispersive liquid-liquid microextraction; SBSE: stir-bar sorptive extraction; PLE: pressurised liquid extraction.
RSD: Relative standard deviation; LOQ: limit of quantification; LOD: limit of detection.
OC: octocrylene; PMDSA: 2-phenylbenzimidazole-5-sulfonic acid; BP-1: 2,4-dihydroxybenzophenone; BP-2: 2,2',4,4'-tetrahydroxybenzophenone; BP-3: 2-hydroxy-4-methoxybenzophenone; BP-4: 2-hydroxy-4-methoxybenzophenone-5-sulphonic acid; BP-6: 2,2'-dihydroxy-4,4'-methoxybenzophenone; BP-8: 2,2'-dihydroxy-4-methoxybenzophenone.

TABLE 3: Determination of flame retardants in environmental samples by LC coupled to mass spectrometry.

Compounds	Samples	Extraction	Determination	Analytical parameters	Reference
38 HFRs	Wastewater	SPE	LC-QqQ APCI (NI) MRM	Recoveries 25–132% LOQs 0.1–5.6 μg·L^{-1}	[105]
TBBPA and five PBDEs	Wastewater, river, and drinking water	LLE	LC-LIT APPI (NI) MRM	Recoveries 43–99% RSD < 17% LOQs 0.2–3.3 ng·L^{-1}	[106]
TBBPA and brominated BPA analogues	Sediment and sludge	Soxhlet	LC-QqQ ESI (NI) MRM	Recoveries 70–105% RSD: 4.9–13.1% LOQs 0.02–0.15 ng·g^{-1}	[107]
HBCD isomers	Suspended sediments from Detroit River	ASE	LC-QqQ, ESI (NI) MRM	—	[109]
HBCD diastereoisomers	Marine sediment	MAE	LC-IT ESI (NI)	Recoveries: 68–91% RSD: 2–11% LOQs: 25–40 pg·g^{-1}	[110]
TBBPA, BPA, Mono-BBPA, Di-BBPA, and Tri-BBPA	Sewage sludge and sediment	Sonication-SPE	LC-QqLIT ESI (NI) SRM	Recoveries: 39–120% RSD < 13% LODs 0.6–2.7 ng·g^{-1}	[111]
HBCDs and TBBPA	Sewage sludge	PLE	UHPLC-MS^2 ESI, APPI, APCI MRM	Recoveries: 65–112% LOQs: 0.005–0.14 ng·g^{-1}	[112]

Abbreviations: LC: liquid chromatography; UHPLC: ultrahigh pressure liquid chromatography; MS: mass spectrometry; QqQ: triple quadrupole; IT: ion trap; LIT: lineal ion trap; ESI: electrospray ionization; APCI: atmospheric pressure chemical ionization; APPI: atmospheric pressure photoionization; NI: negative ion mode of ionisation; PI: positive ion mode of ionisation; MRM: multiple reaction monitoring; SRM: selected reaction monitoring.
SPE: solid phase extraction; ASE: assisted solvent extraction; LLE: liquid-liquid extraction; PLE: pressurised liquid extraction; MAE: microwave-assisted extraction.
RSD: relative standard deviation; LOQ: limit of quantification; LOD: limit of detection.
TBBPA: tetrabromobisphenol A; PBDEs: polybrominated diphenyl ethers; BDE: decabrominated diphenyl ether; HBCD: hexabromocyclododecane; HFRs: halogenated flame retardants.

factors and minimise solvent consumption which avoid the environmental pollution.

Acknowledgment

This work was supported by funds provided by the Spanish Ministry of Science and Innovation, Research Project no. CTQ 2010-20554.

References

[1] C. Baronti, R. Curini, G. D'Ascenzo, A. Di Corcia, A. Gentili, and R. Samperi, "Monitoring natural and synthetic estrogens at activated sludge sewage treatment plants and in a receiving river water," *Environmental Science and Technology*, vol. 34, no. 24, pp. 5059–5066, 2000.

[2] M. J. López De Alda and D. Barceló, "Determination of steroid sex hormones and related synthetic compounds considered as endocrine disrupters in water by liquid chromatography-diode array detection-mass spectrometry," *Journal of Chromatography A*, vol. 892, no. 1-2, pp. 391–406, 2000.

[3] M. Seifert, G. Brenner-Weiß, S. Haindl, M. Nusser, U. Obst, and B. Hock, "A new concept for the bioeffects-related analysis of xenoestrogens: hyphenation of receptor assays with LC-MS," *Fresenius' Journal of Analytical Chemistry*, vol. 363, no. 8, pp. 767–770, 1999.

[4] J. Rose, H. Holbech, C. Lindholst et al., "Vitellogenin induction by 17β-estradiol and 17α-ethinylestradiol in male zebrafish (Danio rerio)," *Comparative Biochemistry and Physiology C*, vol. 131, no. 4, pp. 531–539, 2002.

[5] A. Lagana, A. Bacaloni, G. Fago, and A. Marino, "Trace analysis of estrogenic chemicals in sewage effluent using liquid chromatography combined with tandem mass spectrometry," *Rapid Communications in Mass Spectrometry*, vol. 14, no. 6, pp. 401–407, 2000.

[6] Y. Zhou, Y. Wang, R. Wang, F. Guo, and C. Yan, "Two-dimensional liquid chromatography coupled with mass spectrometry for the analysis of Lobelia chinensis Lour. using an ESI/APCI multimode ion source," *Journal of Separation Science*, vol. 31, no. 13, pp. 2388–2394, 2008.

[7] W. M. A. Niessen, "State-of-the-art in liquid chromatography-mass spectrometry," *Journal of Chromatography A*, vol. 856, no. 1-2, pp. 179–197, 1999.

[8] I. Ferrer and E. M. Thurman, "Liquid chromatography/time-of-flight/mass spectrometry (LC/TOF/MS) for the analysis of emerging contaminants," *Trends in Analytical Chemistry*, vol. 22, no. 10, pp. 750–756, 2003.

[9] K. Wille, H. F. De Brabander, L. Vanhaecke, E. De Wulf, P. Van Caeter, and C. R. Janssen, "Coupled chromatographic and mass-spectrometric techniques for the analysis of emerging pollutants in the aquatic environment," *Trends in Analytical Chemistry*, vol. 35, pp. 87–108, 2012.

[10] Y. Picó, C. Blasco, M. Farré, and D. Barceló, "Analytical utility of quadrupole time-of-flight mass spectrometry for the determination of pesticide residues in comparison with an optimized column high-performance liquid chromatography/tandem mass spectrometry method," *Journal of AOAC International*, vol. 92, no. 3, pp. 734–744, 2009.

[11] C. Soler, B. Hamilton, A. Furey, K. J. James, J. Mañes, and Y. Picó, "Comparison of four mass analyzers for determining carbosulfan and its metabolites in citrus by liquid chromatography/mass spectrometry," *Rapid Communications in Mass Spectrometry*, vol. 20, no. 14, pp. 2151–2164, 2006.

[12] D. Guillarme, *UHPLC in Life Sciencies*, The Royal Society of Chemistry, Cambridge, UK, 2012.

[13] M. Ibáñez, J. V. Sancho, F. Hernández, D. McMillan, and R. Rao, "Rapid non-target screening of organic pollutants in water by ultraperformance liquid chromatography coupled to time-of-light mass spectrometry," *Trends in Analytical Chemistry*, vol. 27, no. 5, pp. 481–489, 2008.

[14] D. Guillarme, J. Schappler, S. Rudaz, and J. L. Veuthey, "Coupling ultra-high-pressure liquid chromatography with mass spectrometry," *Trends in Analytical Chemistry*, vol. 29, no. 1, pp. 15–27, 2010.

[15] L. Nováková and H. Vlčková, "A review of current trends and advances in modern bio-analytical methods: chromatography and sample preparation," *Analytica Chimica Acta*, vol. 656, no. 1-2, pp. 8–35, 2009.

[16] L. Nováková, L. Matysová, and P. Solich, "Advantages of application of UPLC in pharmaceutical analysis," *Talanta*, vol. 68, no. 3, pp. 908–918, 2006.

[17] Y. F. Cheng, Z. Lu, and U. Neue, "Ulltrafast liquid chromatography/ultraviolet and liquid chromatography/tandem mass spectrometric analysis," *Rapid Communications in Mass Spectrometry*, vol. 15, no. 2, pp. 141–151, 2001.

[18] E. Uliyanchenko, P. J. Schoenmakers, and S. van der Wal, "Fast and efficient size-based separations of polymers using ultra-high-pressure liquid chromatography," *Journal of Chromatography A*, vol. 1218, no. 11, pp. 1509–1518, 2011.

[19] M. Ibáñez, J. V. Sancho, F. Hernández, D. McMillan, and R. Rao, "Rapid non-target screening of organic pollutants in water by ultraperformance liquid chromatography coupled to time-of-light mass spectrometry," *Trends in Analytical Chemistry*, vol. 27, no. 5, pp. 481–489, 2008.

[20] D. Vuckovic, X. Zhang, E. Cudjoe, and J. Pawliszyn, "Solid-phase microextraction in bioanalysis: new devices and directions," *Journal of Chromatography A*, vol. 1217, no. 25, pp. 4041–4060, 2010.

[21] M. Lísa, F. Lynen, M. Holčapek, and P. Sandra, "Quantitation of triacylglycerols from plant oils using charged aerosol detection with gradient compensation," *Journal of Chromatography A*, vol. 1176, no. 1-2, pp. 135–142, 2007.

[22] D. Arroyo, M. C. Ortiz, L. A. Sarabia, and F. Palacios, "Determination and identification, according to European Union Decision 2002/657/EC, of malachite green and its metabolite in fish by liquid chromatography-tandem mass spectrometry using an optimized extraction procedure and three-way calibration," *Journal of Chromatography A*, vol. 1216, no. 29, pp. 5472–5482, 2009.

[23] T. Reemtsma and M. Jekel, Eds., *Organic Pollutants in the Water Cycle*, Wiley VCH, Weinheim, Germany, 2006.

[24] T. Reemtsma, S. Weiss, J. Mueller et al., "Polar pollutants entry into the water cycle by municipal wastewater: a European perspective," *Environmental Science and Technology*, vol. 40, no. 17, pp. 5451–5458, 2006.

[25] M. L. Farré, S. Pérez, L. Kantiani, and D. Barceló, "Fate and toxicity of emerging pollutants, their metabolites and transformation products in the aquatic environment," *Trends in Analytical Chemistry*, vol. 27, no. 11, pp. 991–1007, 2008.

[26] A. B. A. Boxall, C. J. Sinclair, K. Fenner, D. Kolpin, and S. J. Maund, "When synthetic chemicals degrade in the environment," *Environmental Science and Technology*, vol. 38, no. 19, pp. 368–375, 2004.

[27] S. Rodriguez-Mozaz, M. J. Lopez de Alda, and D. Barceló, "Advantages and limitations of on-line solid phase extraction coupled to liquid chromatography-mass spectrometry

technologies versus biosensors for monitoring of emerging contaminants in water," *Journal of Chromatography A*, vol. 1152, no. 1-2, pp. 97–115, 2007.

[28] T. Kumazawa, X. P. Lee, K. Sato, and O. Suzuki, "Solid-phase microextraction and liquid chromatography/mass spectrometry in drug analysis," *Analytica Chimica Acta*, vol. 492, no. 1-2, pp. 49–67, 2003.

[29] S. Pedersen-Bjergaard and K. E. Rasmussen, "Bioanalysis of drugs by liquid-phase microextraction coupled to separation techniques," *Journal of Chromatography B*, vol. 817, no. 1, pp. 3–12, 2005.

[30] C. Mahugo-Santana, Z. Sosa-Ferrera, M. E. Torres-Padrón, and J. J. Santana-Rodríguez, "Application of new approaches to liquid-phase microextraction for the determination of emerging pollutants," *Trends in Analytical Chemistry*, vol. 30, no. 5, pp. 731–748, 2011.

[31] S. D. Richardson and T. A. Ternes, "Water analysis: emerging contaminants and current issues," *Analytical Chemistry*, vol. 83, no. 12, pp. 4616–4648, 2011.

[32] D. Barceló and N. Petróvic, "Reducing the environmental risk from emerging pollutants," *Trends in Analytical Chemistry*, vol. 25, no. 3, pp. 191–193, 2006.

[33] K. Kümmerer and K. Kümmerer, *Pharmaceuticals in the Environment*, Springer Verlag, Heidelberg, Germany, 2nd edition, 2004.

[34] T. Kosjek, E. Heath, and B. Kompare, "Removal of pharmaceutical residues in a pilot wastewater treatment plant," *Analytical and Bioanalytical Chemistry*, vol. 387, no. 4, pp. 1379–1387, 2007.

[35] D. Barceló, "Pharmaceutical-residue analysis," *Trends in Analytical Chemistry*, vol. 26, no. 6, pp. 454–455, 2007.

[36] T. A. Ternes, W. Giger, and A. Joss, "Chapter1: introduction," in *Human Pharmaceuticals*, T. A. Ternes and A. Joss, Eds., pp. 1–13, Hormones and Fragrances, IWA Publishing, London, UK, 2006.

[37] T. A. Ternes, "Occurrence of drugs in German sewage treatment plants and rivers," *Water Research*, vol. 32, no. 11, pp. 3245–3260, 1998.

[38] C. D. Metcalfe, B. G. Koenig, D. T. Bennie, M. Servos, T. A. Ternes, and R. Hirsch, "Occurrence of neutral and acidic drugs in the effluents of canadian sewage treatment plants," *Environmental Toxicology and Chemistry*, vol. 22, no. 12, pp. 2872–2880, 2003.

[39] T. A. Ternes, M. Meisenheimer, D. McDowell et al., "Removal of pharmaceuticals during drinking water treatment," *Environmental Science and Technology*, vol. 36, no. 17, pp. 3855–3863, 2002.

[40] K. J. Choi, S. G. Kim, C. W. Kim, and S. H. Kim, "Determination of antibiotic compounds in water by on-line SPE-LC/MSD," *Chemosphere*, vol. 66, no. 6, pp. 977–984, 2007.

[41] L. Chen, X. Zhang, Y. Xu et al., "Determination of fluoroquinolone antibiotics in environmental water samples based on magnetic molecularly imprinted polymer extraction followed by liquid chromatography-tandem mass spectrometry," *Analytica Chimica Acta*, vol. 662, no. 1, pp. 31–38, 2010.

[42] M. J. Martínez Bueno, A. Agüera, M. J. Gómez, M. D. Hernando, J. F. García-Reyes, and A. R. Fernández-Alba, "Application of liquid chromatography/quadrupole-linear ion trap mass spectrometry and time-of-flight mass spectrometry to the determination of pharmaceuticals and related contaminants in wastewater," *Analytical Chemistry*, vol. 79, no. 24, pp. 9372–9384, 2007.

[43] M. Gros, M. Petrović, and D. Barceló, "Tracing pharmaceutical residues of different therapeutic classes in environmental waters by using liquid chromatography/quadrupole-linear ion trap mass spectrometry and automated library searching," *Analytical Chemistry*, vol. 81, no. 3, pp. 898–912, 2009.

[44] A. L. Batt, M. S. Kostich, and J. M. Lazorchak, "Analysis of ecologically relevant pharmaceuticals in wastewater and surface water using selective solid-phase extraction and UPLC-MS/MS," *Analytical Chemistry*, vol. 80, no. 13, pp. 5021–5030, 2008.

[45] K. H. Langford and K. V. Thomas, "Determination of pharmaceutical compounds in hospital effluents and their contribution to wastewater treatment works," *Environment International*, vol. 35, no. 5, pp. 766–770, 2009.

[46] M. Petrovic, M. Gros, and D. Barcelo, "Multi-residue analysis of pharmaceuticals in wastewater by ultra-performance liquid chromatography-quadrupole-time-of-flight mass spectrometry," *Journal of Chromatography A*, vol. 1124, no. 1-2, pp. 68–81, 2006.

[47] J. M. Conley, S. J. Symes, S. A. Kindelberger, and S. M. Richards, "Rapid liquid chromatography-tandem mass spectrometry method for the determination of a broad mixture of pharmaceuticals in surface water," *Journal of Chromatography A*, vol. 1185, no. 2, pp. 206–215, 2008.

[48] B. Kasprzyk-Hordern, R. M. Dinsdale, and A. J. Guwy, "Multi-residue method for the determination of basic/neutral pharmaceuticals and illicit drugs in surface water by solid-phase extraction and ultra performance liquid chromatography-positive electrospray ionisation tandem mass spectrometry," *Journal of Chromatography A*, vol. 1161, no. 1-2, pp. 132–145, 2007.

[49] E. F. Orlando, A. S. Kolok, G. A. Binzcik et al., "Endocrine-disrupting effects of cattle feedlot effluent on an aquatic sentinel species, the fathead minnow," *Environmental Health Perspectives*, vol. 112, no. 3, pp. 353–358, 2004.

[50] P. W. Sorensen, M. Pinillos, and A. P. Scott, "Sexually mature male goldfish release large quantities of androstenedione into the water where it functions as a pheromone," *General and Comparative Endocrinology*, vol. 140, no. 3, pp. 164–175, 2005.

[51] H. Chang, S. Wu, J. Hu, M. Asami, and S. Kunikane, "Trace analysis of androgens and progestogens in environmental waters by ultra-performance liquid chromatography-electrospray tandem mass spectrometry," *Journal of Chromatography A*, vol. 1195, no. 1-2, pp. 44–51, 2008.

[52] M. Huerta-Fontela, M. T. Galceran, and F. Ventura, "Fast liquid chromatography-quadrupole-linear ion trap mass spectrometry for the analysis of pharmaceuticals and hormones in water resources," *Journal of Chromatography A*, vol. 1217, no. 25, pp. 4212–4222, 2010.

[53] E. Pérez-Carrera, M. Hansen, V. M. León et al., "Multiresidue method for the determination of 32 human and veterinary pharmaceuticals in soil and sediment by pressurized-liquid extraction and LC-MS/MS," *Analytical and Bioanalytical Chemistry*, vol. 398, no. 3, pp. 1173–1184, 2010.

[54] J. Antonić and E. Heath, "Determination of NSAIDs in river sediment samples," *Analytical and Bioanalytical Chemistry*, vol. 387, no. 4, pp. 1337–1342, 2007.

[55] A. Göbel, A. Thomsen, C. S. McArdell et al., "Extraction and determination of sulfonamides, macrolides, and trimethoprim in sewage sludge," *Journal of Chromatography A*, vol. 1085, no. 2, pp. 179–189, 2005.

[56] A. M. Jacobsen and B. Halling-Sørensen, "Multi-component analysis of tetracyclines, sulfonamides and tylosin in swine

manure by liquid chromatography-tandem mass spectrometry," *Analytical and Bioanalytical Chemistry*, vol. 384, no. 5, pp. 1164–1174, 2006.

[57] M. Y. Haller, S. R. Müller, C. S. McArdell, A. C. Alder, and M. J. F. Suter, "Quantification of veterinary antibiotics (sulfonamides and trimethoprim) in animal manure by liquid chromatography-mass spectrometry," *Journal of Chromatography A*, vol. 952, no. 1-2, pp. 111–120, 2002.

[58] E. Sagristà, E. Larsson, M. Ezoddin, M. Hidalgo, V. Salvadó, and J. Å. Jönsson, "Determination of non-steroidal anti-inflammatory drugs in sewage sludge by direct hollow fiber supported liquid membrane extraction and liquid chromatography-mass spectrometry," *Journal of Chromatography A*, vol. 1217, no. 40, pp. 6153–6158, 2010.

[59] G. Hamscher, S. Sczesny, H. Höper, and H. Nau, "Determination of persistent tetracycline residues in soil fertilized with liquid manure by high-performance liquid chromatography with electrospray ionization tandem mass spectrometry," *Analytical Chemistry*, vol. 74, no. 7, pp. 1509–1518, 2002.

[60] S. Montesdeoca-Esponda, Z. Sosa-Ferrera, and J. J. Santana-Rodríguez, "Combination of microwave-assisted micellar extraction with liquid chromatography tandem mass spectrometry for the determination of fluoroquinolone antibiotics in coastal marine sediments and sewage sludges samples," *Biomedical Chromatography*, vol. 26, no. 1, pp. 33–40, 2012.

[61] A. M. Jacobsen, B. Halling-Sørensen, F. Ingerslev, and S. H. Hansen, "Simultaneous extraction of tetracycline, macrolide and sulfonamide antibiotics from agricultural soils using pressurised liquid extraction, followed by solid-phase extraction and liquid chromatography-tandem mass spectrometry," *Journal of Chromatography A*, vol. 1038, no. 1-2, pp. 157–170, 2004.

[62] D. Löffler and T. A. Ternes, "Determination of acidic pharmaceuticals, antibiotics and ivermectin in river sediment using liquid chromatography-tandem mass spectrometry," *Journal of Chromatography A*, vol. 1021, no. 1-2, pp. 133–144, 2003.

[63] T. Vega-Morales, Z. Sosa-Ferrera, and J. J. Santana-Rodríguez, "Determination of various estradiol mimicking-compounds in sewage sludge by the combination of microwave-assisted extraction and LC-MS/MS," *Talanta*, vol. 85, no. 4, pp. 1825–1834, 2011.

[64] S. Liu, G. G. Ying, J. L. Zhao et al., "Trace analysis of 28 steroids in surface water, wastewater and sludge samples by rapid resolution liquid chromatography-electrospray ionization tandem mass spectrometry," *Journal of Chromatography A*, vol. 1218, no. 10, pp. 1367–1378, 2011.

[65] N. Gineys, B. Giroud, and E. Vulliet, "Analytical method for the determination of trace levels of steroid hormones and corticosteroids in soil, based on PLE/SPE/LC-MS/MS," *Analytical and Bioanalytical Chemistry*, vol. 397, no. 6, pp. 2295–2302, 2010.

[66] A. Nieto, F. Borrull, E. Pocurull, and R. M. Marcé, "Determination of natural and synthetic estrogens and their conjugates in sewage sludge by pressurized liquid extraction and liquid chromatography-tandem mass spectrometry," *Journal of Chromatography A*, vol. 1213, no. 2, pp. 224–230, 2008.

[67] D. Matejıcek, P. Houserová, and V. Kubáň, "Combined isolation and purification procedures prior to the high-performance liquid chromatographic-ion-trap tandem mass spectrometric determination of estrogens and their conjugates in river sediments," *Journal of Chromatography A*, vol. 1171, no. 1-2, pp. 80–89, 2007.

[68] C. G. Daughton, "Non-regulated water contaminants: emerging research," *Environmental Impact Assessment Review*, vol. 24, no. 7-8, pp. 711–732, 2004.

[69] J. B. Ellis, "Pharmaceutical and personal care products (PPCPs) in urban receiving waters," *Environmental Pollution*, vol. 144, no. 1, pp. 184–189, 2006.

[70] T. A. Ternes, A. Joss, and H. Siegrist, "Scrutinizing pharmaceuticals and personal care products in wastewater treatment," *Environmental Science and Technology*, vol. 38, no. 20, 2004.

[71] Y. Yoon, P. Westerhoff, S. A. Snyder, and E. C. Wert, "Nanofiltration and ultrafiltration of endocrine disrupting compounds, pharmaceuticals and personal care products," *Journal of Membrane Science*, vol. 270, no. 1-2, pp. 88–100, 2006.

[72] M. Carballa, F. Omil, J. M. Lema et al., "Behavior of pharmaceuticals, cosmetics and hormones in a sewage treatment plant," *Water Research*, vol. 38, no. 12, pp. 2918–2926, 2004.

[73] P. Westerhoff, Y. Yoon, S. Snyder, and E. Wert, "Fate of endocrine-disruptor, pharmaceutical, and personal care product chemicals during simulated drinking water treatment processes," *Environmental Science and Technology*, vol. 39, no. 17, pp. 6649–6663, 2005.

[74] S. A. Snyder, S. Adham, A. M. Redding et al., "Role of membranes and activated carbon in the removal of endocrine disruptors and pharmaceuticals," *Desalination*, vol. 202, no. 1–3, pp. 156–181, 2007.

[75] M. Grung, R. Lichtenthaler, M. Ahel, K. E. Tollefsen, K. Langford, and K. V. Thomas, "Effects-directed analysis of organic toxicants in wastewater effluent from Zagreb, Croatia," *Chemosphere*, vol. 67, no. 1, pp. 108–120, 2007.

[76] D. L. Giokas, A. Salvador, and A. Chisvert, "UV filters: from sunscreens to human body and the environment," *Trends in Analytical Chemistry*, vol. 26, no. 5, pp. 360–374, 2007.

[77] M. Schlumpf, P. Schmid, S. Durrer et al., "Endocrine activity and developmental toxicity of cosmetic UV filters - An update," *Toxicology*, vol. 205, no. 1-2, pp. 113–122, 2004.

[78] M. Schlumpf, S. Durrer, O. Faass et al., "Developmental toxicity of UV filters and environmental exposure: a review," *International Journal of Andrology*, vol. 31, no. 2, pp. 144–150, 2008.

[79] P. Y. Kunz, H. F. Galicia, and K. Fent, "Comparison of in vitro and in vivo estrogenic activity of UV filters in fish," *Toxicological Sciences*, vol. 90, no. 2, pp. 349–361, 2006.

[80] S. Chu and C. D. Metcalfe, "Simultaneous determination of triclocarban and triclosan in municipal biosolids by liquid chromatography tandem mass spectrometry," *Journal of Chromatography A*, vol. 1164, no. 1-2, pp. 212–218, 2007.

[81] E. Engelhaupt, "Happy birthday, love canal," *Environmental Science and Technology*, vol. 42, no. 22, pp. 8179–8186, 2008.

[82] M. A. Coogan, R. E. Edziyie, T. W. La Point, and B. J. Venables, "Algal bioaccumulation of triclocarban, triclosan, and methyl-triclosan in a North Texas wastewater treatment plant receiving stream," *Chemosphere*, vol. 67, no. 10, pp. 1911–1918, 2007.

[83] N. Negreira, I. Rodríguez, M. Ramil, E. Rubí, and R. Cela, "Solid-phase extraction followed by liquid chromatography-tandem mass spectrometry for the determination of hydroxylated benzophenone UV absorbers in environmental water samples," *Analytica Chimica Acta*, vol. 654, no. 2, pp. 162–170, 2009.

[84] R. S. Zhao, X. Wang, J. Sun, S. S. Wang, J. P. Yuan, and X. K. Wang, "Trace determination of triclosan and triclocarban in environmental water samples with ionic liquid dispersive

liquid-phase microextraction prior to HPLC-ESI-MS-MS," *Analytical and Bioanalytical Chemistry*, vol. 397, no. 4, pp. 1627–1633, 2010.

[85] D. R. Klein, D. F. Flannelly, and M. M. Schultz, "Quantitative determination of triclocarban in wastewater effluent by stir bar sorptive extraction and liquid desorption-liquid chromatography-tandem mass spectrometry.," *Journal of Chromatography A*, vol. 1217, no. 11, pp. 1742–1747, 2010.

[86] M. Pedrouzo, F. Borrull, R. M. Marcé, and E. Pocurull, "Ultra-high-performance liquid chromatography-tandem mass spectrometry for determining the presence of eleven personal care products in surface and wastewaters," *Journal of Chromatography A*, vol. 1216, no. 42, pp. 6994–7000, 2009.

[87] A. M. Peck, "Analytical methods for the determination of persistent ingredients of personal care products in environmental matrices," *Analytical and Bioanalytical Chemistry*, vol. 386, no. 4, pp. 907–939, 2006.

[88] M. Borremans, J. Van Loco, P. Roos, and L. Goeyens, "Validation of HPLC Analysis of 2-Phenoxyethanol, 1-Phenoxypropan-2-ol, Methyl, Ethyl, Propyl, Butyl and Benzyl 4-Hydroxybenzoate (Parabens) in Cosmetic Products, with Emphasis on Decision Limit and Detection Capability," *Chromatographia*, vol. 59, no. 1-2, pp. 47–53, 2004.

[89] B. Kasprzyk-Hordern, R. M. Dinsdale, and A. J. Guwy, "The occurrence of pharmaceuticals, personal care products, endocrine disruptors and illicit drugs in surface water in South Wales, UK," *Water Research*, vol. 42, no. 13, pp. 3498–3518, 2008.

[90] R. Rodil, J. B. Quintana, P. López-Mahía, S. Muniategui-Lorenzo, and D. Prada-Rodríguez, "Multi-residue analytical method for the determination of emerging pollutants in water by solid-phase extraction and liquid chromatography-tandem mass spectrometry," *Journal of Chromatography A*, vol. 1216, no. 14, pp. 2958–2969, 2009.

[91] R. Rosal, A. Rodríguez, J. A. Perdigón-Melón et al., "Occurrence of emerging pollutants in urban wastewater and their removal through biological treatment followed by ozonation," *Water Research*, vol. 44, no. 2, pp. 578–588, 2010.

[92] Z. Zhang, N. Ren, Y. F. Li, T. Kunisue, D. Gao, and K. Kannan, "Determination of benzotriazole and benzophenone UV filters in sediment and sewage sludge," *Environmental Science and Technology*, vol. 45, no. 9, pp. 3909–3916, 2011.

[93] A. Wick, G. Fink, and T. A. Ternes, "Comparison of electrospray ionization and atmospheric pressure chemical ionization for multi-residue analysis of biocides, UV-filters and benzothiazoles in aqueous matrices and activated sludge by liquid chromatography-tandem mass spectrometry," *Journal of Chromatography A*, vol. 1217, no. 14, pp. 2088–2103, 2010.

[94] A. Nieto, F. Borrull, R. M. Marcé, and E. Pocurull, "Determination of personal care products in sewage sludge by pressurized liquid extraction and ultra high performance liquid chromatography-tandem mass spectrometry," *Journal of Chromatography A*, vol. 1216, no. 30, pp. 5619–5625, 2009.

[95] C. A. de Wit, M. Alaee, and D. C. G. Muir, "Levels and trends of brominated flame retardants in the Arctic," *Chemosphere*, vol. 64, no. 2, pp. 209–233, 2006.

[96] WHO Environmental Health Criteria 192, *Flame Retardants, General Introduction*, World Health Organization, Geneva, Switzerland, 1997.

[97] WHO, *Environmental Health Criteria 172*, WHO/IPCS, Geneva, Switzerland, 1995.

[98] A. Covaci, S. Voorspoels, M. A. E. Abdallah, T. Geens, S. Harrad, and R. J. Law, "Analytical and environmental aspects of the flame retardant tetrabromobisphenol-A and its derivatives," *Journal of Chromatography A*, vol. 1216, no. 3, pp. 346–363, 2009.

[99] A. Covaci, S. Voorspoels, and J. de Boer, "Determination of brominated flame retardants, with emphasis on polybrominated diphenyl ethers (PBDEs) in environmental and human samples - A review," *Environment International*, vol. 29, no. 6, pp. 735–756, 2003.

[100] R. Cariou, J. P. Antignac, L. Debrauwer et al., "Comparison of analytical strategies for the chromatographic and mass spectrometric measurement of brominated flame retardants: 1. Polybrominated diphenylethers," *Journal of Chromatographic Science*, vol. 44, no. 8, pp. 489–497, 2006.

[101] A. Riu, D. Zalko, and L. Debrauwer, "Study of polybrominated diphenyl ethers using both positive and negative atmospheric pressure photoionization and tandem mass spectrometry," *Rapid Communications in Mass Spectrometry*, vol. 20, no. 14, pp. 2133–2142, 2006.

[102] T. Hayama, H. Yoshida, S. Onimaru et al., "Determination of tetrabromobisphenol a in human serum by liquid chromatography-electrospray ionization tandem mass spectrometry," *Journal of Chromatography B*, vol. 809, no. 1, pp. 131–136, 2004.

[103] J. Tollbäck, C. Crescenzi, and E. Dyremark, "Determination of the flame retardant tetrabromobisphenol A in air samples by liquid chromatography-mass spectrometry," *Journal of Chromatography A*, vol. 1104, no. 1-2, pp. 106–112, 2006.

[104] M. Frederiksen, K. Vorkamp, R. Bossi, F. Rigét, M. Dam, and B. Svensmark, "Method development for simultaneous analysis of HBCD, TBBPA, and dimethyl-TBBPA in marine biota from Greenland and the Faroe Islands," *International Journal of Environmental Analytical Chemistry*, vol. 87, no. 15, pp. 1095–1109, 2007.

[105] S. N. Zhou, E. J. Reiner, C. Marvin et al., "Development of liquid chromatography atmospheric pressure chemical ionization tandem mass spectrometry for analysis of halogenated flame retardants in wastewater," *Analytical and Bioanalytical Chemistry*, vol. 396, no. 3, pp. 1311–1320, 2010.

[106] A. Bacaloni, L. Callipo, E. Corradini et al., "Liquid chromatography-negative ion atmospheric pressure photoionization tandem mass spectrometry for the determination of brominated flame retardants in environmental water and industrial effluents," *Journal of Chromatography A*, vol. 1216, no. 36, pp. 6400–6409, 2009.

[107] S. Chu, G. D. Haffner, and R. J. Letcher, "Simultaneous determination of tetrabromobisphenol A, tetrachlorobisphenol A, bisphenol A and other halogenated analogues in sediment and sludge by high performance liquid chromatography-electrospray tandem mass spectrometry," *Journal of Chromatography A*, vol. 1097, no. 1-2, pp. 25–32, 2005.

[108] R. Saint-Louis and E. Pelletier, "LC-ESI-MS-MS method for the analysis of tetrabromobisphenol A in sediment and sewage sludge," *Analyst*, vol. 129, no. 8, pp. 724–730, 2004.

[109] C. H. Marvin, G. T. Tomy, M. Alaee, and G. MacInnis, "Distribution of hexabromocyclododecane in Detroit River suspended sediments," *Chemosphere*, vol. 64, no. 2, pp. 268–275, 2006.

[110] H. H. Wu, H. C. Chen, and W. H. Ding, "Combining microwave-assisted extraction and liquid chromatography-ion-trap mass spectrometry for the analysis of hexabromocyclododecane diastereoisomers in marine sediments," *Journal of Chromatography A*, vol. 1216, no. 45, pp. 7755–7760, 2009.

[111] P. Guerra, E. Eljarrat, and D. Barceló, "Simultaneous determination of hexabromocyclododecane, tetrabromobisphenol A, and related compounds in sewage sludge and sediment
samples from Ebro River basin (Spain)," *Analytical and
Bioanalytical Chemistry*, vol. 397, no. 7, pp. 2817–2824, 2010.

[112] G. Mascolo, V. Locaputo, and G. Mininni, "New perspective
on the determination of flame retardants in sewage sludge
by using ultrahigh pressure liquid chromatography-tandem
mass spectrometry with different ion sources," *Journal of
Chromatography A*, vol. 1217, no. 27, pp. 4601–4611, 2010.

[113] "Commission decision 2002/657/EC," Official Journal of the
European Communities L221/8.

[114] R. Payan, M. A. B. López, R. Fernández-Torres, M. C.
Mochón, and J. L. G. Ariza, "Application of hollow fiber-
based liquid-phase microextraction (HF-LPME) for the
determination of acidic pharmaceuticals in wastewaters,"
Talanta, vol. 82, no. 2, pp. 854–858, 2010.

[115] J. B. Quintana, R. Rodil, and T. Reemtsma, "Suitability of
hollow fibre liquid-phase microextraction for the determination of acidic pharmaceuticals in wastewater by liquid
chromatography-electrospray tandem mass spectrometry
without matrix effects," *Journal of Chromatography A*, vol.
1061, no. 1, pp. 19–26, 2004.

[116] K. Mitani, M. Fujioka, and H. Kataoka, "Fully automated
analysis of estrogens in environmental waters by in-tube
solid-phase microextraction coupled with liquid chromatography-tandem mass spectrometry," *Journal of Chromatography A*, vol. 1081, no. 2, pp. 218–224, 2005.

[117] H.-X. Wang, Y. Zhou, and Q.-W. Jiang, "Simultaneous
screening of estrogens, progestogens, and phenols and their
metabolites in potable water and river water by ultra-performance liquid chromatography coupled with quadrupole
time-of-flight mass spectrometry," *Microchemical Journal*,
vol. 100, no. 1, pp. 83–94, 2012.

[118] S. Montesdeoca-Esponda, Z. Sosa-Ferrera, and J. J. Santana-
Rodríguez, "On-line solid-phase extraction coupled to ultra-
performance liquid chromatography with tandem mass spectrometry detection for the determination of benzotriazole
UV stabilizers in coastal marine and wastewater samples,"
Analytical and Bioanalytical Chemistry, vol. 403, no. 3, pp.
867–876, 2012.

27

Development and Validation of a Stability Indicating LC Method for the Assay and Related Substances Determination of a Proteasome Inhibitor Bortezomib

Kasa Srinivasulu,[1,2] Mopidevi Narasimha Naidu,[1] Kadaboina Rajasekhar,[1] Murki Veerender,[1] and Mulukutla Venkata Suryanarayana[2,3]

[1] Active Pharmaceutical Ingredients, Dr. Reddy's Laboratories Ltd., IPDO, Bachupally, Hyderabad 500072, Andhra Pradesh, India
[2] Department of Chemistry, Osmania University, Hyderabad 500072, Andhra Pradesh, India
[3] Matrix Laboratories Limited, Plot 34 A, Anrich Industrial Estate, Bollaram, Medak District, Jinnaram, Mandal 502 32, Andhra Pradesh, India

Correspondence should be addressed to Kasa Srinivasulu, kasas82003@yahoo.co.in

Academic Editor: Esther Turiel

A novel, simple, sensitive, stability indicating HPLC method was developed and validated for quantification of impurities (process related and degradants) and assay determination of bortezomib. Stability indicating power of the method was established by forced degradation experiments and mass balance study. The chromatographic separation was achieved with Waters SymmetryShield RP18 column using gradient elution using the mobile phase-A consists of a mixture of water-acetonitrile-formic acid (715 : 285 : 1, v/v/v) and the mobile phase-B consists a mixture of methanol-water-formic acid (800 : 200 : 1, v/v/v), respectively. The developed method is validated for parameters like precision, accuracy, linearity, LOD, LOQ, and ruggedness. Central composite experimental design (CCD) was applied to check the robustness of the method. The stability tests were also performed on drug substances as per ICH norms.

1. Introduction

Bortezomib [(1R)-3-methyl-1-[[(2S)-1-oxo-3-phenyl-2-[(pyrazinylcarbonyl)amino]propyl]amino]butyl]boronic acid] is a potent first-in-class dipeptidyl boronic acid proteasome inhibitor [1–6] that was approved in May 2003 in the United States for the treatment of patients with relapsed multiple myeloma where the disease is refractory to conventional lines of therapy. Bortezomib, formerly known as PS-341 [1], bortezomib binds the proteasome via the boronic acid moiety, and therefore, the presence of this moiety is necessary to achieve proteasome inhibition. The proteasome is an interesting new target for cancer therapy, and the proteasome inhibitor PS-341 warrants continued investigation in cancer therapy. One of the potential therapeutic applications of bortezomib is an anticancer agent.

No LC methods were reported in major pharmacopeias like USP, EP, JP, and BP. A few publications are available for bortezomib, some of are available on [7, 8] stability and characterization of bortezomib and metabolites observed in human plasma with the help of MDS sciex API 3000 triple quadruple LC MS using turbo ion spray interface set at 325C and [9] enhanced delivery of cisplatin to international ovarian carcinomas mediated by the effects of bortezomib on human copper transporter and [10] one of it is in human plasma using LC MS. Extensive literature survey reveals there is no stability-indicating LC method for determination of related substances and for quantitative estimation of bortezomib in bulk drugs. An exhaustive study on the stability of bortezomib is demanding as the current International Conference on Harmonisation (ICH) guidelines require that stability analysis should be done by using stability-indicating methods, developed, and validated after stress testing on the drug under a variety of conditions, including hydrolysis (at various pH), oxidation, photolysis, and thermal degradation [11–15]. Moreover, the structural characterization and synthesis of the degradation products allow both to establish the degradation pathways and also their quantitative determination in drug substance. Hence, in the present work, the chemical degradation pathways of bortezomib were established through a forced degradation study and a selective, precise, and accurate LC method for simultaneous estimation of bortezomib and its degradation products was also developed. The validation of the proposed method was also carried out and its applicability was evaluated in commercial form analysis.

2. Experimental

2.1. Chemicals and Reagents.
Methanol HPLC grade and acetonitrile HPLC grade were purchased from Rankem. Formic acid, sodium hydroxide, hydrochloric acid, and hydrogen peroxide was purchased from Merck. HPLC grade water was obtained from Milli-Q water purification system (Millipore, Milford, USA).

Bortezomib drug substance, reference standard, and nine impurities were obtained from Process Research Department of Dr.Reddy's Laboratories, Hyderabad.

2.2. Chemical Names for Bortezomib and Its Impurities

(a) Bortezomib. [(1R)-3-methyl-1-[[(2S)-1-oxo-3-phenyl-2-[(pyrazinylcarbonyl)amino]propyl]amino]butyl]boronic acid.

(b) Imp-A. Pyrazine-2-carboxylic acid (1-carbamoyl-2-phenyl-ethyl)-amide.

(c) Imp-B. (S)-3-Phenyl-2-[(pyrazine-2-carbonyl)-amino]-propionic acid.

(d) Imp-C. 1-[[(2S)-1-oxo-3-phenyl-2-[(pyrazinylcarbonyl)amino]propyl]amino] pentylboronic acid.

(e) Imp-D. [(1R)-3-methyl-1-({(2R)-3-phenyl-2-[(pyrazin-2-ylcarbonyl)amino]propanoyl}amino)butyl]boronic acid. (or) [(1S)-3-methyl-1-({(2S)-3-phenyl-2-[(pyrazin-2-ylcarbonyl)amino]propanoyl}amino)butyl]boronic acid.

(f) Imp-E. Pyrazine-2-carboxylic acid [1-(1-hydroxy-3-methyl-butylcarbamoyl)-2-phenyl-ethyl]-amide.

(g) Imp-F. (S)-3-Phenyl-2-[(pyrazine-2-carbonyl)-amino]-propionic acid methylester.

(h) Imp-G. Pyrazine-2-carboxylic acid [1-(1-hydroxy-3-methyl-butylcarbamoyl)-2-phenyl-ethyl]-amide.

(i) Imp-H. Pyrazine-2-carboxylic acid [1-(3-methyl-butyl-carbamoyl)-2-phenyl-ethyl]-amide.

(j) Imp-I. Pyrazine-2-carboxylic acid {1-[3-methyl-1-(2,9,9-trimethyl-3,5-dioxa-4-bora-tricyclo [6.1.1.02,6]dec-4-yl)-butylcarbamoyl]-2-phenyl-ethyl}-amide.

The structures of Imp-A, Imp-B, Imp-C, Imp-D, Imp-E, Imp-F, Imp-G, Imp-H, Imp-I, and bortezomib were shown in (Figure 1).

2.3. Instrumentation and Software.
Two LC system were used for method development and validation.

> LC 1: Waters make (2695 separation module and a PDA detector 996) with empower software.

> LC 2: Waters make (2695 separation module and a 2487 Dual λ Absorbance detector) with empower software.

2.4. Chromatographic Conditions.
The method was developed using Waters SymmetryShield RP18 5 μm, 4.6 × 250 mm column (Waters, Milford, USA) with mobile phase containing a gradient mixture of mobile phase A and B. Mobile phase-A consists a mixture of water-acetonitrile-formic acid (715 : 285 : 1, v/v/v) and the mobile phase-B consists a mixture of methanol-water-formic acid (800 : 200 : 1, v/v/v), respectively. Exact composition of mobile phase-A preparation is required to achieve resolution between bortezomib and Imp-C ($Rs > 3.0$). The gradient program (T/%B) was set as 0/0, 20.0/0, 35.0/100, 50.0/100, 52.0/0, and 60.0/0. The mobile phase was filtered through a nylon 0.45 μm membrane filter. The flow rate of the mobile phase was 1.0 mL/min. The column temperature was maintained at 35°C and the wavelength was monitored at 270 nm. The injection volume was 10 μL. A mixture of acetonitrile-water (7.5 : 17.5, v/v) was used as diluent.

2.5. LC-MS/MS Conditions.
LC-MS/MS system (Waters 2695 Alliance liquid chromatograph coupled with quattro-micro-mass spectrometer with Mass Lynx software, Waters Corporation, Milford, USA) was used for the unknown compounds formed during forced degradation studies. Waters SymmetryShield RP18 5 μm, 4.6 × 250 mm column (Waters, Milford, USA) was used as stationary phase. Mobile phase-A consists a mixture of water-acetonitrile-formic acid (715 : 285 : 1, v/v/v) and the mobile phase-B consists a mixture of methanol-water-formic acid (800 : 200 : 1, v/v/v) respectively. The gradient program (T/%B) was set as 0/0, 20.0/0, 35.0/100, 50.0/100, 52.0/0 and 60.0/0. The flow rate was 1.0 mL/min. The analysis was performed in positive electro spray positive ionization mode.

The analysis was performed in positive electro spray positive ionization mode. Capillary and cone voltages were 3.5 kV and 25 V, respectively. Source and dissolution temperatures were 120 and 350°C, respectively. Dissolution gas flow was 650 L h^{-1}.

2.6. Preparation of Standard Solutions.
A stock solution of bortezomib standard as well as sample was prepared at 1000 μg/mL for analysis of related substances and for assay determination. A stock solution of impurities (mixture of

Development and Validation of a Stability Indicating LC Method for the Assay and Related Substances Determination of a Proteasome Inhibitor Bortezomib

199

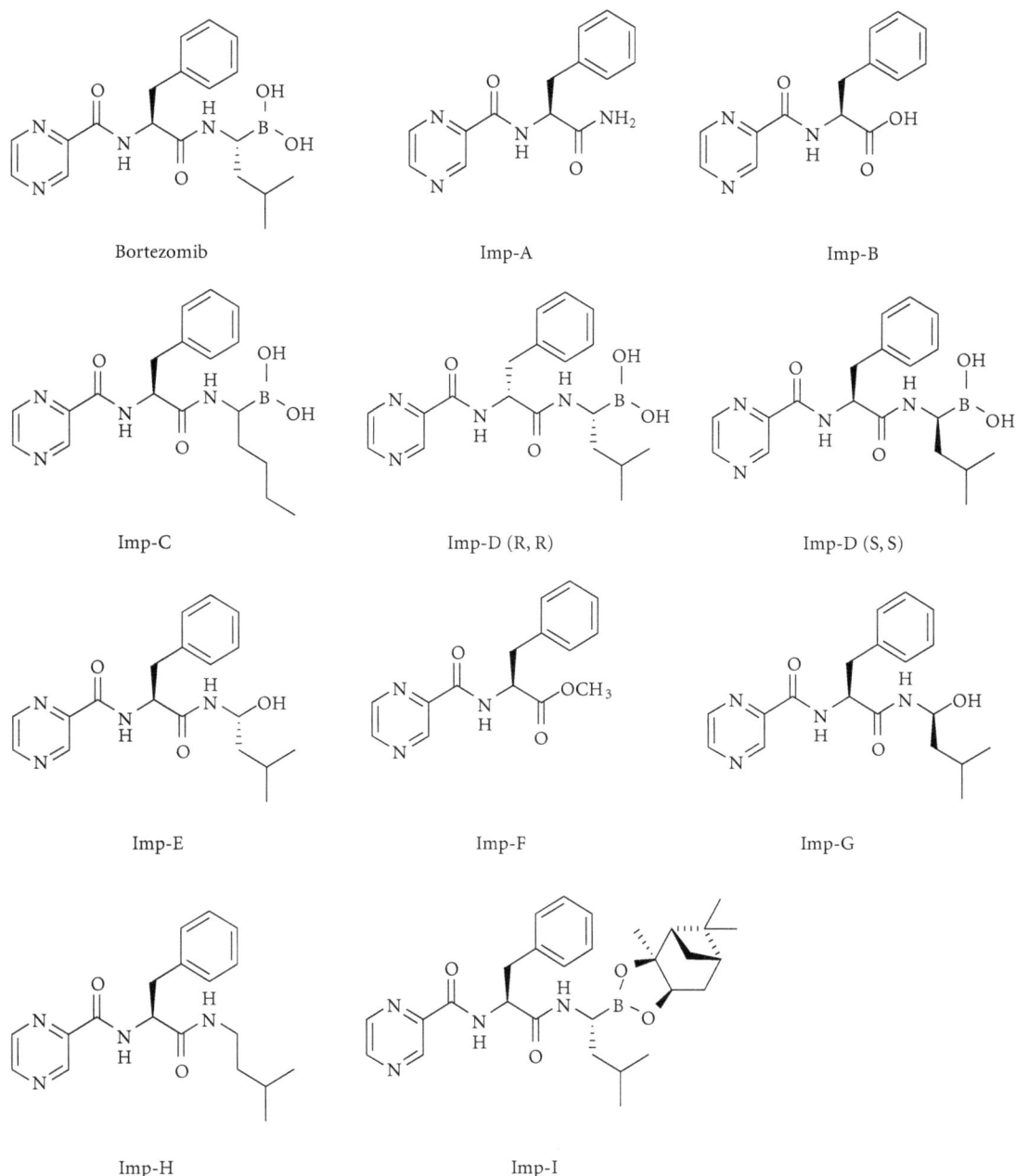

FIGURE 1: Structures of bortezomib and its nine impurities.

Imp-A, Imp-B, Imp-C, Imp-D, Imp-E, Imp-F, Imp-G, Imp-H, and Imp-I) at 100 μg/mL was also prepared in the solvent mixture.

2.7. Method Validation. The proposed method was validated as per ICH guidelines [16].

2.7.1. Solution Stability and Mobile Phase Stability. The stability of bortezomib in solution was determined by leaving test solutions of the sample and reference standard in tightly capped volumetric flasks at room temperature for 48 h

during which they were assayed at 12 h intervals. Stability of mobile phase was determined by analysis of freshly prepared sample solutions at 12 h intervals for 48 h and comparing the results with those obtained from freshly prepared reference standard solutions. The mobile phase was prepared at the beginning of the study period and not changed during the experiment. The % assay of the results was calculated for both the mobile phase and solution-stability experiments.

The stability of bortezomib and its impurities in solution for the related substance method was determined by leaving spiked sample solution in a tightly capped volumetric flask at room temperature for 48 h and measuring the amounts of the nine impurities at every 12 h. The stability of mobile phase was also determined by analysis freshly prepared solution of bortezomib and its impurities at 12 h intervals for 48 h. The mobile phase was not changed during the study period.

2.7.2. Specificity and Mass Balance Study.
Specificity is the ability of the method to measure the analyte response in the presence of its potential impurities. The specificity of the developed LC method for bortezomib was carried out in the presence of its nine impurities. Stress studies were performed at an initial concentration 1000 μg/mL of bortezomib to provide an indication of the stability-indicating property and specificity of the proposed method.

Intentional degradation was attempted to stress condition of Photodegradation as per ICH Q2B (1.2 million lux hours and 200 watt hours/square meter) for 7 days, for thermal degradation, bortezomib was placed in a hot air oven maintained at 105°C for for 7 h. The degradation in acidic condition was done in 5.0 N HCl and the solution was left in dark at 70°C for 72 h. The degradation in basic condition was done in 0.1 N NaOH and the solution was left in dark at 25°C for 7 h. To test the stability in neutral solution bortezomib was dissolved in water with cosolvent and left at 70°C for 72 h.

For oxidative conditions, the degradation was done in 3.0% hydrogen peroxide solution and analyzed immediately. Before LC and LC-MS/MS analyses, acidic and alkaline samples were neutralized and diluted by adding an appropriate volume of diluent. For comparison, an aqueous solution of bortezomib (1000 μg/mL) was prepared, diluted, and analyzed as above.

Peak purity test was carried out for the bortezomib peak by using PDA detector in all stressed samples.

Assay of stressed samples was performed (at 1000 μg/mL) by comparison with qualified reference standard and the mass balance (% assay + % impurities + % degradation products) was calculated. Assay was also calculated for bortezomib sample by spiking all nine impurities at the specification level (i.e., 0.15%).

2.7.3. Linearity.
Linearity test solutions for the assay method were prepared from bortezomib solutions at five concentration levels from 50 to 150% of assay analyte concentration (0.5, 0.75, 1.0, 1.25, and 1.50 mg/mL). The peak area versus concentration data was treated by least-squares linear regression analysis. Linearity test solutions for the related substance method were prepared by diluting stock solutions to the required concentrations. The solutions were prepared at six concentration levels from LOQ to 150% of the specification level (LOQ, 0.0375, 0.075, 0.1125, 0.15, 0.1875, and 0.225%).

2.7.4. Limits of Detection (LOD) and Quantification (LOQ).
The LOD and LOQ for all nine impurities and bortezomib were estimated at a signal-to-noise ratio of 3:1 and 10:1, respectively, [17] and [18], by injecting a series of diluted solutions with known concentration. Precision study was also carried at the LOQ level by injecting six individual preparations of all nine impurities and bortezomib and calculating the % RSD of the area. Accuracy at LOQ level was evaluated in triplicate for the nine impurities by spiking the impurities at the estimated LOQ level to test solution.

2.7.5. Accuracy.
The accuracy of the assay method was evaluated in triplicate using three-concentration levels, that is, 0.5, 1.0, and 1.5 mg/mL in bulk drug sample. The % recoveries were calculated from 1.0 mg/mL of reference standard preparation.

Bulk samples received from Process Research Department of Dr. Reddy's Laboratories show the presence of Imp-A, Imp-B, Imp-C, Imp-D, Imp-E, Imp-F, Imp-G, Imp-H, and Imp-I in between 0.02% and 0.03% levels. Standard addition and recovery experiments were conducted to determine accuracy of the related substance method for the quantification of all nine impurities in bulk drug samples. The study was carried out in triplicate at 0.075%, 0.15%, and 0.225% w/w of the related substances test concentration. The percentages of recoveries for bortezomib and its impurities were calculated.

2.7.6. Precision.
The precision of the method verified by repeatability and by intermediate precision. Repeatability was checked by (waters make 2695 separation module and a PDA detector 996) injecting six individual preparations of bortezomib sample spiked with 0.15% of its nine impurities (0.15% of impurities with respect to 1.0 mg/mL Bortezomib). % RSD of content (%) for each impurity was calculated. The intermediate precision of the method was also evaluated using different analyst and different instrument (waters make 2695 separation module and a 2487 dual λ absorbance detector), and performing the analysis on different days.

Assay method precision was evaluated by carrying out six independent assays of sample of bortezomib at 1.0 mg/mL level against qualified reference standard. The intermediate precision of the assay method was evaluated by different analysts.

2.7.7. Robustness.
As defined by the ICH, the robustness of an analytical procedure describes to its capability to remain unaffected by small and deliberate variations in method parameters [19]. In order to study the simultaneous variation of the factors on the considered responses, a multivariate approach using design of experiments is recommended in robustness testing. A response surface method was performed to obtain more information and to investigate the behavior of the response around the nominal values of the factors. Response surface methodology (RSM) has the following advantages: (a) to allow a complete study where all interaction effects are estimated and (b) to give an approximate description of an experimental region around a center of interest with validity of interpolation [20–22]. Generally, the large numbers of experiments required by standard designs applied in RSM disenchant their use in the validation procedure. However, if an analytical method is fast and requires the testing of a few factors (three or less), a good

Development and Validation of a Stability Indicating LC Method for the Assay and Related Substances Determination of a Proteasome Inhibitor Bortezomib

201

TABLE 1: Factors and level studied for robustness testing.

Factors	Level		
	−1	0	+1
(A) Flow rate (mL/min)	0.8	1.0	1.2
(B) Acetonitrile (%)	28.0	28.5	29.0
(C) Methanol (%)	79.0	80.0	81.0

choice for robustness testing may be the central composite design (CCD) [23], widely employed because of its high efficiency with respect to the number of runs required. In order to study the variables at no more than three levels (−1, 0, +1), the design used in robustness testing of bortezomib was a central composite design (CCD) with D = ±1 [24]. Three factors were considered: flow rate mL min^{-1} (A); acetonitrile % (B) and methanol % (C). The factors and level considered for the study are shown in Table 1.

A precision solution was prepared by spiking the impurities Imp-A, Imp-B, Imp-C, Imp-D, Imp-E, Imp-F, Imp-G, Imp-H, and Imp-I at 0.15% w/w with respect to bortezomib-related substances analysis concentration.

A standard and sample of bortezomib were prepared in assay concentration.

The critical resolution between bortezomib and Imp-C, Imp-E and Imp-F, and % Recovery Assay of bortezomib were studied as response surface.

3. Results and Discussion

3.1. Method Development and Optimization. The main target of the chromatographic method is to get the separation of impurities and degradants generated, from bortezomib. It was also aimed method should be capable of resolving all impurities from each other.

Because alkyl borane compound present in bortezomib was known to be susceptible to oxidation by peroxides, it was speculated that the degradation of Bortetzomib in the presence of peroxide to form major Imp-E and a further degradation to Imp-G and Imp-A [25]. The major degradents Imp-A and Imp-B were observed during the process of basic and acidic catalyzed degradation [25].

These impurities were isolated. The possible degradents and related impurities of bortezomib are very similar to respective drug substance (Figure 1). To obtain a good resolution among the impurities and main drug substances, different stationary phases were tested considering

(a) the feature of stationary phase (RP-C$_8$ and RP-C$_{18}$),

(b) the particle size of the column (3 μm and 5 μm).

Detection was performed at 270 nm, the λ_{max} of bortezomib. The anticipated degradation products as well as all related impurities were expected to absorb at this wavelength and therefore be detected.

These forced degradation samples and all impurities blend solution were injected in Trail-1. In tiral-1, mobile phase A is mixture of water-acetonitrile-formic acid (500 : 500 : 1, v/v/v) and mobile phase B is mixture

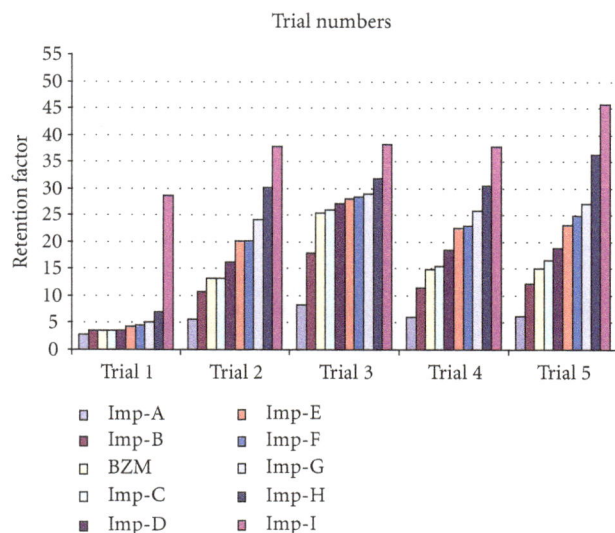

FIGURE 2: Graphical representation of retention factor of components in different trials.

of water-acetonitrile-formic acid (500 : 500 : 1, v/v/v), column temperature was maintained at 25°C, and all other chromatographic conditions adopted were, as described in Section 2.4. In this trial, it is observed that separation between Imp-B, Imp-C and Imp-D was coeluted with the analtye and Imp-E was coeluted with Imp-F. Attempts were given to modify the mobile phase-A and mobile phase-B to increase the resolution and peak symmetry.

It was decided to adopt mobile phase B as water-acetonitrile-formic acid (200 : 800 : 1, v/v/v) and a fixed gradient program mentioned in Section 2.4 for further trials. Organic modifier (mobile phase A) was changed in each trial. The mobile phase A and B used in different trials was given in Table 2. Column temperature was kept at 35°C for trial 5 to reduce the back pressure and to improve the peak symmetry and in all other trials it was maintained at 25°C.

In each trial, the forced degradation and impurity blend samples were injected. Attention was given for the separation of all nine impurities and bortezomib. Retention factor of each impurities and bortezomib obtained with each trial is presented graphically in (Figure 2).

The outcome of each trial is discussed below.

Trial 2. It was observed that the Imp-C co-eluted with bortezomib, Imp-E co-eluted with Imp-F and other impurities were separated from each other.

Trial 3. Imp-C was separated from bortezomib peak but poor resolution was observed. Apart from that, Imp-E, Imp-F, and Imp-G were very closely eluted.

Trial 4. All impurities were separated from each other and with bortezomib, the resolution between bortezomib and Imp-C (*Rs* > 3.0) was increased with satisfactory result, but the resolution between Imp-E and Imp-F (*Rs* < 1.2) were increased but not satisfactory. Few attempts were made

TABLE 2

(a)

Trial Number	Mobile phase A	Composition ratio of mobile phase A
1	Acetonitrile : Water : Formic acid	500 : 500 : 1
2	Acetonitrile : Water : Formic acid	300 : 700 : 1
3	Acetonitrile : Water : Formic acid	250 : 750 : 1
4	Acetonitrile : Water : Formic acid	285 : 715 : 1
5	Acetonitrile : Water : Formic acid	285 : 715 : 1

Mobile phase A used for different trials.

(b)

Trial Number	Mobile phase B	Composition ratio of mobile phase B
1	Acetonitrile : Water : Formic acid	500 : 500 : 1
2	Acetonitrile : Water : Formic acid	800 : 200 : 1
3	Acetonitrile : Water : Formic acid	800 : 200 : 1
4	Acetonitrile : Water : Formic acid	800 : 200 : 1
5	Methanol : Water : Formic acid	800 : 200 : 1

Mobile phase B used for different trials.

to separate those critical impurities by changing the formic acid, buffers in trial-4, but there was no improvement in the resolution between Imp-E and Imp-F.

Trial 5. Result from trial-4 methanol was used as organic modifier in mobile pahse-B. Hence, this trial was conducted with methanol as organic modifier in mobile phase-B and column temperature was kept at 35°C. Imp-E and Imp-F were well resolved (*Rs* > 2.5).

Based on above, it was concluded that trial-5 was highly selective for the quantification of impurities, degradants as well as bortezomib. At these chromatographic conditions, all the impurities and degradants were well separated amongst and also from bortezomib. This gradient program ensured the elution of all other impurities found in crude API. The final chromatographic conditions are concluded on Waters SymmetryShield RP18 5 μm, 4.6 × 250 mm column (Waters, Milford, USA) with mobile phase containing a gradient mixture of mobile phase A and B. Mobile phase-A consists a mixture of water-acetonitrile-formic acid (715 : 285 : 1, v/v/v) and the mobile phase-B consists a mixture of methanol-water-formic acid (800 : 200 : 1, v/v/v), respectively. The gradient program (T/%B) was set as 0/0, 20.0/0, 35.0/100, 50.0/100, 52.0/,0 and 60.0/0. The mobile phase was filtered through a nylon 0.45 μm membrane filter. The flow rate of the mobile phase was 1.0 mL/min. The column temperature was maintained at 35°C and the wavelength was monitored at 270 nm. The injection volume was 10 μL. Under the above conditions, results were as follows: retention time of bortezomib was around 15.0 min, with a tailing factor of 1.1 and % RSD for five replicate injections was 0.1%. Accelerated and long-term stability study results as per ICH Q1A (R2) for bortezomib were generated for 6 months by using the developed LC method and the results were well within the limits, this further confirms the stability indicating of the method.

The typical HPLC chromatograms (Figure 3) represent the satisfactory separation of all components among each other.

3.2. Validation of the Method

3.2.1. Solution Stability and Mobile Phase Stability. Assay (%) of bortezomib during solution stability and mobile phase stability experiments was within ±1%. The variability in the estimation of bortezomib impurities was within ±10% during solution stability and mobile phase experiments when performed using the related substances method. The results from mobile phase stability experiments confirmed that standard solutions and solutions in the mobile phase were stable up to 48 h for assay and related substances analysis.

The results from solution stability experiments confirmed that standard solution was stable up to 48 h and impurities spiked test solution was stable up to 24 h.

3.2.2. Specificity and Mass Balance Study. All forced degradation samples were analyzed at an initial concentration 1000 μg/mL of bortezomib with HPLC conditions mentioned in Section 2.4 using PDA detector to ensure the homogeneity and purity of bortezomib peak. Very significant degradation of bortezomib was observed in oxidative, thermal, acid and base stress conditions leading to the formation of Imp-A, Imp-B, and Imp-E (Figure 4) [25]. This was confirmed by coinjecting impurity standards to these degraded samples and by LC-MS/MS analysis. LC-MS/MS analysis was performed as per Section 2.5. Significant degradation was observed in water hydrolysis and photolytic open stress conditions. bortezomib was found to be stable under photolytic closed stress condition.

Assay studies were carried out for stress samples (at 1000 μg/mL) against bortezomib qualified reference standard. The mass balance (% assay + % sum of all compounds + % sum of all degradants) results were calculated

Development and Validation of a Stability Indicating LC Method for the Assay and Related Substances Determination of
a Proteasome Inhibitor Bortezomib

203

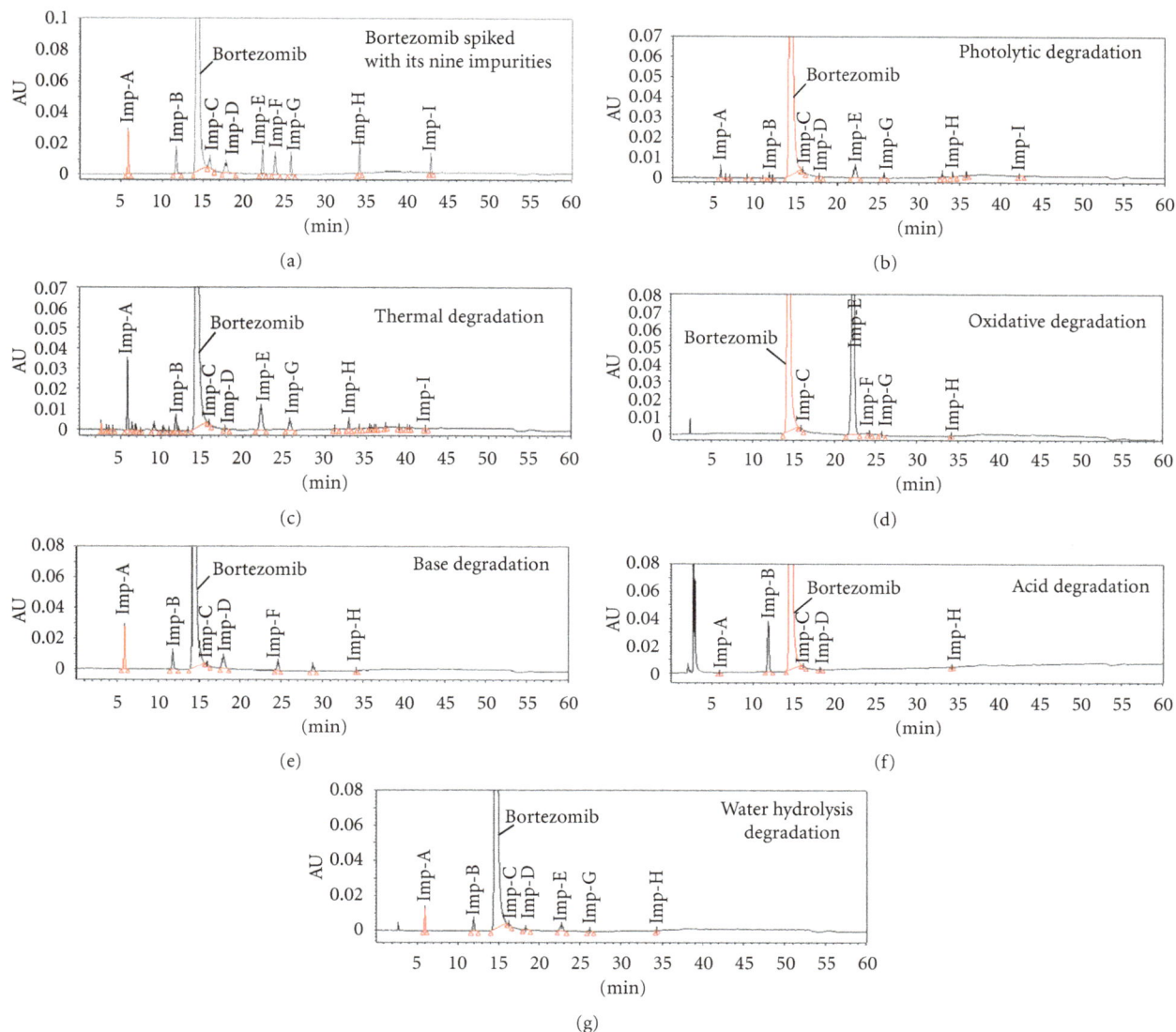

FIGURE 3: Typical chromatograms of bortezomib spiked with its nine impurities and its forced degradation samples.

for all stressed samples and found to be more than 99.2% (Table 3). The purity and assay of bortezomib was unaffected by the presence of its impurities and degradation products and thus confirms the stability-indicating power of the developed method.

3.2.3. Relative Response Factor.

Relative response factor (RRF) was established for Imp-A, Imp-B, Imp-C, Imp-D, Imp-E, Imp-F, Imp-G, Imp-H, and Imp-I as the ratio of slope of impurities and slope of bortezomib. Slope value obtained with linearity calibration plot was used for RRF determination. Established RRF value for Imp-A, Imp-B, Imp-C, Imp-D, Imp-E, Imp-F, Imp-G, Imp-H, and Imp-I are 1.41, 1.45, 0.92, 1.00, 0.94, 1.37, 0.97, 1.09, and 0.67, respectively.

3.2.4. Linearity.

The linearity calibration plot for the assay method was obtained over the calibration ranges tested, and

correlation coefficient obtained was greater than 0.999. The result shows that an excellent correlation existed between the peak area and concentration of the analyte.

Linear calibration plot for the related substance method was obtained over the calibration ranges tested, that is, LOQ to 150% for impurities. The correlation coefficient obtained was greater than 0.999 (Table 4). The above result shows that an excellent correlation existed between the peak area and the concentration of Imp-A, Imp-B, Imp-C, Imp-D, Imp-E, Imp-F, Imp-G, Imp-H, Imp-I, and bortezomib. The linearity established with bortezomib is applicable to unspecified impurities.

3.2.5. Limits of Detection and Quantification.

The determined limit of detection, limit of quantification values for bortezomib, and its nine impurities are reported in Table 4. The method precision for Imp-A, Imp-B, Imp-C, Imp-D, Imp-E, Imp-F, Imp-G, Imp-H, Imp-I, and bortezomib at

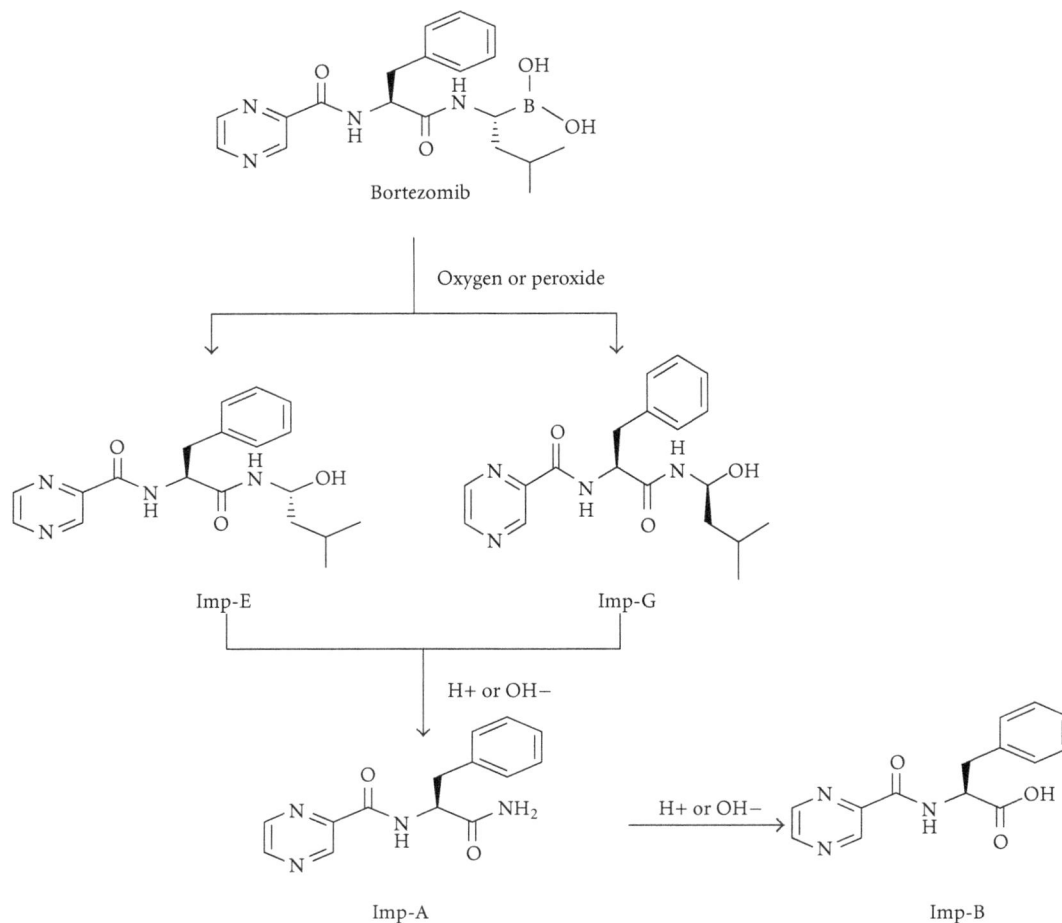

FIGURE 4: Degradation scheme of bortezomib; Imp-E, Imp-G, Imp-A, and Imp-B.

TABLE 3: Mass balance study.

Stress condition	Time	Assay of active substance (% w/w)	Total impurities (% w/w)	Mass balance (assay + total impurities) (% w/w)	Remarks
Thermal treatment (105°C)	7 h	92.4	7.0	99.4	Imp-A, Imp-E and Imp-G were major degradation products
Photo degradation (Open stress)	7 days	96.9	2.4	99.3	Imp-A, Imp-E and Imp-G were major degradation products
Acid hydrolysis (5.0 M HCl, 70°C)	72 h	96.2	3.0	99.2	Imp-B was major degradation product.
Base hydrolysis (0.1 M NaOH, 25°C)	7 h	94.4	5.3	99.8	Imp-A, Imp-B and Imp-D were major degradation products
Oxidation (3% H$_2$O$_2$)	Immediately	81.9	17.9	99.8	Imp-E was major degradation product.
Water hydrolysis (70°C)	72 h	97.4	2.1	99.5	Imp-A, Imp-B and Imp-E were major degradation products

Development and Validation of a Stability Indicating LC Method for the Assay and Related Substances Determination of
a Proteasome Inhibitor Bortezomib

205

TABLE 4: Regression data.

Parameter	Bortezomib	Imp-A	Imp-B	Imp-C	Imp-D	Imp-E	Imp-F	Imp-G	Imp-H	Imp-I
LOD (μg/mL)	0.043	0.012	0.025	0.043	0.043	0.047	0.038	0.045	0.024	0.060
LOQ (μg/mL)	0.174	0.048	0.100	0.173	0.174	0.188	0.152	0.179	0.097	0.239
Regression equation (y) Slope (b)	13289	22831	20852	12009	13934	14688	19330	17640	15812	8765
Intercept (a)	−404	−309	−296	130	45	−386	344	−486	−397	617
Correlation coefficient	0.9995	0.9995	0.9998	0.9993	0.9995	0.9993	0.9994	0.9997	0.9998	0.9996

Linearity range is LOQ-150% with respect to 1.0 mg/mL of bortezomib for impurities;
Linearity range is 50–150% with respect to 1.0 mg/mL of bortezomib for assay.

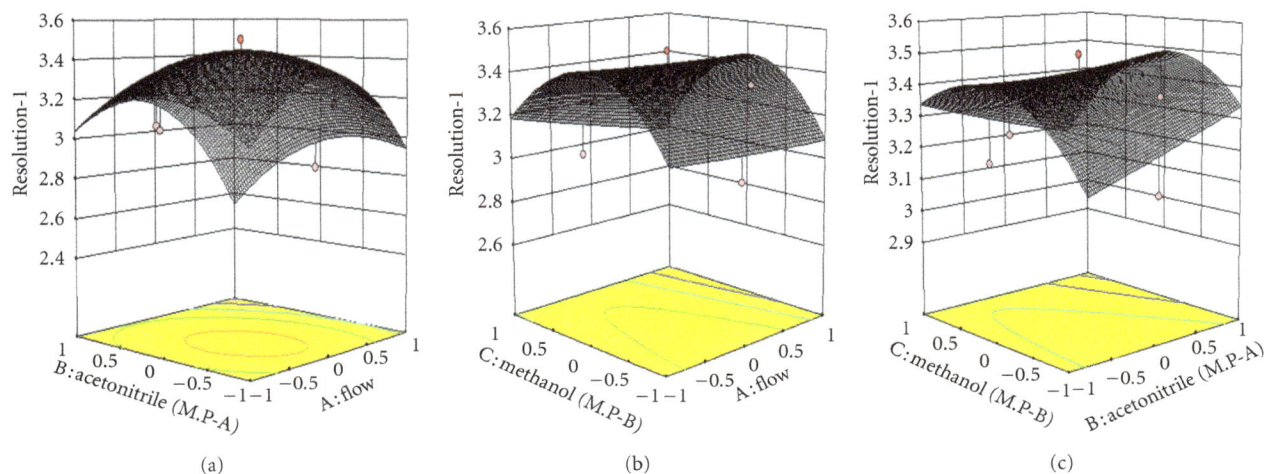

(a)　　　　　　　(b)　　　　　　　(c)

FIGURE 5: Three-dimensional plot of the response surface for Y (found critical resolution-1 between bortezomib and Imp-C), (a) variation of the response Y as a function of A (flow rate) and B (% acetonitrile); fixed factor: C (% methanol) = 80% v/v, (b) variation of the response Y as a function of A (flow rate) and C (% methanol); fixed factor: B (% acetonitrile) = 28.5% v/v, (c) variation of the response Y as a function of B (% acetonitrile) and C (% methanol); fixed factor: A (flow rate) = 1.0 mL min^{-1}.

LOQ level was below 10% RSD. LOD and LOQ established with bortezomib is applicable to unspecified impurities. Recovery at LOQ level for the nine impurities was in the range of 91.1 to 101.7%.

3.2.6. *Accuracy.* The percentage recovery of bortezomib in bulk samples was ranged from 99.6 to 99.9%. The percentage recovery of impurities in bortezomib samples varied from 94.2 to 107.4%. The LC chromatogram of spiked sample of all nine impurities in bortezomib sample is shown in (Figure 3). The % recovery values for bortezomib and impurities are presented in Table 5. The percentage recovery and percentage RSD for three preparations shows that this method is accurate for the determination of assay and related substances of bortezomib.

3.2.7. *Precision.* The % RSD of assay of bortezomib during assay method precision study was well within 0.4%. The %RSD of content (%) of Imp-A, Imp-B, Imp-C, Imp-D, Imp-E, Imp-F, Imp-G, Imp-H, Imp-I in related substance method precision study was within 5%. The % RSD of assay

results obtained in intermediate precision study was within 0.4%, confirming good precision of the method.

3.2.8. *Robustness.* The experimental domain of the selected variables is reported in Table 1. The ranges examined were small deviations from the method settings and the corresponding responses in the resolution and recoveries considered (Y) were observed. A three-factor CCD requires 16 experiments, including two replicates of the center point. Standard run and run order generated by Design Expert software are given in Table 6. By using a fitted full quadratic model (1), a response surface regression analysis for each response factor was performed using coded units. Table 7 shows the values calculated for the coefficients and P-values (P-value is the probability of the null hypothesis). Using a 5% significance level, a factor is considered to affect the response if the coefficients differ from zero significantly and the P-value < 0.050:

$$Y = X_0 + X_A F_A + X_B F_B + X_C F_C + X_{AB} F_A F_B + X_{AC} F_A F_C$$
$$+ X_{BC} F_B F_C + X_{AA} F_{AA} + X_{BB} F_{BB} + X_{CC} F_{CC}, \tag{1}$$

TABLE 5: Accuracy study.

| Added % (n = 3) | Assay | | | Related substances determination | | | | | | | | | | | | | | | | | |
| | Added (in ppm) | % Recovery | % RSD | Added (in ppm) | Imp-A | | Imp-B | | Imp-C | | Imp-D | | Imp-E | | Imp-F | | Imp-G | | Imp-H | | Imp-I | |
					Recovery (%)	RSD (%)	Recovery (%)	RSD (%)	Recovery (%)	RSD (%)	Recovery (%)	RSD (%)	Recovery (%)	RSD (%)	Recovery (%)	RSD (%)	Recovery (%)	RSD (%)	Recovery (%)	RSD (%)	Recovery (%)	RSD (%)
50	500	99.8	0.4	0.75	105.5	1.5	104.8	2.9	104.7	3.1	105.2	2.0	96.5	2.5	102.2	2.2	103.3	1.3	99.7	0.5	99.5	2.5
100	1000	99.6	0.4	1.50	104.8	2.5	102.6	1.1	104.1	1.4	106.4	2.7	105.2	2.2	101.5	1.4	101.4	2.6	97.9	1.1	108.3	2.9
150	1500	99.9	0.3	2.25	106.7	1.2	102.5	1.1	107.4	2.1	94.2	2.3	101.1	0.6	97.3	0.7	101.8	2.3	103.1	0.8	99.8	1.6

Development and Validation of a Stability Indicating LC Method for the Assay and Related Substances Determination of
a Proteasome Inhibitor Bortezomib

207

TABLE 6: Standard run, run order, and response obtained for robustness testing.

| Standard | Run | Level | | | Resolution-1 (Bortezomib and Imp-C) | Resolution-2 (Imp-E and Imp-F) | % w/w |
		Factor A	Factor B	Factor C			Assay
3	1	−1	1	1	3.0	2.3	99.3
1	2	1	1	−1	2.9	2.3	99.1
4	3	−1	−1	−1	3.1	2.4	99.3
5	4	0	0	0	3.5	2.6	99.9
6	5	0	0	0	3.4	2.5	100.1
2	6	1	−1	1	3.0	2.2	99.2
7	7	0	0	0	3.5	2.5	99.9
13	8	0	0	1	3.2	2.4	99.5
12	9	0	0	−1	3.4	2.3	99.6
15	10	0	0	0	3.4	2.6	99.8
14	11	0	0	0	3.5	2.5	99.9
16	12	0	0	0	3.5	2.6	100.1
8	13	−1	0	0	3.1	2.6	99.4
11	14	0	1	0	3.0	2.5	99.6
9	15	1	0	0	2.8	2.4	99.4
10	16	0	−1	0	3.2	2.5	99.8

TABLE 7: Regression coefficients and the associated probability values (P-value) for each response.

| Term | Resolution-1 | | Resolution-2 | | Bortezomib | |
	Coeff.	P-value	Coeff.	P-value	Coeff.	P-value
Constant	3.43	0	2.54	0	99.92	0
Factor A	−0.150	0.1549	−0.100	0.0545	0.000	1.0000
Factor B	−0.100	0.3151	0.000	1.0000	−0.100	0.3763
Factor C	−0.100	0.3151	0.050	0.2668	−0.050	0.6480
A × B	−0.100	0.4039	0.100	0.0968	−0.075	0.5781
A × C	−0.050	0.6677	0.000	1.0000	−0.075	0.5781
B × C	−0.100	0.4039	−0.050	0.3544	0.075	0.5781
A²	−0.344	0.0074	−0.032	0.4038	−0.413	0.0062
B²	−0.194	0.0575	−0.032	0.4038	−0.113	0.2703
C²	0.005	0.9477	−0.182	0.0036	−0.263	0.0343

where Y is the experimental response, X_0 is constant, X_x the coefficients of the factors and interactions and F_x stands for each factor.

The model was validated by the analysis of variance (ANOVA). The statistical analysis is shown in Table 7.

As shown in Figures 5 and 6, the analysis produces three-dimensional graphs by plotting the response model against two of the factors, while the third is held constant at a specified level, usually the proposed optimum.

From Table 7, it can be seen that P values for any of the studied factors are listed. It shows that the method is highly robust for 0.2 mL/min flow rate variation, % acetonitrile variation in the ratio of mobile phase A, and % methanol variation in the ratio of B.

4. Conclusion

A novel, simple, and accurate stability indicating HPLC method for the determination of bortezomib in the presence of degradation products was described for the first time. This method is highly specific for the quantification of degradation products and process-related impurities of bortezomib. The behaviour of bortezomib under various stress conditions was studied and presented. The method was completely validated showing satisfactory data for all the method validation parameters tested. The developed method is stability indicating and can be used for the routine analysis of production samples and also to check the stability of bortezomib samples.

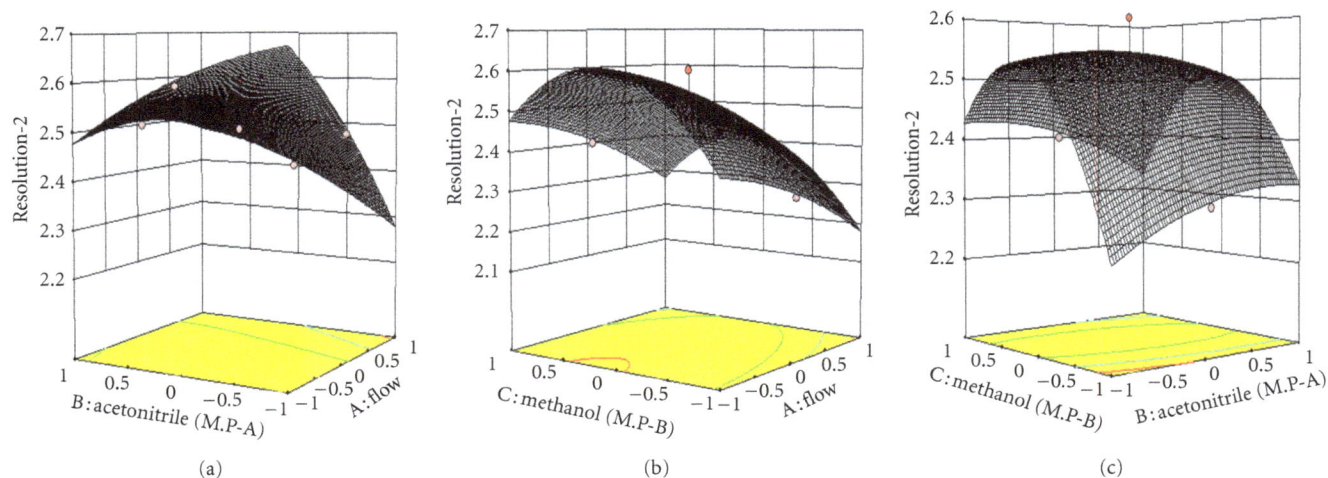

FIGURE 6: Three-dimensional plot of the response surface for Y (found critical resolution-2 between Imp-E and Imp-F), (a) variation of the response Y as a function of A (flow rate) and B (% acetonitrile); fixed factor: C (% methanol) = 80% v/v, (b) variation of the response Y as a function of A (flow rate) and C (% methanol); fixed factor: B (% acetonitrile) = 28.5% v/v, (c) variation of the response Y as a function of B (% acetonitrile) and C (% methanol); fixed factor: A (flow rate) = 1.0 mL min^{-1}.

Acknowledgment

The authors wish to thank the management of Dr. Reddy's Laboratories Ltd. for supporting this work. There is no conflict of interest in the manuscript.

References

[1] J. Adams and R. Stein, "Novel Inhibitors of the Proteasome and Their Therapeutic Use in Inflammation," *Annual Reports in Medicinal Chemistry*, vol. 31, no. C, pp. 279–288, 1996.

[2] J. Adams, "The proteasome: a suitable antineoplastic target," *Nature Reviews Cancer*, vol. 4, no. 5, pp. 349–360, 2004.

[3] J. Adams, Y. Ma, R. Stein, M. Baevsky, L. Grenier, and L. Plamondon, "Boronic ester and acid compounds, synthesis and uses," 1996, US, 1448.012TW01.

[4] P. G. Richardson, T. Hideshima, and K. C. Anderson, "Bortezomib (PS-341): a novel, first-in-class proteasome inhibitor for the treatment of multiple myeloma and other cancers," *Cancer Control*, vol. 10, no. 5, pp. 361–369, 2003.

[5] R. J. Snow and W. W. Bachovchin, "Boronic acid inhibitors of dipeptidyl peptidase IV. A new class of immunosuppressive agents," *Advances in Medicinal Chemistry*, vol. 3, no. C, pp. 149–177, 1995.

[6] European Medicines Agency, "European public assessment report: scientific discussion. The committee for medicinal products for human use," 2004, http://www.emea.eu.int/humandocs/Humans/EPAR/velcade/velcade.htm.

[7] J. S. Daniels, T. Pekol, J. Labutti et al., "Human metabolism of the proteasome inhibitor bortezomib: identification of circulating metabolites," *Drug Metabolism and Disposition*, vol. 33, no. 6, pp. 771–777, 2005.

[8] P. André, S. Cisternino, F. Chiadmi et al., "Stability of bortezomib 1-mg/mL solution in plastic syringe and glass vial," *Annals of Pharmacotherapy*, vol. 39, no. 9, pp. 1462–1466, 2005.

[9] D. D. Jandial, S. Farshchi-Heydari, C. A. Larson, G. I. Elliott, W. J. Wrasidlo, and S. B. Howell, "Enhanced delivery of

cisplatin to intraperitoneal ovarian carcinomas mediated by the effects of bortezomib on the human copper transporter 1," *Clinical Cancer Research*, vol. 15, no. 2, pp. 553–560, 2009.

[10] J. S. Johnston, M. A. Phelps, K. A. Blum et al., "Development and validation of a rapid and sensitive high-performance liquid chromatography-mass spectroscopy assay for determination of 17-(allylamino)-17-demethoxygeldanamycin and 17-(amino)-17-demethoxygeldanamycin in human plasma," *Journal of Chromatography B*, vol. 871, no. 1, pp. 15–21, 2008.

[11] ICH Q1A(R2), "Stability Testing of New Drug Substances and Products," 2000.

[12] I. C. H. Stability, "Testing of New Drug Substances and Products Q1A (R2)," in *Proceedings of the International Conference on Harmonization (IFPMA '03)*, Geneva, Switzerland, 2003.

[13] United States pharmacopoeia, *United States Pharmacopeial Convention*, Rockville, Md, USA, 31st edition, 2008.

[14] T. C. Jens and C. T. Rhodes, *Drug Stability Principles and Practices*, 3rd, Marcel Dekker, New York, NY, USA.

[15] M. Bakshi and S. Singh, "Development of validated stability-indicating assay methods—critical review," *Journal of Pharmaceutical and Biomedical Analysis*, vol. 28, no. 6, pp. 1011–1040, 2002.

[16] ICH Q2 (R1), "Validation of Analytical Procedures: Text and Methodology," 2005.

[17] G. A. Shabir, "Validation of high-performance liquid chromatography methods for pharmaceutical analysis: understanding the differences and similarities between validation requirements of the US Food and Drug Administration, the US Pharmacopeia and the International Conference on Harmonization," *Journal of Chromatography A*, vol. 987, no. 1-2, pp. 57–66, 2003.

[18] M. Ribani, C. B. Grespan Bottoli, C. H. Collins, I. C. S. Fontes Jardim, and L. F. Costa Melo, "Validation for chromatographic and electrophoretic methods," *Quimica Nova*, vol. 27, no. 5, pp. 771–780, 2004.

[19] Y. Vander Heyden, A. Nijhuis, J. Smeyers-Verbeke, B. G. M. Vandeginste, and D. L. Massart, "Guidance for robustness/

Development and Validation of a Stability Indicating LC Method for the Assay and Related Substances Determination of
a Proteasome Inhibitor Bortezomib

209

ruggedness tests in method validation," *Journal of Pharmaceutical and Biomedical Analysis*, vol. 24, no. 5-6, pp. 723–753, 2001.

[20] D. K. Lin, "Discussion on papers by Box and Liu, Box, and Myers," *Journal of Quality Technology*, vol. 31, pp. 61–66, 1999.

[21] K. K. Hockman and D. Berengut, "Design of experiments," *Chemical Engineering*, vol. 102, pp. 142–148, 1995.

[22] H. Fabre, "Robustness testing in liquid chromatography and capillary electrophoresis," *Journal of Pharmaceutical and Biomedical Analysis*, vol. 14, pp. 1125–1132, 1996.

[23] B. Singh, R. Kumar, and N. Ahuja, "Optimizing drug delivery systems using systematic "design of experiments." Part I: fundamental aspects," *Critical Reviews in Therapeutic Drug Carrier Systems*, vol. 22, pp. 27–105, 2004.

[24] S. Pinzauti, P. Gratteri, S. Furlanetto, P. Mura, E. Dreassi, and R. Phan-Tan-Luu, "Experimental design in the development of voltammetric method for the assay of omeprazole," *Journal of Pharmaceutical and Biomedical Analysis*, vol. 14, no. 8-10, pp. 881–889, 1996.

[25] S. Wu, W. Waugh, and V. J. Stella, "Degradation pathways of a peptide boronic acid derivative, 2–Pyz–(CO)–Phe–Leu–B(OH)2," *Journal of Pharmaceutical Sciences*, vol. 89, pp. 758–765, 2000.

Permissions

The contributors of this book come from diverse backgrounds, making this book a truly international effort. This book will bring forth new frontiers with its revolutionizing research information and detailed analysis of the nascent developments around the world.

We would like to thank all the contributing authors for lending their expertise to make the book truly unique. They have played a crucial role in the development of this book. Without their invaluable contributions this book wouldn't have been possible. They have made vital efforts to compile up to date information on the varied aspects of this subject to make this book a valuable addition to the collection of many professionals and students.

This book was conceptualized with the vision of imparting up-to-date information and advanced data in this field. To ensure the same, a matchless editorial board was set up. Every individual on the board went through rigorous rounds of assessment to prove their worth. After which they invested a large part of their time researching and compiling the most relevant data for our readers. Conferences and sessions were held from time to time between the editorial board and the contributing authors to present the data in the most comprehensible form. The editorial team has worked tirelessly to provide valuable and valid information to help people across the globe.

Every chapter published in this book has been scrutinized by our experts. Their significance has been extensively debated. The topics covered herein carry significant findings which will fuel the growth of the discipline. They may even be implemented as practical applications or may be referred to as a beginning point for another development. Chapters in this book were first published by Hindawi Publishing Corporation; hereby published with permission under the Creative Commons Attribution License or equivalent.

The editorial board has been involved in producing this book since its inception. They have spent rigorous hours researching and exploring the diverse topics which have resulted in the successful publishing of this book. They have passed on their knowledge of decades through this book. To expedite this challenging task, the publisher supported the team at every step. A small team of assistant editors was also appointed to further simplify the editing procedure and attain best results for the readers.

Our editorial team has been hand-picked from every corner of the world. Their multi-ethnicity adds dynamic inputs to the discussions which result in innovative outcomes. These outcomes are then further discussed with the researchers and contributors who give their valuable feedback and opinion regarding the same. The feedback is then collaborated with the researches and they are edited in a comprehensive manner to aid the understanding of the subject.

Apart from the editorial board, the designing team has also invested a significant amount of their time in understanding the subject and creating the most relevant covers. They scrutinized every image to scout for the most suitable representation of the subject and create an appropriate cover for the book.

The publishing team has been involved in this book since its early stages. They were actively engaged in every process, be it collecting the data, connecting with the contributors or procuring relevant information. The team has been an ardent support to the editorial, designing and production team. Their endless efforts to recruit the best for this project, has resulted in the accomplishment of this book. They are a veteran in the field of academics and their pool of knowledge is as vast as their experience in printing. Their expertise and guidance has proved useful at every step. Their uncompromising quality standards have made this book an exceptional effort. Their encouragement from time to time has been an inspiration for everyone.

The publisher and the editorial board hope that this book will prove to be a valuable piece of knowledge for researchers, students, practitioners and scholars across the globe.

List of Contributors

Hiroyuki Tanaka
Faculty of Pharmaceutical Science, Kyushu University, 3-1-1 Maidashi, Higashiku, Fukuoka 812-0855, Japan

Waraporn Putalun
Faculty of Pharmaceutical Science, Khon Kaen University, Khon Kaen 40002, Thailand

Yukihiro Shoyama
Faculty of Pharmaceutical Science, Nagasaki International University/2825-7 Huis Ten Bosch, Sasebo, Nagasaki 859-3298, Japan

Joaquim Chico
Departament de Quimica Anal´ıtica, Facultat de Quimica, Universitat de Barcelona, c/Mart´ı i Franques 1, 08028 Barcelona, Spain

Frederique van Holthoon and Tina Zuidema
RIKILT-Institute of Food Safety, Wageningen University and Research Center, Akkermaalsbos 2, 6708 WB, P.O. Box 230, 6700AE Wageningen, The Netherlands

Nitin Dubey
College of Pharmacy, IPS Academy, Indore, India

Nidhi Dubey
School of Pharmacy, Devi Ahilya Vishwavidyalaya, Indore, India

Rajendra Mehta
A. R .College of Pharmacy, Vallabh Vidyanagar, India

Daya L. Chothani
Pharmacognosy Department, Pioneer Pharmacy Degree College, Ajwa-Nimeta Road, Sayajipura, Vadodara 390019, India

M. B. Patel and S. H. Mishra
Herbal Drug Technology Lab, Pharmacy Department, Faculty of Technology and Engineering, The M. S. University of Baroda, Vadodara 390 001, India

Mieczysław Sajewicz, Dorota Staszek, Michał S.Wrobel and Teresa Kowalska
Institute of Chemistry, University of Silesia, 9 Szkolna Street, 40-006 Katowice, Poland

Monika Waksmundzka-Hajnos
Department of Inorganic Chemistry, Faculty of Pharmacy, Collegium Pharmaceuticum, Medical University of Lublin, 4a Chod´zki Street, 20-093 Lublin, Poland

Singaram Kathirvel
Department of Pharmaceutical Analysis, Hindu College of Pharmacy, Amaravathi Road, Guntur 522002, India

Suggala Venkata Satyanarayana
Department of Chemical Engineering, JNTU College of Engineering, Anantapur 515002, India

Garikapati Devalarao
Department of Pharmaceutical Analysis, KVSR Siddhartha College of Pharmaceutical Sciences, Vijayawada 520008, India

Mareike Perzborn, Christoph Syldatk and Jens Rudat
Institute of Process Engineering in Life Sciences, Section II: Technical Biology, Karlsruhe Institute of Technology (KIT), 76131 Karlsruhe, Germany

Steingrimur Stefansson
Fuzbien Technology Institute, 9700 Great Seneca Hwy, Suite 302, Rockville, MD 20850, USA

Daniel L. Adams and Cha-Mei Tang
Creatv Microtech Inc., 11609 Lake Potomac Drive, Potomac, MD 20854, USA

A. A. Shirkhedkar, J. K. Rajput, D. K. Rajput and S. J. Surana
Department of Pharmaceutical Chemistry, R.C. Patel Institute of Pharmaceutical Education and Research, Karwand Naka, Shirpur, Dhule District 425 405, India

Maria Rambla-Alegre
Area de Quimica Anal'ıtica, QFA, Universitat Jaume I, 12071 Castello, Spain

Ramakrishna Kommana and Praveen Basappa
Department of Pharmaceutical Analysis and Quality Assurance, Gokaraju Rangaraju College of Pharmacy, Hyderabad 500090, India

Mei-Liang Chin-Chen, Samuel Carda-Broch, Josep Esteve-Romero and Juan Peris-Vicente
QFA, ESTCE, Universitat Jaume I, 12071 Castello, Spain

Maria Rambla-Alegre
Department of Organic Chemistry, Ghent University, 9000 Ghent, Belgium

Pinak M. Sanchaniya, Falgun A. Mehta and Nirav B. Uchadadiya
Indukaka Ipcowala College of Pharmacy, Beyond GIDC, P.B. No. 53, Vitthal Udyognagar, Gujarat 388121, India

Batuk Dabhi, Bhavesh Parmar, Nitish Patel, Yashwantsinh Jadeja, Madhavi Patel, Hetal Jebaliya and A. K. Shah
Department of Chemistry, Saurashtra University, Rajkot-360 005, Gujarat, India

Denish Karia
Arts Commerce and Science College, Borsad 388540, Gujarat, India

Yan-yan Jia, Song Ying, Chen-tao Lu, Jing Yang, Li-kun Ding, Ai-dong Wen and Yan-rong Zhu
Department of Pharmacy, Xijing Hospital of the Fourth Military Medical University, Xian 710032, China

Najma Memon, Huma I. Shaikh and Amber R. Solangi
National Centre of Excellence in Analytical Chemistry, University of Sindh, Jamshoro, Sindh 78060, Pakistan

Nirav Uchadadiya, Falgun Mehta and Pinak Sanchaniya
Department of Pharmaceutical Chemistry and Analysis, Indukaka Ipcowala College of Pharmacy, New Vallabh Vidyanagar, Gujarat 388121, India

Łukasz Ciesla
Department of Inorganic Chemistry, Medical University of Lublin, Chodzki 4a, 20-093 Lublin, Poland
Department of Biochemistry, Institute of Soil Science and Plant Cultivation, State Research Institute, Czartoryskich 8, 24-100 Puławy, Poland

Artem U. Kulikov
Laboratory of Pharmacopoeial Analysis, Scientific and Expert Pharmacopoeial Centre, Astronomicheskaya Street 33, Kharkov 61085, Ukraine

Jozef Rzepa, Mieczysław Sajewicz, Patrycja Gorczyca, Magdalena Włodarek and Teresa Kowalska
Institute of Chemistry, University of Silesia, 9 Szkolna Street, 40-006 Katowice, Poland

Tomasz Baj, Patrycja Gorczyca, Magdalena Włodarek and Kazimierz Głowniak
Department of Pharmacognosy with Medicinal Plant Unit, Faculty of Pharmacy, Medical University of Lublin,1 Chodzki Street, 20-093 Lublin, Poland

Monika Waksmundzka-Hajnos
Department of Inorganic Chemistry, Faculty of Pharmacy, Medical University of Lublin, Collegium Pharmaceuticum, 4a Chodzki Street, 20-093 Lublin, Poland

Josilene Chaves Ruela Correa and Herida Regina Nunes Salgado
School of Pharmaceutical Sciences, Universidade Estadual Paulista, Rodovia Araraquara-Ja´u km 1, 14801-902 Araraquara, SP, Brazil

Cristina Duarte Vianna-Soares
Department of Pharmaceutical Products, Federal University of Minas Gerais, Avenida Antonio Carlos 6627, 31270-901 Belo Horizonte, MG, Brazil

Monika L. Jadhav and Santosh R. Tambe
Pharmaceutical Chemistry Department, M.G.V's Pharmacy College, Panchavati, Nashik 422003, India

Cristina A. Diagone, Renata Colombo, Fernando M. Lancas and Janete H. Yariwake
Instituto de Quimica de Sao Carlos, Universidade de Sao Paulo, Caixa Postal 780, 13560-970 Sao Carlos, SP, Brazil

Irena Malinowska and Katarzyna E. Stepnik
Chair of Physical Chemistry, Department of Planar Chromatography, Faculty of Chemistry, Maria Curie Skłodowska University, M. Curie—Skłodowska Sq. 3, 20-031 Lublin, Poland

Md. Saddam Nawaz
Quality Assurance Department, ACI Ltd., Narayanganj 1400, Bangladesh

Zoraida Sosa-Ferrera, CristinaMahugo-Santana and Jose Juan Santana-Rodrıguez
Departamento de Quimica, Universidad de Las Palmas de Gran Canaria, 35017 Las Palmas de Gran Canaria, Spain

Mopidevi Narasimha Naidu, Kadaboina Rajasekhar and Murki Veerender
Active Pharmaceutical Ingredients, Dr. Reddy's Laboratories Ltd., IPDO, Bachupally, Hyderabad 500072, Andhra Pradesh, India

Kasa Srinivasulu
Active Pharmaceutical Ingredients, Dr. Reddy's Laboratories Ltd., IPDO, Bachupally, Hyderabad 500072, Andhra Pradesh, India
Department of Chemistry, Osmania University, Hyderabad 500072, Andhra Pradesh, India

Mulukutla Venkata Suryanarayana
Department of Chemistry, Osmania University, Hyderabad 500072, Andhra Pradesh, India
Matrix Laboratories Limited, Plot 34 A, Anrich Industrial Estate, Bollaram, Medak District, Jinnaram, Mandal 502 32, Andhra Pradesh, India